# An Institute for an Empire

*An Institute for an Empire* is the first scholarly study of one of the world's foremost scientific institutions, the Physikalisch-Technische Reichsanstalt (PTR) in Imperial Germany. The Reichsanstalt stood at the forefront of institutional innovation in science and technology during the late nineteenth and early twentieth centuries, uniting diverse practitioners and representatives of physics, technology, industry, and the state. During its first three decades of activity (1887–1918), the Reichsanstalt demonstrated how physics and industrial technology could help build a modern society and a modern nation-state. Moreover, it encouraged and helped inaugurate the era of Big Science. For many in the scientific, technological, and industrial worlds, the Reichsanstalt in Berlin came to symbolize, in another sphere, the young German Reich's newly acquired political power and authority. The Reichsanstalt was an institute for an empire.

Using a wealth of archival and printed sources, David Cahan pursues two interrelated themes – the institutionalization of science (and technology) and measurement as the "essence" of the Reichsanstalt – to understand the Reichsanstalt's founding, its development into a scientific bureaucracy, its relation to other scientific, industrial, and political institutions in the German Reich and beyond, and the reasons for its scientific successes and limitations. Professor Cahan also discusses the Reichsanstalt's leaders and scientists, including Werner von Siemens and Hermann von Helmholtz, as well as its scientific and technological work. Among the Reichsanstalt's many accomplishments were contributions to the new quantum physics, development of physical standards and measuring instruments for science, industry, and the state, and testing work for a variety of German industries.

FOR MY WIFE

# Contents

# Illustrations

# Tables

# Acknowledgments

This study could not have been carried out without the cooperation of numerous institutions and their staffs, who kindly assisted me in the course of my research. It is a pleasure to thank the following individuals and institutions for allowing me to use their holdings and for permission to cite parts of their collections or to reprint photographs: the Milton S. Eisenhower Library, Special Collections, The Johns Hopkins University, Baltimore; the Bancroft Library, University of California, Berkeley; Akademie-Verlag, Berlin; Amerika-Gedenkbibliothek Berliner Zentralbibliothek, Berlin; Bildarchiv Preussischer Kulturbesitz, Berlin; Physikalisch-Technische Bundesanstalt-Institut Berlin; Staatsbibliothek Preussischer Kulturbesitz, Handschriftenabteilung, Berlin; Technische Universität, Berlin; Universitätsbibliothek, Hochschularchiv, Berlin; Zentrales Archiv der Akademie der Wissenschaften der DDR, Historische Abteilung, Berlin; Physikalisch-Technische Bundesanstalt (in particular Dr. Wilfried Hauser), Braunschweig; Cambridge University Library; Massachusetts Institute of Technology, the Libraries, Institute Archives and Special Collections, Cambridge, Massachusetts; National Bureau of Standards, Gaithersburg, Maryland; Handschriftenabteilung, Niedersächsische Staats- und Universitätsbibliothek, Göttingen; the Albert Einstein Papers, Department of Manuscripts and Archives, the Hebrew University of Jerusalem; Bundesarchiv, Koblenz; Emil Warburg Nachlass, Dr. Peter Meyer-Viol, Lanaken, Belgium; Zentrales Staatsarchiv der DDR, Merseburg; Deutsches Museum, Sondersammlungen, Munich; Siemens Museum, Munich; American Institute of Physics, Niels Bohr Library, Center for History of Physics (including the Archive for History of Quantum Physics), New York; American Philosophical Society, Philadelphia; Zentrales Staatsarchiv der DDR, Potsdam; the

John Hay Library, Archives, Brown University, Providence, Rhode Island; Deutsches Röntgen-Museum, Remscheid; the Nobel Archives, Kgl. Vetenskapsakademien, Stockholm; Stockholms Universitetsbibliotek; Archives du Bas-Rhin, Strasbourg; Carnegie Institution of Washington; National Academy of Sciences, Washington, D.C.; the National Archives, Washington, D.C.; and (for the Henry Rowland Papers) Mrs. Jean Rowland, West Lebanon, New Hampshire. I also thank the University of California Press for permission to use parts of two articles that I published previously in *Historical Studies in the Physical Sciences 12:2* (1982): 253-83, *15:2* (1985): 1-65, ©1982 and 1985 by the Regents of the University of California.

This study first took shape as a dissertation, and it is a pleasure to thank the Fulbright Commission for an award that allowed me to spend a year in Germany doing research, and Robert H. Kargon and Russell McCormmach for their counsel and reading of my dissertation. I benefited equally from the comments and criticisms of John L. Heilbron, Robert H. Kargon, Robert J. Richards, and Eugene Rudd, all of whom read the final version of the book manuscript. Finally, by far my greatest debt, both intellectual and personal, is to Jean Axelrad Cahan, who from first to last took time from her own work to listen to my ideas about the Reichsanstalt and gently to criticize my writing about it. Any mistakes of fact and interpretation are, of course, my own.

# Abbreviations

AIP    American Institute of Physics, Niels Bohr Library, Center for History of Physics, New York

*AP*    *Annalen der Physik (und Chemie*, until 1900)

*APAWB*    *Abhandlungen der Preussischen Akademie der Wissenschaften zu Berlin*

BA    Bundesarchiv, Reichsanstalt Papers held in the archive of the PTB-IB

BAK    Bundesarchiv, Koblenz

CBP    Carl Barus Papers, John Hay Library, Archives, Brown University, Providence, Rhode Island

DM    Deutsches Museum, Handschriftenabteilung, Munich

*DMZ*    *Deutsche Mechaniker-Zeitung* (supplement to *ZI* [1898])

*DSB*    *Dictionary of Scientific Biography*, ed. Charles Coulston-Gillispie, 16 vols. (New York, 1970-80)

EFP    Emil Fischer Papers, Bancroft Library, University of California, Berkeley

*EZ*    *Elektrotechnische Zeitschrift*

*GAFK*    *Gesammelte Abhandlungen von Friedrich Kohlrausch*, ed. Wilhelm Hallwachs, Adolf Heydweiller, Karl Strecker, and Otto Wiener, 2 vols. (Leipzig, 1910-11)

*HEDR*    "Anlage IV. Reichsamt des Innern," in *Haushalts-Etat des Deutschen Reichs: Reichshaushalts-Etat für das Etatsjahr [auf das Rechnungsjahr] . . . nebst Anlagen* (title varies) (Berlin, 1872ff.)

*HSPS*    *Historical Studies in the Physical and Biological Sciences*

*NTM*    *NTM-Schriftenreihe für Geschichte der Naturwissenschaften, Technik und Medizin*

*NWN*    *Die Naturwissenschaften*

*PB*    *Physikalische Blätter*

PPI      *Physiker über Physiker: Wahlvorschläge zur Aufnahme von Physikern in die Berliner Akademie 1870 bis 1929 von Hermann v. Helmholtz bis Erwin Schrödinger,* ed. Christa Kirsten and Hans-Günther Körber (Berlin, 1975)

PPII    *Physiker über Physiker II: Antrittsreden, Erwiderungen bei der Aufnahme von Physikern in die Berliner Akademie, Gedächtnisreden von 1870 bis 1929,* ed. Christa Kirsten and Hans-Günther Körber (Berlin, 1979)

PTB-IB   Physikalisch-Technische Bundesanstalt-Institut Berlin

PTR     Physikalisch-Technische Reichsanstalt

PZ      *Physikalische Zeitschrift*

RI       Reichsamt des Innern, 1879-1919; thereafter, Reichsministerium des Innern

SAA     Siemens-Archiv-Akten, Siemens Museum, Munich

SAC     Svante Arrhenius Collection, Stockholms Universitetsbibliotek, Stockholm

SB       *Sitzungsberichte der Preussischen Akademie der Wissenschaften zu Berlin*

SBVR    *Stenographische Berichte über die Verhandlungen des Reichstages* (title varies)

SPK     Staatsbibliothek Preussischer Kulturbesitz, Handschriftenabteilung, Berlin

TAIEE   *Transactions of the American Institute of Electrical Engineers*

VDPG   *Verhandlungen der deutschen Physikalischen Gesellschaft*

WAPTR *Wissenschaftliche Abhandlungen der Physikalisch-Technischen Reichsanstalt*

ZI       *Zeitschrift für Instrumentenkunde*

ZStAP   Zentrales Staatsarchiv, Dienststelle Potsdam

ZTP     *Zeitschrift für technische Physik*

# German Institutions, Organizations, and Agencies

*Allgemeine Elektrizitätsgesellschaft* (General Electricity Company)
*Artillerie- und Ingenieurschule* (Artillery and Engineering School)
*Astrophysikalisches Observatorium* (Astrophysical Observatory)
*Bauakademie* (School of Architecture)
*Beleuchtungstechnische Gesellschaft* (Society for Illumination Technology)
*Berlin-Charlottenburger Strassenbahngesellschaft* (Berlin-Charlottenburg Streetcar Company)
*Berliner (Deutsche) Physikalische Gesellschaft* (Berlin [German] Physical Society)
*Berliner Elektrizitätswerke* (Berlin Electrical Works)
*Chemisch-Technische Reichsanstalt* (Imperial Institute of Chemistry and Technology)
*Chemisch-Technische Versuchsanstalt* (Chemical-Technical Experimental Testing Station)
*Deutsche Gesellschaft für Mechanik und Optik* (German Society for Precision Mechanics and Optics)
*Deutsche Gesellschaft für Technische Physik* (German Society for Technical Physics)
*Deutsche Seewarte* (German Marine Observatory)
*Deutscher Verein für Gas- und Wasserfachmänner* (German Association of Gas and Water Specialists)
*Deutscher Verein für Rübenzucker-Industrie* (German Association of the Beet Sugar Industry)
*Eidgenössische Technische Hochschule* (Federal Institute of Technology, Zurich)

*Elektrotechnischer Verein zu Berlin* (Berlin Electrical Engineers Association)

*Gesellschaft Deutscher Naturforscher und Aerzte* (Society of German Natural Scientists and Physicians)

*Gesellschaft für drahtlose Telegraphie* (Wireless Telegraphy Company)

*Gesellschaft für Linde's Eismaschinen* (Linde's Freezer Company)

*Gewerbeakademie* (Industrial School)

*Gewerbeschule* (Trade School)

*Göttinger Vereinigung zur Förderung der angewandten Mathematik und Physik* (Göttingen Association for the Advancement of Applied Mathematics and Physics)

*Internationale Beleuchtungskommission* (International Illumination Commission)

*Kaiserliche Normal Eichungskommission* (Imperial Institute for Weights and Measures)

*Kaiser-Wilhelm-Gesellschaft* (Kaiser William Society)

*Königliche Preussische Mechanisch-Technische Versuchsanstalt* (Royal Prussian Mechanical-Technical Experimental Testing Station)

*Kriegsakademie* (War College)

*Kunstakademie* (Art Academy)

*Landesaufname* (Ordnance Survey)

*Landestriangulation* (State Triangulation Service)

*Lichtnormalien-Kommission* (Light Standards Commission)

*Max-Planck-Gesellschaft zur Förderung der Wissenschaften* (Max Planck Society for the Advancement of the Sciences)

*Physikalisch-Technische Bundesanstalt-Institut Berlin* (Federal Institute of Physics and Technology, Berlin)

*Physikalisch-Technische Reichsanstalt* (Imperial Institute of Physics and Technology)

*Polytechnische Gesellschaft zu Berlin* (Berlin Polytechnical Society)

*Preussische Akademie der Wissenschaften* (Prussian Academy of Sciences)

*Preussisches Kultusministerium* (*Ministerium der geistlichen, Unterrichts- und Medizinal- Angelegenheiten*, Prussian Ministry of Culture)

*Union-Elektrizitätsgesellschaft* (Union Electrical Company)

*Verband Deutscher Elektrotechniker* (Association of German Electrical Engineers)

*Verein zur Beförderung des Gewerbefleisses* (Association for the Advancement of Industry)

*Verein Deutscher Ingenieure* (Association of German Engineers)

*Vereinigung der Elektrizitätswerke* (Alliance of Electrical Works)

*Zentraldirektorium des Vermessungswesens* (Central Survey Office)

# Introduction

In 1887 a private philanthropist and the imperial German government combined resources to establish an institution, the Physikalisch-Technische Reichsanstalt – in English, the Imperial Institute of Physics and Technology – which was to represent the best of both pure science and industrial technology. The Reichsanstalt's scientific studies, like that on blackbody radiation, which helped usher in the new quantum physics; its institutional commitment to physical metrology, for example in the development of electrical standards needed by science and industry; and its testing and certification work on scientific instruments, measuring apparatus, and materials – these and other activities demonstrated some of the intellectual and social benefits that science and technology could contribute to modern society and to the modern nation-state. By the turn of the century, the Reichsanstalt stood at the forefront of institutional innovation in science and technology, uniting diverse practitioners and representatives of physics, technology, industry, and the state. For many in the scientific, technological, and industrial worlds, the Reichsanstalt in Berlin came to symbolize, in another sphere, the young German Reich's newly acquired political power and authority. The Reichsanstalt was an institute for an empire.

Success soon brought imitators. In 1898, barely a decade after the Reichsanstalt's founding, the mathematician Felix Klein, with the aid and support of the Prussian Kultusministerium (Ministry of Culture) and a group of German industrialists, established the Göttinger Vereinigung zur Förderung der angewandten Mathematik und Physik (Göttingen Association for the Advancement of Applied Mathematics and Physics). Like the Reichsanstalt, it helped unite and advance physics and technology; but unlike the Reichsanstalt, it also trained advanced students in these disciplines.[1]

German chemists also viewed the Reichsanstalt as a model institution for advancing their discipline. During the early years of the twentieth century, several leading German academic chemists – including Emil Fischer, Walther Nernst, and Wilhelm Ostwald – and chemical firms – with representatives from the chemical giants Agfa, BASF, and Bayer – espoused the idea of a Chemisch-Technische Reichsanstalt (Imperial Institute of Chemistry and Technology) that they hoped would do for academic and industrial chemistry what the Physikalisch-Technische Reichsanstalt was already doing for physics and its allied industries. The proposed Chemische Reichsanstalt was established in 1921. In 1910, however, it formed the core idea of a proposal that led to the establishment of the most prestigious scientific institutes in twentieth-century Germany: the Kaiser-Wilhelm-Gesellschaft (Kaiser William Society), founded in 1911 and known since 1946 as the Max-Planck-Gesellschaft zur Förderung der Wissenschaften (Max Planck Society for the Advancement of the Sciences). Not surprisingly, the Gesellschaft's first three constituent institutes were devoted to chemistry.[2]

The Reichsanstalt had imitators abroad as well as at home. In 1895, Douglas Galton, the reigning president of the British Association for the Advancement of Science, began promoting an idea suggested in 1891 by Oliver Lodge: the construction of a British national physical laboratory to help advance measuring physics and precision instrumentation, just as the new German Reichsanstalt was doing. Galton complained that British scientists had to resort to assistance from Paris and Berlin in order to standardize their precision instruments. To make Britain more independent in metrological matters and to help advance science-based industry there, he called for "the extension of the Kew Observatory, in order to develop it on the model of the Reichsanstalt," going so far as to display detailed construction plans of the Reichsanstalt and to analyze its organization.[3] Galton and a stellar crew of British scientists – including Lord Rayleigh, Joseph Lister, Henry Roscoe, and Michael Foster – called for a scientific institution "which should do for the United Kingdom that which the Reichsanstalt did for Germany."[4] By 1898 British scientists and industrialists had persuaded the Treasury to approve their plans; a year later the new National Physical Laboratory in Teddington began operations.[5]

American scientists, engineers, and manufacturers were also well acquainted with the Reichsanstalt's achievements in standardization and in conducting research in science-based technology aimed at helping industry.[6] In 1897, Henry Pritchett, the director of the American government's five-man Office of Standard Weights and Measures, hired a physicist, Samuel W. Stratton, to review the office's work "and to recommend . . . a plan for its enlargement into a more efficient bureau of standards, which might perform in some measure for this country the work carried on by the Reichsanstalt in Germany."[7] Pritchett and Stratton's campaign for an American equivalent of the Reichsanstalt found strong support from physicists, among others. George Ellery Hale, astrophysicist and scientific empire builder, advised Stratton that American engineering industries required an American standards and testing laboratory that would provide the same sort of "guarantee of excellence furnished by the authority of the Reichsanstalt." [8] Hale's distinguished colleague, Joseph S. Ames, professor of physics at The Johns Hopkins University, argued similarly that both American science and American industry would receive untold benefits from an American institute like the Reichsanstalt. Ames claimed that "everyone" agreed "that the tremendous advance in all branches of manufactures in Germany, in the last ten years, is due largely, if not entirely, to the work of the Reichsanstalt."[9] Pritchett and Stratton soon found that, in addition to the warm support of physicists like Hale and Ames, they also had the backing of major congressional and government figures. In 1901, their efforts came to fruition: The United States Congress established a National Bureau of Standards. Less than a year later, the bureau's first president, Samuel W. Stratton, was in Berlin, as *Science* magazine reported, "studying the Reichsanstalt with a view to the buildings to be erected at Washington for the newly established Bureau of Standards."[10]

The Reichsanstalt retained its preeminence at least until the outbreak of the First World War, providing the model, further, for Japan's Institute of Physical and Chemical Research.[11] But by 1914 the compliment of imitation had turned into the threat of competition. The National Physical Laboratory and the Bureau of Standards as well as other institutional circumstances within Germany challenged the Reichsanstalt's hegemony in metrology and its previously unmatched research opportunities.[12]

The Reichsanstalt's scientific and technological accomplishments, its importance for German industry, society, and the state, and its role in

inducing others to establish complementary or similar institutions in
Germany and abroad make it a natural object for historical study. Yet
apart from a number of studies of the Reichsanstalt's origins,[13] and
apart from three celebratory chronicles by the Reichsanstalt itself and
by its post-World War II successor, the Bundesanstalt,[14] there has been
to date no scholarly study of the Reichsanstalt in Imperial Germany.[15]
Two reasons help explain this historiographical gap: First, Allied
bombing of Berlin during World War II destroyed about 40 percent of
the Reichsanstalt's buildings, whose contents would doubtless have
greatly aided any would-be historian.[16] Valuable Reichsanstalt archival materials were also lost due to combat in Berlin.[17] Second, the
sheer size of the institution and the volume of its activities make it difficult to organize the study of the Reichsanstalt. In short, documentary
source material and a thematic approach are needed for a scholarly and
manageable history of the Reichsanstalt in imperial Germany.

The present study of the Reichsanstalt is the evolutionary product
of two previous studies that addressed these needs. In the first, a doctoral dissertation, I located and organized a good deal of previously un-
exploited documentary source material.[18] Three sets of archival holdings proved particularly valuable for that (and this) study. At the
Siemens Museum in Munich, I found materials on Werner von Siemens and, even more important, copies of the minutes of the meetings
of the Reichsanstalt's Board of Directors (Kuratorium), minutes that
had been destroyed or lost during the bombing of Berlin in World War
II. At the Zentrales Staatsarchiv in Potsdam (in the German Democratic
Republic), I found extensive Reichsanstalt materials, particularly concerning personnel matters. Finally, access to Reichsanstalt materials in
the archive of the Physikalisch-Technische Bundesanstalt-Institut Berlin (Federal Institute of Physics and Technology, Berlin) proved to be
equally vital. In the second of my two predecessor studies, this one intended for a German-reading audience on the occasion of the Reichsanstalt's centennial, I had an opportunity to rework my doctoral dissertation, paying less attention to day-to-day details and attempting to
discern general trends and drawing more general conclusions.[19] In the
present study I have further revised my empirical findings and, more
important, have attempted to elaborate more fully themes that were
relatively undeveloped or merely implicit in my previous work.

The present study of the Physikalisch-Technische Reichsanstalt in
imperial Germany seeks to provide an understanding of the workings

and results of this model scientific institute for the new German Reich and the means by which the Reich helped advance, and at times hindered, the enterprises of science and technology. In so doing, it pursues two interrelated themes: the institutionalization of science (and technology) and measurement as the "essence" of the Reichsanstalt. In examining the first of these, it analyzes not only how the Reichsanstalt was founded but also how it developed both internally and in relation to other scientific and political institutions. In addressing the second of these, it argues that the common element of the Reichsanstalt's diverse contributions to the German scientific, industrial, and political communities was the creation of measuring methods, instruments, standards, and data. It was its commitment to advancing the art and science of physical measurement that gave the institution its character and coherence. Whether Reichsanstalt scientists and technicians were helping at a given moment to form new ideas about the nature of radiation or to establish safety standards for boilers, they were engaged in the field of physical metrology – the creation or improvement of measuring methods and instruments and the actual making of measurements.

The Reichsanstalt originated and functioned in multiple contexts: at once political, scientific, and industrial. Chapter 1 discusses several of the social roles of physics and technology in imperial Germany, including instruction in physical measurement, aid to science-based industries (above all the optical and electrical), and the enhanced cohesiveness of German political life through the establishment of nation-wide physical units and standards. Moreover, it discusses the need for state support of new physics institutes and shows the revolutionary change that took place in the institutional bases of physics between 1865 and 1918. In so doing, it sets the stage for the emergence of the Reichsanstalt after 1871 and for one of the central, intractable problems that confronted the Institute after 1900: competition from academic physics institutes.

Chapter 2 pursues the theme of institutionalization by arguing that during the process of the Reichsanstalt's establishment in the 1880s a struggle occurred between two groups of individuals over the future institute's purposes and course. In particular, it shows that the industrial scientist Werner von Siemens – the central figure in the Reichsanstalt's founding – and his scientist allies were interested principally in building an institute devoted to pure scientific research, yet one that would also address the long- and short-term needs of technology. Siemens and his allies wanted something more than an institution devoted to the more immediate needs of technology; they wanted a state-supported, nonteaching physics research institution, one that would fill what they perceived to be certain shortcomings of German academic science. By contrast, Siemens's adversaries – instrumentmakers, professors of engineering, and interested professional associations of engineers and artisans – wanted an institute that would more directly pursue work that was immediately relevant to their interests. The struggle between these two groups resulted in a rift in the Institute's foundations, a rift that was to be only partially overcome and that deeply marked the Institute's subsequent development.

Chapter 3 also pursues the theme of institutionalization, now by relating how Hermann von Helmholtz, Germany's foremost scientist during the last third of the nineteenth century, built the new Institute. It shows how he used his understanding of the needs of physics, technology, and the state and his widespread contacts in German scientific life to take the Institute through its "heroic" period (1887-1894). The aging but still charismatic Helmholtz began the process of shaping an institutional structure that could flourish without him – by, among other devices, establishing a governing board of directors, developing formal rules of authority within a hierarchically ordered institute, and securing loyal organs of publication. Chapter 3 argues that there was a causal relationship between the buildup of the workspaces under Helmholtz and the pace and type of work done at the Institute during his reign. The need and desire first to construct buildings for scientific research delayed pure scientific research under Helmholtz, while the need to show its adversaries that it could contribute to German industry led the Institute to occupy available rooms at the nearby Technische Hochschule Charlottenburg in order to achieve some immediate technological results of use to industry.

When he died in 1894, Helmholtz left behind a small, partially completed institute, a nascent scientific bureaucracy. His successor, Friedrich Kohlrausch, completed the institutionalization of the Reichsanstalt and oversaw its development into a full-scale bureaucracy. Chapter 4 shows that Kohlrausch's presidency (1895-1905) brought unprecedented growth to the Reichsanstalt, making it a scientific institute sui generis and one of the first institutional embodiments of Big Science.

Chapter 4 also presents a discussion of the Institute's most dramatic scientific results: its measurements of blackbody radiation, work that served as the experimental foundations of Max Planck's new quantum theory and that helped bring the Reichsanstalt's Willy Wien as well as Planck the Nobel Prize in physics for 1911 and 1918, respectively. Building on Hans Kangro's outstanding historical study of the background to and experimental foundations of Planck's law of blackbody radiation,[20] Chapter 4 argues, against Kangro, that the practical needs of the German illumination industry – better temperature measurements and better understanding of the economy of heat and light radiation – provided the institutional justification and motivation for the Reichsanstalt's blackbody work. To be sure, individual Reichsanstalt physicists were principally concerned with finding (as Kangro and others have so well shown) a spectral-energy distribution law for blackbody radiation. But they worked in an institute that faced multiple, complex demands and that supplied the finest material resources then available for conducting physical measurements. In short, Chapter 4 argues that utilitarian as well as purely scientific interests led to the Reichsanstalt's blackbody-radiation work and that a full appreciation of the timing and setting of the origins of quantum physics requires attention to both types of interests.

At the same time, Chapter 4 discloses that scientific and industrial interests at the Reichsanstalt sometimes clashed with one another. In particular, the ever increasing demands of the electrical industry for more testing work by the Reichsanstalt and for the right to run electrical streetcars in front of it threatened to, and in part did, damage the Reichsanstalt's capacity to conduct scientific research. Conversely, industry's unwillingness to allow the Reichsanstalt full control over the implementation of legal electrical standards and the testing of electrical meters further contributed to a sometimes antagonistic relationship between the industry and the Reichsanstalt.

By 1905, when Emil Warburg succeeded Kohlrausch as the Reichsanstalt's third president, the Reichsanstalt stood at the height of its scientific fame. It had become the world's foremost center for physical measurement – for both scientific and industrial purposes. It had also become a large scientific bureaucracy, greatly in need of reform. Chapter 5 analyzes the nature and causes of a series of institutional problems facing the Reichsanstalt after 1905, along with Warburg's attempted reforms.

The outstanding and paradoxical symptom of these problems was that, despite its contributions to blackbody-radiation work between 1893 and 1901, by the start of Warburg's presidency (1905-1922) the Reichsanstalt was but little engaged in pursuing topics at the forefront of contemporary research in pure physics. The very successes of its leading scientists during the 1890s enabled many of them to leave the Reichsanstalt for academic positions at the new physics institutes of the German universities and the older, expanded institutes of the Technische Hochschulen (institutes of technology). After 1900, the expanded research facilities of German academic institutes and the greater attractions of academic life for a number of Reichsanstalt physicists meant that academic physics institutes could now more than compete with the Reichsanstalt in conducting physical research. As Chapter 5 shows, in comparison with academic institutions the Reichsanstalt offered its scientists inadequate professional rewards and opportunities. Moreover, the enormous testing burdens on the Reichsanstalt constantly forced its scientists to forgo research in pure physics and to concentrate instead on research for industry or on conducting merely routine testing work. Largely as a consequence of these two forces – the increased research capacities of academic institutes and the enormous testing burdens on the Reichsanstalt's resources – the Reichsanstalt had evolved from the original vision into an institution devoted to the one area of physics that the academic institutes could not sufficiently address and that industry increasingly required: measuring physics.

Warburg sought to overcome the Reichsanstalt's institutional inadequacies and to meet the demands of contemporary science and technology through a series of reforms, ranging from a search for increased financial support to a major internal restructuring of the Institute. But Warburg's reforms had barely any opportunity to be tested. The outbreak of World War I made a shambles of his reforming efforts, crippling the Institute simultaneously with the demise of the Reich itself.

In pursuing the two themes of institutionalization and measurement, I have attempted to combine a chronological and analytical approach to the Reichsanstalt's history during Germany's imperial era. Like all institutional historians, I have had to confront the problem of faithfully relating the institution's achievements without letting the study degenerate into a mere catalogue of personnel and events. Indeed, in the case at hand the sheer number of scientists and technicians employed at the Reichsanstalt – about 150 – and the innumerable observations, experiments, tests, and other scientific activities conducted by them between 1887 and 1918 prohibit any linear, year-by-year chronology. In order to deal with this methodological problem, I have focused my analysis of the Reichsanstalt's scientific and technical work on three years – 1893, 1903, and 1913 – which typify its work under each of its first three presidents: Helmholtz, Kohlrausch, and Warburg. By adopting this cross-sectional mode of analysis I have avoided, I hope, some of the common pitfalls of institutional history and provided a sense of both continuity and change at the Reichsanstalt. However, I have not hesitated to abandon this format when reporting important results that were achieved in other years or that stretched over a number of years, as in the case of blackbody radiation. Readers interested in learning more about the Reichsanstalt's activities in the intervening years can consult the Institute's (nearly) annual reports (*Tätigkeitsberichte*), which appeared in the *Zeitschrift für Instrumentenkunde* and which form the basis of this study's discussion of the Reichsanstalt's work.

# 1

## Physics and Empire

### *"The Flag of Science"*

On 19 January 1871, Alexandre Regnault, a young but already highly acclaimed Parisian artist, was killed by a Prussian bullet at the battle of Buzenval during the Franco-Prussian War. The battle was a decisive one for France and for the new German Reich, founded only the day before in the Hall of Mirrors at Versailles, where the Germans had recently set up their military headquarters. It sealed the fate of Paris, and soon brought an armistice and the surrender of France to Germany, leaving only the Parisians to destroy one another during the coming months. It sealed the fate, too, of the young artist's father, Henri Victor Regnault, one of France's leading scientists, a physicist known particularly for his creation of new heat-measuring methods, instruments, and data.

After his son's funeral, Regnault sought refuge in his laboratory at the nearby Sèvres porcelain works, of which he was director. Thanks to an agreement that he had reached with the Prussian crown prince, his barricaded laboratory in occupied Sèvres stood under official protection, and entrance to it was expressly prohibited. Upon entering the laboratory, however, Regnault discovered that all of his precision instruments and apparatus – thermometers, barometers, manometers, balances, and so on – had been smashed to pieces or bent beyond repair and that his manuscripts and data-recording notebooks, containing the results of a decade's worth of the painstaking empirical research for which he was so renowned, had been burned. Not a lock had been forced nor a window broken by the Prussian soldier who had entered the superbly equipped laboratory and destroyed its contents. The destruction, Regnault declared, was the work "of a true connoisseur."[1] The victimized physicist never recovered from this double catastrophe:

His son's death and his ruined laboratory robbed him of any further desire to work. Before the decade had ended, Regnault was dead, and French physics, which since 1830 had been in slow but steady decline, had reached its nineteenth-century nadir.[2]

While the perpetrators of political violence destroyed a bit of European scientific culture, German academics did their share to support and encourage their military and political leaders in the war effort. In August 1870, for example, Emil DuBois-Reymond, physiologist, scientific politico, and current dean of the University of Berlin, delivered a rabid, hate-filled, anti- French speech to the university's faculty.[3] Six months later he again took to the podium, this time to exhort his fellow members of the Prussian Akademie der Wissenschaften (Academy of Sciences) in Berlin "to uphold the flag of science" in the academies and universities while their students, sons, and brothers upheld the national flag of Germany on the battlefield.[4] DuBois-Reymond went on to declare that German and French scientists belonged to the same, unified scientific community.[5] These thin words of solidarity were spoken barely a week after the destruction in Sèvres. Nearly a half-century later, as World War I burst upon them, German scientists like Max Planck would continue to be torn between their love of fatherland and their internationalist spirit.[6]

Upon the founding of the new Reich, scientists were quick to stake out areas for scientific enterprise in the new society. At the 1871 meeting of the Gesellschaft Deutscher Naturforscher und Aerzte (Society of German Natural Scientists and Physicians), for example, Rudolf Virchow, medical scientist and politician, declared that science had two sorts of contributions to make to the Reich: It could contribute to its material welfare and to its ideal values. Though Virchow considered the former the less important of the two, he nonetheless believed that "the meaning of the natural sciences" lay "essentially in the material benefit that it produces, in the use that it creates." Medicine, trade, commerce, mining, agriculture, navigation, transportation, the military, even *"Küche und Keller"* (food and drink) – all these areas of life, Virchow said, were now being influenced by and benefited from science. As to the ideal values, Virchow argued that it was the task of the natural sciences to unite the German peoples into one spiritual and cultural unit. That was the "national meaning" of all scientific work.[7] Science and politics were to go hand and hand in the new Reich.

## The Empire Needs Physics and Technology

Physics, second perhaps only to chemistry and medicine among the
sciences, indeed had much to offer the Reich. Its offerings are best
classified into three (closely related) categories: the science and art of
physical measurement per se; new cognitive developments in physics
that led to the emergence and development of science-based industry
and that brought new technologies to civic life and the military; and instruction in laboratory physics for future professionals and industrial
managers and scientists. These categories, to use Virchow's language,
together constituted the contributions of physics to the Reich's material
welfare and ideal values.

Starting in the 1830s a new branch of physics known as measuring
or precision physics arose in Germany and elsewhere. Its main preoccupation was, naturally enough, greater precision in physical measurement. But in order to achieve that goal, measuring physicists had to
create new physical instruments, methods, and workspaces with which
to conduct their highly delicate observational and experimental work.
New or improved, and increasingly specialized, instruments for precision measurement appeared in (and beyond) the scientific laboratory:
telescopes, microscopes, barometers, thermometers, calorimeters, pyrometers, thermocouples, galvanometers, voltammeters, manometers,
photometers, polarimeters, spectroscopes, along with geodetical, nauti-
cal, and meteorological instruments, to name but a few. For the generation of physicists who matured professionally after 1840, measurement
became the shibboleth of scientific progress. And precision technology
became, to take the high philosophical road, "the foundation of every
art of the measurement of space and time."[8]

Basic to measurement is a universally agreed upon set of physical
units and standards – a subject pioneered by German physicists. In the
early 1830s Carl Friedrich Gauss, with the assistance of Wilhelm
Weber, introduced the idea of a system of absolute units of measurement – that is, he selected distance, time, and mass as the (mechanical)
basis for measuring all phenomena, including the nonmechanical phenomena of electricity and magnetism. This fundamental step in
metrology permitted, in principle, measurements of the properties of a
body or of any physical phenomenon to be reproduced anywhere
without first having a precalibrated instrument. In the 1830s, Gauss and
Weber devoted much time to measuring magnetic phenomena; in the
1840s and 1850s, Weber and Rudolph Kohlrausch extended this new

metrological method to electricity; in the 1860s and beyond, Rudolph's son, Friedrich, would continue this work at several institutions, including the Reichsanstalt.[9]

By the 1870s, as each of the scientifically advanced European nations moved toward establishing its own electrical standards and as the fledgling electrical industry began to show its economic potential, it became clear that scientists and industrialists alike needed a set of international metrological units and standards. In 1875, seventeen European nations banded together to form the Bureau international des poids et mesures (International Bureau of Weights and Measures). They reached the first international agreement on weights and measures; however, they could not reach an agreement on electrical standards. Their difficulties issued not only from the era's nationalistic fervor – in particular, from the fear that one nation might dominate metrological matters and so injure the interests of rival powers – but also from the scientific and technical uncertainties involved in attaining a trustworthy standard. No nation and no scientist could then offer any universally acceptable electrical standard.

Partial progress toward the achievement of such standards was made six years later, in 1881, when twenty-two nations sent 250 delegates – including many of the world's foremost electrical scientists – to Paris for the first International Electrical Congress. The participants agreed that the CGS system of units – that is, a system of measurement based on the centimeter, gram, and second – should constitute the basis for an absolute system, to which were tied the electrical units of ohm for resistance, volt for electromotive force, and ampere for current. However, there remained widespread disagreement on the actual standards – in particular, for resistance and luminous intensity – and on the methods and materials to be used in reproducing those standards locally. Additional international electrical congresses had to be held in Chicago (1893), St. Louis (1904), and London (1908) before an agreed upon set of standards was attained.[10] In time, international standards for mechanical, thermal, and optical phenomena were also reached. By prosecuting this standardization work, the culture of physics gradually helped unite diverse political entities, from the individual German states to the nation-states of Europe and beyond.

Before the establishment of the Reichsanstalt many academic physicists devoted much of their time and laboratory space trying to achieve one or more of the desired universal standards and building and testing precision instruments. Indeed, with the establishment of a series of new laws of optics, electromagnetism, and thermodynamics during the nineteenth century, many physicists after 1870 had come to believe that the fundamental laws of physics had already been discovered and that, therefore, their major tasks lay in the refinement of physical laws, standards, constants, and functions to the highest degree of precision possible or in the discovery of new phenomena through the extension of precision measurement into unexplored areas.[11]

By the 1870s, and especially after 1890, a new group of science-based, "high-technology" industries had begun to develop: refrigeration, steel, metal extraction, advanced railway and ship construction, automobile (particularly with its internal-combustion engine), aeronautics, precision mechanical (particularly machine tooling), chemical (organic and synthetics), optical and glass, and electrical (including telegraphy, telephony, and radio). These new science-based industries pulled Western Europe and the United States out of the (so-called) Great Depression (1873-1896) and sustained growth until 1914; they were crucial to the rapid process of industrialization in Germany during this period and to Germany's ascension to economic preponderance by 1914. Economic historians have aptly termed these events a Second Industrial Revolution.[12] Two of the industries – the optical (along with the closely associated glass) and the electrical – merit special attention, for their efforts to be technologically innovative, to generate and maintain national and international standards, and to have their products tested and certified made them particularly influential in setting the Reichsanstalt's course.

Between the late 1860s and the late 1880s the German optical and glass industries were transformed from economically limited industries based on traditional trial-and-error techniques to highly profitable ones based on newly acquired physical and chemical knowledge. The central figure in the transformation of the German optical industry was Ernst Abbe, a physicist trained under Weber at Göttingen. In 1863, two years after having received his doctorate, Abbe became a lecturer at Jena University. One of his first concerns was to build up Jena's physical facilities, above all to furnish it with a proper set of physical instruments for teaching and research.

To improve his understanding of the scientific basis of microscopes and telescopes, Abbe chose optical theory as his area of specialization. In particular, he sought to reconsider the theoretical basis of microscope objectives and, with the aid of the Jena optical firm Karl Zeiss & Co., to standardize the manufacture of microscope lenses. Abbe soon saw that the key to maximizing resolving power lay in understanding the concept of numerical aperture – a term he introduced – and its influence on the objective. In 1873 he published a revolutionary theory of microscopic imaging, the essence of which was that greater resolution required lenses of high aperture and the use of oblique light. To this theoretical work Abbe, stimulated by the English microscopist J. W. Stephenson, added a practical component: the use of homogeneous immersion lenses for microscope objectives in order to increase the numerical aperture. Abbe's theory and the lenses of Zeiss & Co., of which Abbe became a partner in 1876, brought unprecedented improvements to microscopy. By the late 1870s the name "Zeiss" signified precision instrumentation of the highest order.[13]

As successful as Abbe and Zeiss's work of the 1870s had been, they and others were (still) confronted with another major difficulty: the correction of chromatic aberration. The difficulty lay in the composition of the glass used in manufacturing lenses. The crown and flint glasses then in use failed to disperse light rays of different wavelengths in just the right way so as to correct the chromatic aberration produced by each type of glass; hence it was impossible to bring differently colored rays to the same focal point. The result was a penumbra of colored fringes around the image. To solve this problem, Abbe turned in 1879 to Otto Schott, an expert glassmaker who appreciated Abbe's theoretical advances in optics, understood the chemistry of glass, and yearned to make a breakthrough in glass manufacture. With the combined financial support of Zeiss & Co. and the Prussian government, Schott undertook large-scale glass-melting experiments. By the end of the 1880s, Schott's Jena glassworks was producing more than eighty types of glass, all of which met the theoretical specifications of Abbe's imaging theory and all of which were composed of the proper chemical constituents needed to construct the lenses – so-called apochromatics – used to eliminate chromatic aberration.[14]

The entire production of Zeiss optical instruments had now been made dependent on and was controlled by scientific theory. As a result of the theoretical discoveries of Abbe and the chemical experimentation of Schott, there was an enormous improvement in the quality of precision optical instruments and breathtaking growth in the German optical and glass industries. New types of or radically improved ap-

paratus like thermometers, microscopes, telescopes, and photographic
equipment made the German optical industry one of the most dynamic
sectors of the economy. Whereas in 1870 the German optical industry
had imported from France and England, by 1900 it exported at least 20
million marks' worth of optical goods to the world market; the total
production of optical products in 1900 ranged somewhere between 30
and 35 million marks. The young German industry set the world's pace
in both quantity and quality, in both ordinary consumer goods and advanced scientific products.[15] And the glass industry experienced a similar turnabout in its fortunes: It too changed from an importing to an ex-
porting industry. In 1897, for example, it produced glass products
worth more than 115 million marks, of which more than one-third was
exported. Between 1861 and 1895 it grew from 664 factories employing about 13,000 workers to more than 3,000 factories employing
nearly 60,000 workers.[16] Moreover, the German optical and glass industries helped supply physicists, astronomers, chemists, biologists,
medical doctors, and others with the finest types of optical and glass
products – from ordinary beakers for chemical experiments to high-
precision objectives for microscopes, telescopes, pyrometers, and so
on.

   The emergence of the electrical industry during the second half of
the nineteenth century, to turn to the second of the science-based industries so crucial to the Reichsanstalt's development, represented a
new historical phenomenon: Unlike any other industry that had
emerged before 1900, the electrical owed its existence to science. With
the discovery of several laws of electromagnetism, above all that of induction by Michael Faraday, electrical telegraphy became (in the
1840s) the new means of communication. Indeed, physicists like
Faraday, Georg Simon Ohm, Gauss, Weber, James Prescott Joule, William Thomson, and James Clerk Maxwell made major contributions to
electrical science and quickly came to see the technological possibilities of their discoveries.

   From the late 1860s to the early 1890s the combined drive of increased scientific understanding of electromagnetic phenomena and a
series of technological innovations transformed the industry from one
dominated by low-voltage technology to one dominated by high voltage. To cite some of the main developments of this period: In the late
1860s Werner von Siemens and others invented the dynamo; toward
the end of the 1870s both the telephone and electric-arc lighting were

improved and became commercially profitable; in 1879 Thomas Edison discovered in carbon a high-resistance filament that made the electric light bulb of practical use; in 1882 Edison opened the Pearl Street Station in New York City, thereby demonstrating the feasibility of an electric lighting system; in the late 1880s the electrical streetcar was introduced and brought important growth to the industry during the 1890s; and by the early 1890s the invention of polyphase, alternating-current systems allowed the transmission of high-voltage electrical power from a central source to distant regions. These and numerous other innovations rendered electricity an economically profitable source of illumination, communication, transportation, and power.[17]

From the 1870s to 1914, the German electrical industry dramatically increased its employment, production, and corporate investment. Between 1875 and 1907 the number of factories increased from 88 to more than 1,000 and the number of employees from slightly less than 1,300 to more than 91,000. In 1890 the total value of production was about 30 million marks; in 1900 about 210 million; in 1913 about 104 billion.[18] The number of electric lamps in use, to look at another index of growth, increased from 450,000 in 1891 to 32 million in 1908 and to 75 million in 1914.[19] By 1900 the German electrical industry exported more than 23 million marks' worth of goods (and imported about 6.5 million marks' worth).[20] The growth in capitalization was equally impressive: In 1895 there were 35 public firms capitalized at about 156 million marks; by 1900 there were 131 at about 891 million.[21] In short, the industry's growth was sensational. Only the industry-wide "crisis" between 1901 and 1903 temporarily slowed this economic whirlwind.

The whirlwind had its epicenter in Berlin, where Werner von Siemens's firm of Siemens & Halske and Emil Rathenau's Allgemeine Elektrizitätsgesellschaft (General Electricity Company, or AEG) were headquartered. Together Siemens the inventor and Rathenau the market strategist mustered the resources to dominate the German market. Then, in turn, Germany led the European electrical industry and, in conjunction with the United States, dominated the international market.[22]

The burgeoning electrical industry required more than scientific laws, technological innovations, and entrepreneurial risk takers. It also needed a set of electrical units and standards of measurement. From the 1870s on, scientists, industrialists, and government officials had a common, pressing need to establish trustworthy measures for a score of electrical phenomena: for example, the amount of light radiated, the luminous intensity, the energy consumption and light-energy distribution of an illuminating source; the amount of resistance in a wire; the electromotive force in a dynamo; and the electrostatic capacity and inductivity of a cable. To obtain such measures required trained measurers, who issued mostly from academic physics institutes. It soon became clear that such measuring work was at once too demanding and too routine for the resources of an academic institute; some sort of non-academic institution would be needed as well.

The new products and processes in science-based industries naturally brought changes to the general population and to ordinary life. Inexpensive yet accurate medical thermometers; diagnostic X-ray units and therapeutic radiation sources and technology; tuning forks (for musical instruction and performance); electrical and gas meters; security apparatus for hot-water heaters; saccharimeters (for measuring sugar content) – these and numerous other technological devices required the establishment of government-regulated standards, testing, and certification before the consumer might feel assured of the reliability of the products and services being offered. To the technological changes in the home and workplace should be added those in transportation, in particular the appearance in the 1890s of electrical streetcar systems in the big (and sometimes not so big) cities of Germany. Here, too, standardization and regulation were required.

Finally, new technologies in transportation, communication, and weaponry changed the military in Germany (and elsewhere). Helmut von Moltke, chief of the German general staff from 1857 to 1887 and a lifelong student of physics, made use of the German railroads and electrical telegraph system a prime feature of his military strategy.[23] Moreover, thanks to the army's need for geodetic and optical instruments, he advanced the cause of precision technology in Prussia.[24] He and his successors introduced new, more accurate, and more powerful types of weapons (e.g., breech-loading rifled firearms and machine guns) that were increasingly dependent on precision engineering, and new means of communication (in particular, electrical signal equipment, field telephones, and radio) that issued from results in electrical science and technology. By World War I, even acoustical theory had found military application when the techniques of sound ranging and sonar were developed to locate enemy positions on land and in the sea.[25] As for the German navy, its traditional needs for hydrographical, meteorological, and astronomical instrumentation had long made it highly dependent on precision technology. With the navy's aggressive expansion after 1898, a still greater call on physics and technology was made: for example, to overcome the effects of magnetic disturbances to the compasses on board its new armored ships and to develop new forms of seaborne communication for its new submarines as well as for other parts of the fleet.[26]

Physics had a third type of offering for the new Reich: instruction in laboratory measuring skills for future professionals and industrial managers and scientists. From the mid-1860s on, university students made ever heavier demands on physicists for laboratory instruction. Between the mid-1860s and 1914 there was nearly a tenfold increase in the number of university students of natural science and a sixfold increase in the number of students of medicine.[27] Those enrolling in laboratory physics courses at a steadily increasing rate consisted primarily of medical, pharmaceutical, and chemical students and, especially after 1876, future secondary-school science and mathematics teachers. (At a number of universities, the resident physicist had to introduce special laboratory sections solely for these groups of students.) Moreover, the need to accommodate students from new interdisciplinary fields – astrophysics, biophysics, geophysics, physical chemistry, and electrical engineering – made new teaching demands on academic physicists. So, too, did new requirements in virtually every field of natural science as well as in mathematics, engineering, law, forestry, and military science. The demographic pressure represented by this increased number of students became the principal reason for building new or expanding older physics institutes in Germany.[28]

Nearly all German (and many foreign) students learned their laboratory skills by doing the exercises prescribed in Friedrich Kohlrausch's *Leitfaden der Praktischen Physik,* which reached the open market in 1870 and underwent eleven editions between then and 1910. Kohlrausch's manual, soon translated into English as *An Introduction to Physical Measurements,* was the first of its kind in Germany, if not elsewhere; its appearance in the late 1860s and its quick, widespread adoption by German physicists was itself a recognition of the need to expand and reorganize elementary laboratory instruction and of the increased student demand for instruction in the art of making physical measurements. From aspiring pharmacists to aspiring electrical engineers to aspiring theoretical physicists, German students of science and technology turned to Kohlrausch's *Leitfaden* for instruction in elementary and advanced measuring work. His text dominated the market and became the bible of all measuring physicists between 1870 and 1914.[29]

## Physics Needs the State

Physics (and physics-based technology) thus contributed in a number of ways to the Reich's material welfare and ideal values: It provided physical standards and measurements for the nation-state and industry; it facilitated the emergence of science-based industry and brought new technologies to the nation; and it provided instruction in laboratory physics for an unprecedented number of future professionals and industrial managers and scientists. To make these contributions, it in turn needed something from the state: new physical workspaces.

Before the 1860s, physical workspaces, which were almost exclusively located in universities, were known as physical cabinets. At its simplest, the cabinet was a collection of instruments that, when not in use, were contained in a large, glass-enclosed cabinet or case. The more complex cabinets also included a lecture hall, workrooms, and a workshop. From the 1830s onward, German university cabinets changed markedly in nature and size: They acquired laboratories for teaching and research, and their staffs and advanced students expanded in number and adopted an ethos of research as a natural, complementary task to their traditional responsibility of teaching. The result was that by 1870 the cabinet had evolved into the institute.[30]

Between 1865 and 1914 the individual German states constructed new independent buildings to house the institutes that had slowly developed after 1830 (Table 1.1). As Figure 1.1 shows, the 1880s and 1890s were the great building decades for German physics institutes; six institutes were built during each, making a total of more than half of the twenty-three physics institutes built between 1872 and 1915. The 1870s and 1900s saw the building of four institutes each. From 1900 on, growth began to slow; by 1915, it had ended.[31]

Responsibility for educational and (for the most part) scientific affairs lay with the individual German states (*Länder*), not the Reich. Prussia was by far the largest and most powerful German state; it contained ten of the twenty-one German universities and built ten of the twenty-three German physics institutes constructed during the imperial

Table 1.1. *German physics institutes: date and cost of establishment of new independent buildings, 1870–1920*

| Institute | State | Period of construction | Year opened | Cost (marks) |
|---|---|---|---|---|
| Berlin | Prussia | 1873–78 | 1878 | 1,542,578 |
| Bonn | Prussia | 1911–13 | 1913 | 436,700 |
| Breslau | Prussia | 1898–1900/1 | 1900/1 | 363,900 |
| Erlangen | Bavaria | 1892–94 | 1894 | 211,500 |
| Freiburg | Baden | 1888–90 | 1891 | 193,162 |
| Giessen | Hesse | 1897–1900 | 1900 | 350,000 |
| Göttingen | Prussia | 1903–5 | 1905 | 429,900 |
| Greifswald | Prussia | 1889–91 | 1891 | 204,500 |
| Halle | Prussia | 1887–90 | 1890 | 296,240 |
| Heidelberg | Baden | 1912–13 | 1913 | 791,000 |
| Jena | Thuringia | 1882–84 | 1884 | 65,000 |
|  |  | 1900–2 | 1902 | 175,000 |
| Kiel | Prussia | 1899–1901 | 1901 | 237,600 |
| Königsberg | Prussia | 1884–88 | 1888 | 339,924 |
| Leipzig | Saxony | 1872–73/74 | 1873/74 | 151,200 |
|  |  | 1901–4 | 1904/5 | 1,363,000 |
| Marburg | Prussia | 1912–15 | 1915 | 419,330 |
| Münster | Prussia | 1898–1901 | 1901 | 127,400 |
| Munich | Bavaria | 1893–94 | 1894 | 430,000 |
| Rostock | Mecklenburg | 1909–10 | 1910 | 229,200 |
| Strasbourg | Alsace | 1879–83 | 1882 | 583,542 |
| Tübingen | Württemberg | 1886–88 | 1888 | 260,000 |
| Würzburg | Bavaria | 1878–79 | 1879 | 162,650 |

*Note:* These figures include the total cost of construction (i.e., foundations and the building itself; heating, water, gas, and electrical connections, etc.; director's residence [in most cases]; additional plant; and external plant, furnishings, and initial equipment).
*Source:* David Cahan, "The Institutional Revolution in German Physics, 1865–1914," *HSPS 15:2* (1985), 1–65, on 15–18, 63–65.

era. Moreover, Prussia's size and pattern of growth set the pace for the other states. By 1915, all the German states together had spent about 9.4 million marks to build and furnish twenty-three new physics institutes, with the average institute costing about 407,000 marks. The most expensive institute was Berlin's (costing more than 1.5 million marks),

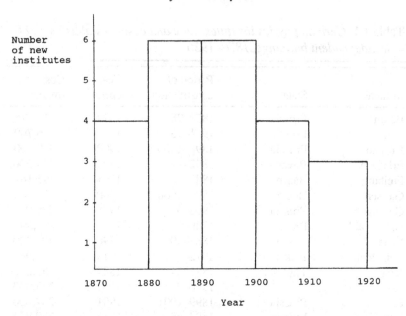

Figure 1.1. Number of new physics institutes under construction per decade, 1870-1920. Adapted from David Cahan, "The Institutional Revolution in German Physics, 1865-1914," *HSPS 15:2* (1985): 20.

built to attract Hermann von Helmholtz there in 1871 and to help symbolize Berlin as the new scientific and cultural capital of Germany.[32]

The imperial era witnessed an institutional revolution in German physics. Whereas before 1865 Germany had only a group of largely antiquated physical cabinets housed in limited quarters built for other purposes, by 1914 it had an entire set of new physics institutes. Whereas before 1865 research in physics was carried out by only a limited number of physicists, by 1914 it pervaded every institute. Whereas before 1865 only a handful of private laboratories existed, by 1914 every university had its own public laboratory, and many of these were capable of instructing a hundred or more students. Whereas before 1865 attendance at physics lectures rarely reached more than 50 students per class, by 1914 it was not uncommon to find 300 or more students listening to lectures on experimental physics. Whereas before 1865 there was never more than one assistant in physics at any one university, by 1914 many institutes had three or four or more assistants. And whereas before 1865 cabinets were either without budgets or provided only

minimal financial help, by 1914 every institute had a substantial
budget, sometimes running to tens of thousands of marks per year.

All physicists in imperial Germany recognized that a radical institutional transformation had taken place in their discipline after 1865.
Two of the most active participants in and articulate commentators on
that transformation were Emil Warburg and Friedrich Kohlrausch.
Both were among Germany's outstanding experimental physicists;
both were two of its most important pedagogues in physics; and both
became president of the Reichsanstalt. On the occasion of the formal
opening of his new institute in Freiburg in 1891, Warburg argued that
it was science's increased utility that had led to its increased state support:

> As far as physics is concerned, the so-called rise of the natural
> sciences, which characterizes modern times, lies not in the
> number and significance of discoveries or principles of re-
> search. It is due much more to the greatly increased effect that
> this science exerts on civic life and on the branches of tech-
> nology dependent on it. And, as we must add, to the counteref-
> fects that hereby result.[33]

Five years later, in his inaugural address before the Prussian Akademie
der Wissenschaften, Kohlrausch echoed Warburg's analysis:

> In the last few decades physics has grown out of its formerly
> outwardly very modest place into a recognized position of the
> first rank. It owes this, on the one hand, to its teaching, whose
> manifold power of cultivating thought and creativity no longer
> can be denied by anyone. However, if one were to analyze the
> motives which have led governments and legislatures to ap-
> prove many millions for physical institutes, then the effective
> motive would prove to be the connection of physical research
> with life, with technology. . . . Physics has come to its rich re-
> sources through its interaction with cultural development, ac-
> cording to the modern principle, that one must risk capital in an
> enterprise that promises to be of use.[34]

An opportune political situation in Germany and an entrepreneurial spirit among German physicists; a burgeoning science ripe
for technological applications; an economy in search of new productive
industries; and a booming student population, all led to an institutional

revolution in German physics between 1865 and 1914. As Germany transformed itself from an agricultural to an industrial society, it simultaneously transformed and enormously expanded its education system, an increasingly important part of which was devoted to instruction in the natural sciences. The pre-1870 physical cabinet became to the post-1870 physical institute what the cottage industry became to the factory system. In short, the institutional revolution in German physics after 1865 was due principally to the social needs of modernizing German society.

## *False Start in Charlottenburg*

No sooner had the institutional revolution in German physics begun and the Franco-Prussian War ended than a number of Prussian scientists formed a group aimed at improving the state of precision technology in Prussia. Operating with the support of the emperor and the crown prince, the group consisted of DuBois-Reymond; Helmholtz; Heinrich Bertram, the Berlin school commissioner; Wilhelm Foerster, of whom more presently; Carl Adolf Paalzow, professor of physics at the Bauakademie (School of Architecture) and at the Kriegsakademie (War College) in Berlin; and Karl Schellbach, professor of mathematics and physics at the Friedrich-Wilhelms-Gymnasium and at the Kriegsakademie in Berlin and a leading figure in reforming physics and mathematics instruction in Berlin and in building scientific and technical institutes there.[35]

Foerster was the group's dominant figure. He was well known to German scientific, military, and industrial leaders as a promoter and director of scientific and technological institutions. His responsibilities as professor of astronomy at the University of Berlin, director of the Berlin Observatory, cofounder of the Astrophysikalisches Observatorium (Astrophysical Observatory) at Potsdam, director of the Kaiserliche Normal Eichungskommission (Imperial Institute for Weights and Measures), member of the Bureau international des poids et mesures and ex officio member of the Prussian Zentraldirektorium des Vermessungswesens (Central Survey Office) gave him a special interest in precision technology.

In July 1872 the group issued the so-called Schellbach Memorandum outlining the state of precision technology in Prussia and calling for state support in building an institute or museum devoted to scientific and precision-mechanical studies, including the construction of new apparatus. The memorandum was submitted to the Kultusministerium, which forwarded it to the Akademie der Wissenschaften for review. Although the Akademie recognized the need to advance experimental studies in general and precision mechanics in particular, it held the proposed institute to be an inappropriate means for meeting the problems of such studies. Berlin's scientists, it claimed, would be "essentially better served through an increase in the endowment of their [academic] institutes." The Akademie saw the proposed institute as being "in competition" with the extant academic institutes. As for the need to improve the state of precision mechanics, it argued that if the academic institutes were provided with more money they could increase their orders for precision instruments with Berlin's instrument-makers and so help improve the latter's economic condition.[36] The Akademie, known for conservatism and hostility toward technology, categorically rejected the establishment of the proposed institute.[37]

Although the Schellbach Memorandum failed to convince the Akademie, it did win support from the military, in particular from Moltke. Moltke was well aware of both the scientific and military importance of precision instrumentation and of the current state of Prussian precision technology: He had regularly attended Foerster's public lectures on astronomy during the mid-1860s; as we have seen, he made the electric telegraph a principal feature of his military strategy; and, as the first director of the Zentraldirektorium des Vermessungswesens, he knew the importance of having the best measuring instruments to advance the army's topographical needs.[38]

In reaction to the Akademie's rejection of the Schellbach Memorandum, Foerster turned to the Prussian Landestriangulation (State Triangulation Service) for help. Like the Zentraldirektorium des Vermessungswesens, the Landestriangulation, along with its director Brigadier-General Otto von Morozowicz, was under Moltke's direct command. In the fall of 1873 Foerster explained to Morozowicz that there were scientific, political, and economic reasons for improving the state of precision technology in Germany (and in particular in Berlin). Foerster regarded the state as the sole means of improving precision technology and called upon Morozowicz as head of the Landestriangulation to form a committee aimed at doing so.[39] Morozowicz's office immediately proceeded to investigate the state of precision technology in Prussia.[40] Foerster had found a secure ally in the military.

The Morozowicz committee met in December 1873.[41] It concurred with Foerster's view that only a state-supported institute could improve precision technology. Siemens proposed that the state establish an institute similar to France's Conservatoire des arts et métiers where artisans and instrumentmakers could conduct work that they could not do elsewhere. Schellbach and several others supported Siemens's proposal. Characteristically, it was the long-term solution to the problems of precision technology that interested Siemens. On the other hand, Karl Wilhelm Gallenkamp, director of the Berlin Gewerbeschule (Trade School); Georg Neumayer, the navy's chief hydrographer and the founding director in 1876 of the Deutsche Seewarte (German Marine Observatory); and Franz Reuleaux, director of the Berlin Gewerbeakademie (Industrial School) and one of Germany's foremost engineering professors – all wanted aid for precision technology's immediate problems. Though the committee reached no specific resolution on whether to aim for a long- or short-term solution to the problem, it did call for a state-supported, nonprofit mechanical institute. But it emphasized that the state itself was not to become involved in a profit-making enterprise.[42] Moreover, at the instigation of DuBois-Reymond and Siemens the committee agreed to substitute the phrases "scientific mechanics" and "institute for scientific mechanics" for "precision mechanics" and "technical institute," respectively.[43] This was the first sign that the institute might have a scientific, as opposed to an industrial, orientation. Here, as later, it was principally Siemens who sought to ensure a scientific basis.

In January 1874 the committee promulgated its report. It called for a well-equipped mechanical institute to help advance precision mechanics, astronomy, geodesy, physics, chemistry, and other disciplines. The institute was to be a nonacademic technological research institute. Apart from a few passing references, there was no mention of pure scientific research. The committee argued that the state's duty toward scientific mechanics was similar to its duty toward general science: namely, to prepare the foundations ("the source of true productivity") of these fields and not to worry about "immediate," practical results.[44] The committee was thus proposing a new type of state obligation: the support of industrial technological research that would someday yield economic gain for German industry.

The report called for a director to lead the institute's daily business and a board to oversee its long-term welfare. The institute was to consist of two sections: the first, a museum for the display of instruments and products and the holding of lectures; the second, a workshop for the development of new products and the training of new mechanics. Upon the director's approval, the museum would be open to all young artisans and instrumentmakers who wished to work there.[45]

The report formed the basis for all future plans to establish an institute to advance precision technology in Prussia. In July 1875, the Kultusministerium recommended attaching the proposed mechanical institute to a new building being planned for the Gewerbeakademie.[46] The Prussian Landtag, however, postponed voting the necessary funds and, instead, debated the nature of higher technical education and the desirability of forming a new, unified organizational structure, a polytechnicum.[47] It decided to unite the Gewerbeakademie and the Bauakademie; in 1879 the two merged to form the Technische Hochschule Charlottenburg. As of November 1882, however, construction of the new building for the Technische Hochschule had not progressed very far, and the mechanical institute had yet to be allotted space or funds.[48]

During this seven-year period of delay (1876-1882) important changes occurred in precision technology. First, for some unknown reason the state of Prussian precision technology improved from the mid-1870s onward. The need for a mechanical institute was now called into question.[49] Second, new scientific and industrial developments in electricity occurred that had important bearings on precision technology. As we have seen, the needs of academic physics and industry alike now led governments and the world physics community to seek the establishment of a set of electrical units and standards. At the International Electrical Congress convened in Paris in 1881, the members of the German committee – led by Foerster, Helmholtz, Kohlrausch, and Gustav Wiedemann – harbored serious doubts about the ability of the congress to reach trustworthy decisions and about the feasibility of establishing an international electrical standards laboratory in Paris. Moreover, they naturally opposed the French desire for hegemony in metrology. The committee and the German government recognized that, if they were to satisfy their own scientific and industrial needs in electrical metrology and if they wanted to prevent French hegemony in this field, then they must soon provide institutional support for such work in Germany. Indeed, the Germans themselves had hopes of becoming the supreme power in electrical metrology.[50]

The delays in securing quarters for the mechanical institute, the improvement of the state of precision technology, the rise of the electrical industry and its accompanying need to establish secure foundations for electrical units and standards, and the French threat of hegemony presented new conditions and problems for establishing a German mechanical institute. By the end of 1882 the institute's promoters had begun to reconsider the purpose and feasibility of their proposed mechanical institute. They had made a false start in Charlottenburg. A new beginning was needed, one that was oriented to the national political as well as the scientific needs of the young German Reich.

# 2

# Rift in the Foundations

As the promoters of the mechanical institute reconsidered their plans to unite physics and technology for the benefit of the Reich, they slowly came to realize that the proposed union was a far more complex affair than they had originally anticipated. It took them nearly five years (1882-1887) to overcome a rift between those who wanted a pure-science-oriented research institution and those who wanted a practice-oriented technological institution. That they overcame this rift at all, that they managed to establish the Reichsanstalt, was due principally to the vision, largess, and political skill of Werner von Siemens. After the Reichsanstalt's establishment in 1887, however, it would become evident that the rift had been only temporarily overcome; the various sources of tension connected with the Reichsanstalt's fundamental purposes reemerged after a few years, transforming the Institute's development and Siemens's ideals.

## Toward a Prussian Physical-Mechanical Institute

In light of the delays and changing conditions for establishing an institute for scientific mechanics, Moltke and the head of the Prussian Kultusministerium, Gustav von Gossler, convened a committee in November 1882 to reconsider the feasibility of establishing such an institute.[1] During the committee's six-month existence, its scientists and technologists showed themselves to be divided over the question of the institute's purposes.[2] Despite the common concern over the welfare of precision technology, the committee was split by a chasm of interests.

The rift became the main issue of dispute at the committee meetings. At the initial meeting, for example, the astronomer Foerster, with the support of Helmholtz and Siemens, proposed that the new, improved circumstances of precision technology required the committee to eliminate the institute's training function for mechanics; instead, Foerster (along with Helmholtz and Siemens) wanted the institute to investigate and test materials and, in general, to give support to individual artisans and instrumentmakers. Helmholtz added that the institute should be "an administrative authority that granted funds for precision instruments, just as the Akademie der Wissenschaften did for pure scientific purposes."[3] Franz Reuleaux opposed their plan to subsidize individual artisans and instrumentmakers; instead, he wanted the institute to be a commercially oriented experimental testing station (*Versuchsstation*) attached to the Technische Hochschule Charlottenburg, where he taught engineering. To this the instrumentmaker Rudolf Fuess added a call for the creation of an instrument collection and the establishment of a fund for mechanics.[4]

Fuess's proposal was an old one; it had already been made in the original Schellbach Memorandum of 1872. In response, Siemens recommended a fundamental change in the institute's purposes. He claimed that science formed the foundations of precision mechanics, that it was "the higher point of view." He thus proposed establishing an institute for "great scientists who are forced to waste [time] in instruction and administration as well as for talented younger scientists [*Gelehrte*] . . . where, undisturbed by other duties, they can serve science alone."[5] Believing that the natural sciences formed the foundation of technology, Siemens simultaneously sought to secure that foundation and to support science itself by providing both older, accomplished and younger, promising scientists with a well-equipped institute for advanced scientific research, one free from the burdens of teaching and administration.

The response of the committee members to Siemens's proposal followed professional lines. The scientists Foerster and Helmholtz shared Siemens's views and supported his goal. Helmholtz confessed that physicists in his physics institute at the University of Berlin lacked the time and facilities to conduct systematic physical research – this despite the fact that the institute was less than four years old and far larger and far more costly than any other in Germany. He, too, wanted an institute where full-time scientific research could be done. The engineer Reuleaux, by contrast, complained that Siemens, Foerster, and Helmholtz unjustly denigrated the diverse experimental testing stations and that the latter could well form the basis of an institute for precision mechanics. Foerster and Helmholtz rejoined that the stations were practically and commercially oriented. "The new institute," Foerster added, "has higher duties to undertake that cannot be solved with the means of the present experimental testing stations." The tense debate revealed an unmistakable difference in views between the engineers and instrument-makers, on the one hand, who favored an institute devoted exclusively to the immediate needs of technology, and the scientists, on the other, who favored a combined scientific and technological institute where scientific developments would eventually be applied to the solution of technological problems. By the end of the second meeting, the committee, noting the limited funds that would be available for the proposed institute, recommended that a pure scientific function not be added to the institute and instead voted to establish an institute for scientific mechanics to be attached to the Technische Hochschule Charlottenburg. It relegated Siemens's proposal for a pure scientific institute to a subcommittee.[6] The technologists had won the first round.

Their victory was only a temporary one, however. By the time of its final report, in June 1883, the committee came to prefer the scientists' (including Siemens's) vision of the new institute to that of the technologists. The report called for the establishment of a physical-mechanical institute devoted to "the experimental advancement of exact natural scientific research and precision technology." It claimed that, in order for technology and the economy to flourish, science itself required more attention and care; that meant that the state had an "interest in maintaining and developing the deeper scientific foundations of precision technology." The committee was thus now calling for an institution with a broader purpose than that of simply advancing technology itself:

> Besides providing for precision technology, it is even more necessary that there be set up in Prussia, through the establishment of a physical laboratory, at least the nucleus of an institute for the entire field of exact natural [scientific] research. Through this institute the recognized shortcomings of the existing institutes [insufficient laboratory space, equipment, personnel, etc.] will receive timely relief before more severe disadvantages in the form of intellectual decline proceed and before the necessary strength required for the considerable intellectual efforts needed to create and maintain special institu-

tions serving only free experimental research and the higher development of precision technology begins to fail.[7]

To the argument of domestic scientific need the committee added the danger of foreign economic competition. It observed that in France, Great Britain, and Russia the state aided both pure science and technology and, hence, the economy. The implied threat was clear: France, Britain, and Russia might become more economically competitive than Germany by supporting national institutions devoted to scientific research and precision technology. The report called for an institute devoted to scientific and technological research in optics, electricity, mechanics, metallurgy, and other fields, as well as to general testing and certification of all types of physical instruments, materials, and products.[8]

The report also painstakingly differentiated the institute's research and testing responsibilities from those of academic and industrial institutions. First, regarding its scientific (and related technological) research, the institute was to conduct only work that could not be done by academic scientists in their small, teaching-oriented laboratories or by commerical engineers and instrumentmakers who lacked the economic wherewithal for such work. To put the matter positively, the institute was to conduct costly, long-term observational and experimental measuring work that required teams of scientists and technicians working on a full-time basis. Second, it was to conduct only those scientific and technical investigations that private industry could not perform. Finally, it was to purchase its equipment from private industry and to construct only those instruments and process only those materials that private industry proved incapable of providing. The physical-mechanical institute, in short, was to support and complement, not compete with, academic science and private industry.[9] It was to do what the physics institutes at the universities and polytechnics could not do: full-time scientific and technological research and testing.

Accompanying this report were three statements in favor of the proposed institute. The second and third of these, by Helmholtz and Siemens, respectively, merit extended attention, because they express the views and hopes of the dominant figures in the Reichsanstalt's founding and early years. By contrast, the first statement, by Lieutenant-Colonel Schreiber, chief of the Trigonometric Section of the Prussian Landesaufnahme (Ordnance Survey), merely concerned the latter's dependence on progress in the field of precision technology;

it showed the military's unflagging interest in a precision-mechanical institute.

Helmholtz's statement addressed the institute's scientific tasks, both those pertaining directly to pure science and those contributing indirectly to the advancement of precision technology. In arguing for a state-supported pure scientific institute, Helmholtz claimed "that in the end every serious scientific work yields its practical application where one had previously least expected it." He cited navigation, time calculation, practical optics, and geodetic measurements as examples of practical economic concerns that were dependent on pure science. He enumerated a list of unsolved or only partially solved problems in pure physics that the institute should undertake, including the exact determination of gravitational intensity, the absolute measurement of gravitation and the velocity of light, the establishment of electrical units and standards, and gas-thermodynamic measurements. These problems belonged to pure and applied physics: Geodesy, electrical technology, and heat technology as well as pure physics sought solutions to them. Finally, as if concerned that his remarks on the economic importance of pure physics might not be entirely convincing, Helmholtz added a touch of nationalistic rhetoric: "It is obviously unworthy of a nation, which through its power and intelligence occupies and defends one of the foremost positions among civilized peoples, to relinquish to other nations or to the casual dilettantism of private, well-to-do individuals the care for the creation of fundamental knowledge." As a matter of honor if not of commerce, Germany must provide for the research needs of its scientists if it was not to abandon them and the entire nation to foreign dependency or serendipity.[10]

The third statement, that of Siemens, deserves equally close attention, for from the spring of 1883 onward Siemens's planning and strivings to found a physical-mechanical institute were to prove decisive in the realization of the Reichsanstalt. Siemens argued that if Germany wanted to remain economically competitive with France, England, and America, and sought honor and prestige from scientific research, then it would have to underwrite the establishment of a natural scientific research institute. "The state," he claimed, "has applied its power with undoubted success to the advancement of scientific instruction . . . . Scientific research, however, is not a professional activity within the state structure; it is only a tolerated, private activity of scientists beside their profession – teaching." He regarded the state academies of sciences and the experimental testing stations as inadequate for dealing with the problems facing contemporary German science, industry, and the nation. The university laboratories, he correctly noted, were basically set up for instruction and were "usually . . . inappropriate for finer and more extensive scientific investigations." The lack of proper support for research, he added, was particularly evident in experimental physics. He wanted the state to do for physics what the chemical industry was doing for chemistry. He feared that, without the proposed institute, the efforts of the French, English, and Americans to advance scientific instruction and research would prove economically crippling for Germany. He wrote:

> Recently England, France, and America, those countries which are our most dangerous enemies in the struggle for survival, have recognized the great meaning of scientific superiority for material interests and have zealously striven to improve natural scientific education through pedagogical improvements and to create institutions that promote scientific progress.

He thus sought an organization that might fulfill a double purpose: namely, "to advance natural scientific research in general and to support industry through the solution of the scientific-technical problems that essentially condition its development." Pure physics and technical applications were to go "hand in hand" at this state-supported scientific research institute.

Like Helmholtz and the committee as a whole, Siemens specified the types of physical problems that the envisaged institute should undertake. He proposed extensive experimental investigations on the chemical composition, the physical properties, and the manufacturing methods of glass. He recommended a parallel set of investigations for metals and their alloys. And he pointed to the scientific and financial imperatives for establishing electrical units and standards: "Already these burning questions of electrical measuring units," he warned, "make it an absolute necessity to create as quickly as possible an organization for scientific investigation with the appropriate facilities and equipment."[11] In Siemens's view only a state-supported physical laboratory, not an academic or industrial one, could afford to conduct fundamental, systematic, and long-term research. He saw its intellectual and economic benefits to Germany as far outweighing its costs.

## Werner von Siemens: An Industrial Scientist

Siemens's views of the relations between science, technology, the economy, and the state and his plan for an institute in which both pure physics and technology would flourish to their mutual benefit were the crystallization of a lifetime's experience in the scientific, technological, and industrial worlds. From 1835 to 1838 he had attended the Berlin Artillerie- und Ingenieurschule (Artillery and Engineering School), dedicating his free time to studying his *"Lieblingswissenschaften,"* mathematics, physics, and chemistry. "The love for these sciences," he later wrote, "has remained with me throughout my life and has formed the foundation of my later success."[12] He believed his technological achievements and success in business were due to his scientific studies.

Siemens built a large part of his business on contracts with the Prussian military and diverse European governments. He owed his first successes to electrochemical technology: While still in the Prussian army he developed explosives that helped win military victory for Prussia and promotion for himself.[13] In 1847 he and the artisan J. G. Halske established a firm for developing and erecting electric telegraph systems. By the 1860s Siemens's multinational enterprises had erected telegraph lines from London to Calcutta. With his brother Sir William, Werner was also among the main contractors and participants in the laying of the Atlantic cable in the 1860s. His electrical wires, as the physicist August Kundt dramatically put it, "spanned continents and seas."[14] They allowed peoples and nations to communicate with one another, and they brought Siemens a thriving business.

It was the invention and commercial development of the generator from 1867 onward that transformed his thriving business into an industrial empire. Exploiting Faraday's discovery of electromagnetic induction, he achieved a new power source that helped usher in new areas of industry (e.g., electric lighting and electric traction systems) and brought prodigious industrial growth to his firm and to Germany. During the 1880s, while he fought for the Reichsanstalt, his firm grew faster than ever before, thanks largely to its new business in high-voltage technology.[15] By the time of his retirement in 1889, the firm of Siemens and Halske dominated the low-voltage branch of the German electrical industry and, with Emil Rathenau's Allgemeine Elektrizitätsgesellschaft, dominated the high-voltage branch as well.[16] By the 1880s, Siemens had become the Bismarck of industry (Figures 2.1 and 2.2).

Figure 2.1. Sketch of Werner von Siemens in 1887, drawn by Ismael Gentz. Courtesy of Siemens Museum, Munich.

Two other dimensions of Siemens's career – his attitude toward and accomplishments in pure physics, and his patriotism – shaped his motivations for establishing the Reichsanstalt. Siemens stated repeatedly that science, not technology, was his highest cultural value. He wrote "My love always belonged to science as such, while my work and accomplishments lay mostly in the field of technology."[17] And with love went understanding, and support, of the ideology of pure science and of pure scientists:

> Scientific research should not be the means to the end. The German scientist [*Gelehrte*] has from time immemorial distinguished himself by conducting science for its own sake, for the satisfaction of his thirst for knowledge. In this sense I have al-

Figure 2.2. Werner von Siemens: the successful industrialist late in life. Reprinted from *Personal Recollections of Werner Siemens,* trans. W. C. Coupland (London, 1893), frontispiece.

ways counted myself more among scientists than technologists, since the expectation of application has not directed me, or certainly only in special cases, in the choice of my scientific work. Entrance into the intimate circle of the greatest scientists greatly inspired me and stimulated me to scientific activity.[18]

His most valued friends were not businessmen, as one might expect, but rather scientists, especially physicists: Wilhelm Beetz, Ernst Brücke, Rudolph Clausius, DuBois-Reymond, Helmholtz, Gustav Kirchhoff, Karl Hermann Knoblauch, Karl Ludwig, and Gustav Wiedemann. These were his closest friends, the regular visitors to his home. In 1845 he joined with several of these men to found the Berlin (later Deutsche) Physikalische Gesellschaft (Physical Society). The award of an honorary doctorate from the University of Berlin in 1860, his selection in 1874 as the first technologist invited to join the exclusive Prussian Akademie der Wissenschaften, and the marriage in 1884 of his eldest son Arnold to Helmholtz's daughter Ellen were great sources of satisfaction and honor for him and established further, if largely symbolic, ties to the German scientific community.[19]

His ties were substantative ones, too. He was, in fact, a highly accomplished experimental physicist. Among his outstanding contributions to pure physics were his work toward establishing absolute electrical units and standards of measurement (including those for
resistance, current strength, electromotive force, and luminosity); his
studies on electrostatic induction, on the optimal conditions for the
flow of electrical currents and the performance of magnets, on the conductivity of metals and the effect of temperature on conductivity; his
codiscovery of the dynamo; and his development of the galvanometer
and other scientific instruments.[20] These accomplishments were of a
piece with his practical work. The physicist Gustav Mie rightly judged
that "Siemens would not have been able to become the creator of a new,
vast branch of technology, if he simultaneously had not been an excellent physicist." [21] Kundt declared Siemens to have been "one of the
best and noblest of those who took part in our communal work." [22] The
German physics community regarded Siemens as one of their own.[23]

Siemens always felt and expressed strong patriotic, even nationalistic, sentiments and supported a united Germany. Patriotism is one of
the leitmotifs of his autobiography: "From earliest youth on," he wrote,
"the disunion and powerlessness of the German nation pained me." Siemens sought a united and powerful German nation through Prussia. He
regarded himself, his family, and his class – the industrial capitalists –
as the new leaders of a united Germany. He believed that neither the
workers and social democrats, on the one hand, nor the large land-
owners, the nobility, the military, and government officials, on the
other, could preserve the German nation and lead it to greatness.
Rather, he thought that "in the dawning natural scientific age the large
technical business firms" will secure for each country called upon "the
appropriate position in the great contest of the civilized world."[24] With
the aid of their scientific and technological compatriots, Siemens
thought, the leaders of industrial capitalism were alone capable of con-
serving and developing a modern, industrialized Germany.

Siemens spent only a small portion of his career in formal politics.
From 1862 to 1866, as a member of the liberal Progressive Party, he
represented the district of Solingen-Remscheid in the Prussian Landtag.[25] Though he at first supported the revolution of 1848, he quickly
joined the reaction; the prospect of a workers' state was distasteful to
him, no matter how much he despised the German ruling class. Until
the 1860s he opposed Bismarck, but after the defeat of Austria in 1866

he changed his views and became a supporter, at least of Bismarck's foreign policy. Like most German liberals, he marveled at Bismarck's stunning military successes between 1864 and 1871 and put his longing for a strong, united Germany ahead of his liberalism.

## A Gift for the Reich

Establishing the Reichsanstalt capped Siemens's career and perpetuated his scientific and industrial interests. In 1883 he took the matter of founding a physical-mechanical institute into his own hands. In a letter to von Gossler, he reminded the minister of culture of their recent efforts to establish "a state institute for natural scientific experimental investigations" free from all teaching obligations. He noted that the institute, to be housed in rooms of the Technische Hochschule Charlottenburg, would be so organized that, in addition to precision technological research, "also general scientific investigations could be performed." But he was worried that the institute, particularly in the future, would lack the necessary space and financial wherewithal "for the pure Scientific Section." He therefore offered to give Prussia 12,000 square meters of land as a site for a pure scientific institute. Siemens thought that the land, which was located in the neighborhood of the Technische Hochschule, would be an appropriate place for establishing a "state institute for experimental physics" that would contribute to the "essential advancement of science itself and, thereby, also to the technology closely bound to it." It was to be a place to undertake "exact scientific experiments."[26] The technological research was to be done in rooms at the Technische Hochschule; the scientific in the new institute nearby.

In return for his gift of land, Siemens expected that "the state [would] build, equip, and continually maintain at public cost an institute . . . for experimental investigations with the necessary laboratories and a residence for the director." His gift, he presumed, would be "tax free." His "sole motive" in offering the gift, he said, was to help realize the establishment of a scientific workplace, "a goal to which I ascribe great importance." The source of his motive, he claimed, was his lifelong success at using scientific and technical results for business purposes.[27] His letter to von Gossler was at once a statement of his support for pure physics, a pronouncement of his views on the relations of pure science and technology, a demonstration of his patriotic senti-

ments, and a concrete business offer. With it he gave momentum and
breadth to an inert and narrowly conceived project, opening up the
possibility of constructing a nonteaching, pure scientific research institute. The enlarged purpose and facilities of the physical-mechanical institute were a far cry from the original plan of 1872 for an institute for
aiding precision mechanics.

The government responded quickly and positively. Von Gossler
acknowledged Siemens's offer as another example of his "patriotic and
unselfish disposition to contribute to the state for the well-being of the
fatherland and in the interest of science." He warned Siemens,
however, that two difficult barriers had to be overcome before his plan
could be realized: the agreement of the Prussian minister of finance and
that of the Prussian Landtag.[28]

The death of Siemens's brother William in November 1883 provided the occasion and the means to speed up the slow-moving,
bureaucratic process. In January 1884 Siemens renewed and enlarged
his proffered gift to Prussia. To accelerate the project, he offered his inheritance from his brother's estate for the construction of the scientific
building, which he and his coplanner Helmholtz intended to be the institute's principal facility for scientific work, and he asked von Gossler
to allow Helmholtz and the government architect Paul Spieker to design the institute.[29] Von Gossler promised to expedite the matter as
best he could, but observed that negotiations with the minister of finance had still to take place.[30] In the meantime, Siemens told Spieker
that he expected an inheritance of some 250,000 marks and that he was
prepared, if necessary, to spend up to 300,000 marks for the construction of the scientific building.[31]

In March Siemens again changed his offer: He now transferred it
from Prussia to the Reich. His letter to the Reich minister of the interior, Heinrich von Boetticher, sums up his motives and his hopes:

> For today's crucial natural scientific experiments there must be
> well-planned rooms protected from external disturbances, ex-
> cellent and costly instruments, and the complete devotion of
> the scientists, equipped with all [available] knowledge for the
> solution of the problem undertaken. The teachers and laborato-
> ries of the universities and pedagogical schools are not appro-
> priate for the purpose; neither are the professors employed at
> them. The more active these latter are and the more they have
> proved themselves to be pathbreaking researchers, the more

they are overburdened by their teaching obligations and the extra duties bound up with them. Besides lacking spare time for deepening the intellectual foundations of their research work, they also lack the appropriate locale and means for making the necessary instruments and equipment. . . .

From the planned natural scientific workplace, both material and ideal advantages of great importance would accrue to the Reich. In the current vigorous struggles of peoples, the country that opens new paths and newly creates or enlivens important branches of industry has a decisive superiority. . . . The discovery of a new scientific fact normally becomes technologically applicable only after its complete systematic processing, i.e., usually after a long time. Thus scientific progress may not be made dependent on material interests. Modern culture rests on the rule of man over natural forces, and each newly discovered natural law expands this rule and thereby the highest excellence of our race! . . .The patronage of natural scientific research is thus in eminent degree an advancement of the country's material interests! . . . It thus appears as a task of the Reich, and not of the individual [state] governments, to supply the necessary equipment for bringing this scientific production to the highest level and for maintaining it. . . .

Finally, I note that in offering 1/2 million marks in real estate or capital for the founding of the planned institute my purpose is only to perform a service for my fatherland and to prove my love for science, to which I exclusively thank my prosperity in life.[32]

Von Gossler recorded that Siemens turned to the Reich because of the institute's "national meaning" and in the hope that he could get more money. He enthusiastically supported Siemens's plan to develop a "center for the experimental research of all German physicists . . . whose investigations would [also] advance the economic interests of the entire Reich." Reich officials, for their part, responded cautiously. The minister of finance, though sympathetic to the project, expressed doubts about its financial feasibility. The minister of the interior, under whose ministry the Reichsanstalt would fall, was partial to Siemens's offer and asked the minister of the treasury for his cooperation in persuading Bismarck to accept it.[33]

In July the ministers of the interior and of the treasury informed

Bismarck of the history behind and nature of Siemens's offer, its purposes, its scientific and economic importance for the Reich, and their
own and von Gossler's support for the project. They noted, too, that the
minister of finance was not opposed to such plans. And they alerted
Bismarck to the two main problems entailed in accepting Siemens's
gift: First, the government needed to win the approval of the Bundesrat
and the Reichstag. Second, since the Reich had no legal competence in
matters of scientific research (except in the representation of all the
German states vis-à-vis foreign countries), the issue of the legality of
the proposed plan for a physical-mechanical institute required resolution; Bismarck, the ministers suggested, should therefore consider
whether the institute should be a Prussian or an imperial one or some
mixture of the two. Bismarck unhesitatingly replied that the project
should be undertaken "*solely for the Reich*" and adduced as legal warrant the Reich's authority to operate an "experimental testing station."
His choice of language suggests that he may not have fully understood
that Siemens proposed to establish a *scientific research* institute, a task
that traditionally and legally belonged to the individual German states.
Alternatively, and more likely, it suggests that he favored promoting
Reich scientific institutions no matter what their denomination or legality might be. The Reichsanstalt was doubtless a welcome, if unplanned,
addition to the efforts of Bismarck and his senior staff to make Berlin
the center of Germany's political, economic, cultural, and academic
life. Boetticher informed Siemens of Bismarck's acceptance of Siemens's offer and of the chancellor's request that Siemens, together with
Reich officials, specify the institute's purposes more precisely and prepare an organizational and financial plan aimed at convincing the
legislature of the institute's necessity and appropriateness.[34] Siemens
had easily won the backing of the top Reich officials. Others were to
show a different disposition.

## Allies and Adversaries

Following Bismarck's request, Siemens and several Reich officials
formed an advisory committee.[35] Between October 1884 and December 1885 the committee held five full and thirteen subcommittee sessions. Its members discussed in extensive detail the proposed institute's
objectives and finances; the responsibilities, qualifications, and re-
munerations of its employees and guests; the function, location, size,

layout, and equipping of the various buildings; and the nature and responsibilities of the board of directors.[36] Nonetheless, this was not an institute planned by a committee, for Siemens and, to a lesser degree, Helmholtz dominated these committee meetings.

An illustration of Siemens's leadership in planning the institute can be seen from a subcommittee meeting held in November 1884. Siemens here called for the institute to be bifurcated into a Scientific (*Physikalische*) Section and a Technical (*Technische*) Section, with both sections under the control of a single president. The latter, in cooperation with the board of directors, was to decide upon the scientific and technological problems to be undertaken; to hire the personnel; to set the budget; and, in general, to administer the institute. Siemens anticipated these prerogatives by listing the general types of problems to be investigated by the Scientific Section, including the search for "fundamental scientific findings," the "experimental determination of unresolved questions," and "experimental research work toward the furthering of our natural knowledge." His characterization of the section's work was vague yet it made clear that he intended the section to do pure scientific research. For the Technical Section he recommended a division into five highly specific subsections: materials testing and the determination of physical constants; precision mechanics; optics (including work on glass and its constants, etc.); thermometry; and electrical technology (including the establishment and testing of electrical units and standards). Siemens recognized that during its first years the institute could conduct only limited operations. Nevertheless, he thought it important to set out its program in full in order to instruct the German Parliament about its function and scope and to convince it of its importance. He accurately predicted it would require a decade before the institute could operate at full capacity.[37]

Helmholtz, for his part, was, in effect, Siemens's right-hand man, a point best illustrated by the fact that Helmholtz acted as Siemens's plenipotentiary when the latter could not attend committee meetings. At Siemens's request, Helmholtz assumed primary responsibility for the Scientific Section. Moreover, he was particularly influential in selecting the board members and in determining the pay levels for Reichsanstalt employees. Foerster, the other most active member at these meetings, assumed primary responsibility for the Technical Section, again at Siemens's request. By contrast, the technologists at these meetings, Bamberg and Fuess, played minor roles, as did government

Figure 2.3. The proposed Reichsanstalt, drawn sometime between 1884 and 1887. Courtesy of National Bureau of Standards, Gaithersburg, Maryland.

officials. It was Siemens's intent during these meetings to determine how best to use his and the Reich's assets (Figure 2.3).

Parallel to these committee meetings, Siemens initiated three sets of events during 1884-1885 that were crucial for the Reichsanstalt's birth. First, to avoid further delays, he authorized construction of the scientific building. He did this at his own expense, knowing that Parliament might refuse to approve the Reichsanstalt. Second, he sought to ensure that Helmholtz would be the institute's first president. Indeed, the institute was in part designed for and by Helmholtz, who was the only candidate for the first presidency. DuBois-Reymond, who knew both Siemens and Helmholtz intimately, reported that he and others were aware that Siemens had "always regretted" that Helmholtz had to devote so much time to teaching instead of to research. "And we saw," DuBois-Reymond continued, "that he [Siemens] had intended the position of president of that institution for Helmholtz, a position that would free him from every nonscientific activity."[38] Helmholtz, however, refused to accept the presidency until the minister of the treasury accepted his enormously high salary request: circa 24,000 marks per year. The issue, as we shall see in Chapter 3, required three years before it was resolved. Third, Siemens initiated a campaign to gain parliamentary support. In September 1884 he wrote the mayor of Cologne:

> I have strived to set up an imperial institute dedicated exclusively to natural scientific work, not to teaching. I shall make considerable personal sacrifice for it. Bismarck, however, currently still holds science for a type of sport without practical meaning; and this view is still rather widespread. Through the lecture [in Cologne] I intend to introduce an effective journalistic propaganda [campaign] in order to spread knowledge of the great social meaning of natural scientific research.[39]

However, as knowledge of the proposed institute became public, Siemens began to worry about its public image. In particular, he was concerned that some individuals might think that the institute was meant only for Berlin scientists and artisans. He wrote Helmholtz:

> At previous discussions we had already definitely agreed that it is absolutely necessary to give the institute the specific Berlin character, [that is,] that through its organization it generally appear as a place of work open to all outstanding German scientists. If this does not happen, then the rumor which is already widely spread – scientific development should be monopolized for a Berlin institute and its workers – will seriously spread everywhere, especially to the Bundesrat and Reichstag. That must be avoided at all costs.[40]

Siemens had good reasons to be worried. For in late 1885 his campaign met its first real adversaries: the German legislature and parts of the German technological community.

During 1886 Bismarck, the legislature, and technological groups all presented reasons for slowing down the development of the Reichsanstalt or eliminating its Scientific Section. Siemens's ostensible ally, Bismarck, now refused to seek financial support for the Reichsanstalt. The Iron Chancellor was apparently concerned that the legislature might block his plan to have a liquor monopoly (in which he held a financial interest) passed if at the same time he promoted the Reichsanstalt. Although Bismarck was not opposed to the Reichsanstalt as such, he did have higher financial and political priorities than its legislative approval. His unhurried attitude naturally irritated Siemens. "It is of course in our interest," Siemens wrote to a relative, "to attend to the pri-

orities and thus not bring our project, which has already been under ne-
gotiation for many years, to the public too soon. Unfortunately, through
Bismarck's obstinancy another construction year is lost." [41]

To spur the recalcitrant chancellor into pushing for legislative approval, Siemens intensified his lobbying campaign. He turned to the
Elektrotechnischer Verein zu Berlin (Berlin Electrical Engineers Association), an organization that represented the new subdiscipline of electrical engineering and with which he had close ties: He had helped
found it in 1879 and had been its first president. Siemens informed
Richard Rühlmann, the editor of the Verein's journal, that Bismarck
had postponed asking for legislative approval of the Reichsanstalt. He
feared that further postponement might lead to fading interest in the
Reichsanstalt and to financial loss to the German electrical industry. He
pointed out that contracts involving millions of marks might be
threatened because they did not rest on legal electrical standards:
"There is a danger at hand that this lawlessness in regard to electrical
standards will lead to great legal uncertainty and many court cases."
Siemens asked Rühlmann to write an article for the Verein's journal
calling on Bismarck to accelerate approval of the Reichsanstalt's fi-
nancing and construction. He added that most countries were now trying to establish electrical units and standards and that "it will be dis-
graceful for German science and technology, which had led in this
field, if it should now remain behind its foreign sisters." [42]

Siemens expressed similar concerns to Friedrich Hammacher, a
director of several large mining enterprises and a leader of the National
Liberals in the Reichstag. He disclosed that the minister of the interior
had confidentially told him that he would like to see the issue raised in
the Reichstag, so that he could force Bismarck to support the Reichsanstalt. Siemens asked Hammacher for his help: "If the bill [for the
Reichsanstalt] comes from a member of a party from the middle and is
supported by other parties, that would give the imperial chancellor a
reason to designate the issue as pressing, and simultaneously serve as a
guarantee that the Reichstag is well disposed to it. Now my request,
therefore, is that you will agree to introduce the bill." Siemens added
that Rudolf Virchow, Ernst Haeckel, "and other scientific people"
would all certainly support the Reichsanstalt. A day later he wrote
again to report that he intended to bring the issue to the newspapers to
motivate parliamentary discussion and to gain Bismarck's support in fi-
nancing the Reichsanstalt. [43]

Word of the proposed institute now also reached the German technological community. The Verein zur Beförderung des Gewerbefleisses (Association for the Advancement of Industry) held a discussion with Siemens about the Reichsanstalt and proved to be a friendly supporter of it.[44] In contrast, the Verein Deutscher Ingenieure (Association of German Engineers), the single most powerful and influential agent of German engineering interests, was highly skeptical about Siemens's undertaking. In October 1886 the Verein's leadership wrote to the Bundesrat about their objections to the Reichsanstalt, which issued from the fact that the Verein represented primarily mechanical engineers. The Verein's leadership conceded that the general idea and function of the Reichsanstalt were useful; at the same time, they disregarded the Scientific Section, which they recognized as Siemens's special project. Their principal concern was with the Technical Section, which they saw as too narrow and misdirected in its purposes. They considered the Technical Section's emphasis on precision mechanics to be largely irrelevant to their concerns in applied mechanics. They wanted the Reichsanstalt to conduct research in non-precision technology, for example in fluid and solid mechanics. Such investigations, they claimed, had not been included in the programs of either section of the Reichsanstalt, because such research did not require precision apparatus. "Nonetheless, such investigations are also of great importance for scientific and practical technology." The Verein's leaders asked the Bundesrat to see that the Reichsanstalt's Technical Section became more oriented toward non-precision engineering; to increase the number of engineers on the Reichsanstalt's board beyond the current two; and to appoint a suitable director for the Technical Section. To make sure that the Bundesrat felt the full force of their concerns, they published their letter in the Verein's trade journal.[45] In sum, German (mechanical) engineers registered their complaints and wants with the government and legislature and sought to wrest partial control of the emerging Reichsanstalt from Siemens and company.

So, too, did the Deutsche Gesellschaft für Mechanik und Optik (German Society for Precision Mechanics and Optics), an organization devoted to precision mechanics and optics and one that stood to gain much from the Reichsanstalt. The Gesellschaft's leaders, among whom were none other than the technologists Bamberg and Fuess of Siemens's planning committee, told Bismarck that the Reichsanstalt's Scientific Section would have "only an indirect influence" on their industry; like the Verein Deutscher Ingenieure, they peremptorily discounted its value for their interests. Their real concern was not with the Scientific Section but with the Technical Section, which they believed would become devoted solely to testing and certification work, "while precision technological researches, with which the earlier planned mechanical institute [was] supposed to be concerned, will now move into the background." They wanted Bismarck to ensure the Technical Section a "just" position alongside the Scientific Section and its "necessary connection with the practice of mechanics and the auxiliary trades that stand near to it." They urged that the Technical Section publish in a journal read by practicing mechanics, by which they meant their *Zeitschrift für Instrumentenkunde*. Finally, they called for the director of the Technical Section to be someone who "has grown together with our industry, who knows its goals and our aspirations . . . , [who] offers an open mind and a warm heart. Only such a man can maintain the necessary connection with the different branches of precision mechanics and optics and himself select those assistants who can offer the solutions to numerous difficulties [and] be equal to the practical and scientific assignments." They wanted to have their man in the number two post at the Reichsanstalt, and they added that they would be able to find the appropriate individual – they had in mind Leopold Loewenherz, the eventual director – if Bismarck would only kindly allow them the opportunity.[46] Bamberg and Fuess, having witnessed firsthand the power of Siemens and his allies on the committee to control the direction of the future institute, thus tried to compensate for their powerlessness at the committee meetings through a personal appeal to Bismarck. What they had failed to achieve as individuals in Siemens's committee, they sought to gain directly from the chancellor through the vehicle of their political pressure group, the Deutsche Gesellschaft für Mechanik und Optik.

At the end of 1886 a governmental committee informed Bismarck that the technologists' complaints about the Technical Section's alleged shortcomings were unfounded and that some of their demands had already been met. The committee noted that the Reichsanstalt lacked the space and the money to expand the scope of its program. It pointed out, furthermore, that those technological problems that the Verein Deutscher Ingenieure had enumerated but that the Reichsanstalt did not plan on undertaking should be addressed by the experimental testing stations operated by the individual states. However, the committee did concede that engineering interests should receive increased representation on the board of directors.[47]

As the technologists protested, Parliament acted. In June 1886, the Bundesrat consented to a first payment of 300,000 marks for the Reichsanstalt. At the Reichstag Budget Commission hearings in November, however, Siemens's efforts suffered a sharp setback, in part owing to the protests of the technologists. The Liberals opposed the Reichsanstalt as an unwarranted governmental incursion into the economic realm; the Conservatives as "extravagant pampering of industry"; and the Center Party as an infringement on the exclusive right of the states to conduct scientific research.[48] In addition, Budget Commission members objected to the amount of money requested by the government; they questioned the appropriateness of a *Reichsanstalt* as opposed to a *Landesanstalt;* and they doubted that the Scientific Section was truly necessary for the Technical Section to meet its goals. Only the Social Democrats were favorably disposed toward the Reichsanstalt. The arguments and interests against Siemens's version of the Reichsanstalt prevailed: The Budget Commission approved the 160,000 marks requested for the Technical Section but voted to eliminate altogether the Scientific Section, including the president's salary.[49] In two days of sessions the commission nullified three years of work by Siemens and his allies. The technologists had won this battle.

The calamitous Budget Commission recommendations led Siemens, Foerster, and the crown prince – who was a close friend of Siemens's and who would ennoble him in 1888[50] – to redouble their lobbying efforts. In December, Siemens and a group of electrical technologists published an appeal urging members of the Elektrotechnischer Verein to lobby for passage of the full government funding request, including that for the Scientific Section. The crown prince instructed Foerster to prepare a pamphlet justifying the need for the Scientific Section and allaying fears of interference by the Reichsanstalt with research at the German universities and Technische Hochschulen.[51] The prince also personally lobbied opponents of the Reichsanstalt's Scientific Section to reconsider their views. According to Foerster, it was "undoubtedly" the crown prince's "earnestness . . . for the good cause" that led to a change of minds among Budget Commission and Reichstag members.[52]

In January 1887 the government took its case before the full (yet poorly attended) Reichstag. It sought above all to get the legislators to approve the Scientific Section. Several legislators expressed uncertainty over the section's purpose and functions. However, Representatives Karl Schrader (Progressive) and Fritz Kalle (National Liberal) argued for it, pointing to its expected contributions both to the Technical Section and to other Reich institutes. Bruno Geiser (Social Democrat), by contrast, considered the Scientific Section's assignment to be the advancement of pure science for its own sake, though he too hoped for and expected technological and, hence, economic benefits as well. Rudolf Virchow (Progressive) favored supporting a Scientific Section both for its contributions to pure science and for its technological and economic implications. Representative Witte (National Liberal), who reported on the bill to the Reichstag, emphasized that the Reichsanstalt would not conduct all types of scientific work, but only "scientific work that is eminently connected to practice and, indeed, to commercial fields that are of the most decisive importance to German industry and trade and [Germany's] position in the world market." The Reichsanstalt, as von Gossler told the Reichstag, was to be part and parcel of "a German-national character." "We are forming in the German Reich . . . a unitary circle of interests in the field of science, trade, industry, and in many other fields as well."[53] It was the potential, if long-term, economic benefits along with the threat of economic competition from abroad that provided the decisive argument for legislative (and governmental) support for the Reichsanstalt's Scientific Section. Nationalistic appeals helped, too. Neither the government nor the legislature ever considered supporting a Scientific Section dedicated exclusively to pure science.

The most outspoken opponent of the Reichsanstalt, and especially its Scientific Section, was Georg von Hertling, then a professor of philosophy at the University of Munich and the Center Party's leader of the southern and conservative wing. Hertling alluded to "a rather marked antipathy against the planned Reichsanstalt" and claimed that his (unnamed) scientific friends viewed the Scientific Section as an infringement upon the rights of the universities and Technische Hochschulen. The philosophy professor declared that he was no opponent of science as such. However, he and his Bavarian constituents feared that the Reichsanstalt would become "a planned central institute that would stand over all other existing scientific institutes."[54] Hertling and other adversaries of the Reichsanstalt from Bavaria sought to maintain the traditional and legal rights of the individual German states to the exclusive conduct of general scientific research: They opposed the Scientific Section as an instance of Berlin's cultural imperialism. Unlike the Reichsanstalt's backers in Berlin and throughout Prussia, Hertling and his supporters in southern Germany did *not* want an institute for an empire. They wanted, instead, to maintain local (and traditional) control of scientific institutes. Their message was registered and ignored: Although the Reichstag denied the government's full funding request, it did approve 250,000 marks for the scientific building and 100,000 marks for the Technical Section. Bismarck, however, dissolved the Reichstag before a final reading of the bill could take place, and the fate of the proposed institute was left hanging.[55]

The final reading in the new Reichstag gave the technologists another opportunity to express their concern that the Reichsanstalt would not address their needs. Their spokesman, Representative Witte, related that they still feared that the Technical Section would be a mere testing and certification institute or that it would be subservient to a dominant Scientific Section. He reminded the government of the technologists' petition for a voice in choosing the director of the Technical Section and their plea for adequate representation of their interests on the board of directors. The minister of the interior replied that the Technical Section would not be treated as a "stepchild" and that the technologists would be sufficiently represented on the board.[56] These last-minute governmental assurances satisfied the Reichstag. At the bill's final reading (28 March 1887), the newly elected Reichstag, which unlike its two predecessors (1881-1887) fully supported Bismarck's legislative proposals, approved 700,432 marks for the Reichsanstalt, including 100,432 marks for wages, salary, housing allowances, and so on; 120,000 marks as a first installment for the Technical Section; and 480,000 marks toward the construction of and furnishings for the Scientific Section's scientific building.[57]

Siemens and his allies had mounted an effective lobbying campaign that brought him his Scientific Section and that, in the public realm, overcame his adversaries' concerns about the direction of the Technical Section. With the exception of Helmholtz, Siemens had campaigned without the help of Germany's physicists. For most of the latter the Reichsanstalt came as a surprising fait accompli. A letter to Siemens from his friend Friedrich Kohlrausch indicates the depth of their ignorance as late as April 1887:

I heard in Berlin that plans for the physical Reichsanstalt are al-

ready underway. And I also accidentally read of this in a news-
paper. All that I know about the Institute is limited to news-
paper reports or hearsay. It would interest me very much to
learn something more definite.[58]

Kohlrausch, then professor of physics at Würzburg and at the center of
the German physics profession, wrote this letter six months before the
Reichsanstalt's official opening. (Three months later he would be
named a member of the newly created Reichsanstalt Board of Directors.) As noted, Helmholtz was the only physicist involved in planning
and founding the Reichsanstalt, a fact that lends substance to the ac-
cusation by technologists and southern Germans that Siemens established the Reichsanstalt, and in particular its Scientific Section, as a gift
to his friend Helmholtz. The establishment of the Reichsanstalt occurred independently of the academic physicists and constituted an implicit critique of the German academic system's ability to conduct
scientific research on a full-time basis. German science had begun to
find a new setting beyond the universities and the Technische Hochschulen.

No more was the Reichsanstalt the product of German industrial
groups, either large or small. To be sure, Siemens, governmental officials, and, eventually, the legislature hoped to achieve economic gain
and national prestige through the Reichsanstalt's operation. But there
were no industrial interests or circles immediately involved in its
origin, apart from those – like the Verein Deutscher Ingenieure and the
Deutsche Gesellschaft für Mechanik und Optik – which complained
that their interests were being ignored or insufficiently represented.
Nor did the Prussian or Reich governments initiate the efforts to establish the Reichsanstalt, though, with the exceptions of Bismarck and his
finance ministers, they sought to expedite it. Neither Prussian nor
Reich "science policy," if, as is most doubtful, any such policy existed,
played a role in the establishment of the Reichsanstalt.

Though the desire to advance precision technology had provided
the occasion for Siemens's interest in a mechanical institute, it was his
interest in pure science, his desire to provide scientific foundations for
technology, and his patriotic sentiments that motivated him to establish
the Reichsanstalt. His electrical firm stood to benefit from the Reichsanstalt's work, as did the entire electrical industry; and Germany's
strong nationalistic spirit and her unmet scientific, technological, and
industrial needs would in part be fulfilled by the creation of such an in-

stitute. But these advantages were at best necessary, not sufficient, conditions. It was the vision, the largess, and the political skill of Siemens (and his allies) that proved essential to the founding of the Reichsanstalt.

## Compromises

Although Siemens and his allies had won parliamentary approval for their version of the Reichsanstalt, their victory remained incomplete. At the inaugural meetings of the Reichsanstalt's Board of Directors, held less than six months after Siemens and his allies had won public approval, their adversaries again attempted to set the Technical Section's character and direction. The latter now sought to do so in private board meetings, as the Reichsanstalt's directors met to establish the Institute's formal purposes and organization along with a set of guidelines and policies for the institution as a whole. In the course of these meetings, the rift in the Reichsanstalt's foundations resurfaced.

Since the technologists had recognized all along that the Scientific Section was Siemens's favorite, they simply remained silent in board discussions about it. The board easily agreed on two formal purposes for the Scientific Section. First, it was to undertake physical research, especially of the measuring sort, that demanded more time, money, instrumentation, and materiel "than [could] normally be provided by private individuals or institutes of instruction."[59] This first (and central) responsibility seemingly allowed the section to investigate all physical problems, including those of a broad theoretical nature. However, its precise intent and effect was to avoid competition between the Reichsanstalt, on the one hand, and private industry, the universities, and the Technische Hochschulen, on the other. Indeed, as early as 1884 Helmholtz had declared that the section's work should include only such problems "that cannot be executed in the other laboratories [i.e., those of the universities]."[60] Moreover, the existence of a number of sister institutes – including the Königliche Preussische Mechanisch-Technische Versuchsanstalt (Royal Prussian Mechanical-Technical Experimental Testing Station), the Chemisch-Technische Versuchsanstalt (Chemical-Technical Experimental Testing Station), and the Normal Eichungskommission – and the recent establishment of a number of engineering laboratories at the Technische Hochschulen further restricted the range of the Reichsanstalt's scientific research.[61] The latter was to be limited to those problems that industry and higher educational institutions proved unable to investigate.[62] The board provided the section with a niche among Germany's scientific institutions.

The Scientific Section's second purpose was to undertake work that the Technical Section could not complete without its help.[63] As we shall see in the remainder of this study, this charge provided an opening for the Technical Section to impinge decidedly upon the Scientific Section's resources.

The renewed disputes between Siemens and his adversaries again concerned the Technical Section's purposes and its relationship to the Scientific Section. The instrumentmaker J. A. Repsold, for example, questioned the point of having two separate sections and predicted that the Technical Section would function "more as a physical than as a technical" division. Foerster replied that two sections had been created in order "to relegate scientific research exclusively to the first section and, in particular, to keep it far away from involvement with vested interests [i.e., industry]." Helmholtz added two further points: First, the Reichstag was particularly interested in the Technical Section and hoped that it would quickly provide some results; and second, rooms for the Technical Section had long been held in reserve at the Technische Hochschule Charlottenburg.[64] Building a physics institute, Helmholtz knew, would take several years. By promptly establishing a separate Technical Section in the available quarters of the nearby Technische Hochschule, he hoped the Reichsanstalt could quickly demonstrate its value to German industry and so resolve any lingering doubts among those Reichstag representatives who had questioned its utility.

Repsold was satisfied with these replies, but Franz Grashof, professor of mechanical engineering at the Technische Hochschule Karlsruhe, was not. Grashof declared that "engineering circles" – doubtless meaning, above all, the Verein Deutscher Ingenieure, of which he was one of the founders and leading figures – wanted the Reichsanstalt to include in its mission "a more direct prosecution and furtherance of the engineering disciplines." Grashof, together with Repsold and the Deutsche Gesellschaft für Mechanik und Optik, was concerned about who would become the director of the Technical Section. He wanted the latter to be independent (i.e., not dependent on Helmholtz and Siemens) and to have had a broad, advanced "higher scientific education, . . . someone who, in particular, fully understands all the engineering disciplines." [65] He wanted, in other words, strong representation for the interests of German engineers at the Reichsanstalt.

His demands elicited sharp replies. Carl Paalzow, professor of physics at the Technische Hochschule Charlottenburg, wanted to know precisely which branches of engineering Grashof thought the Reichsanstalt should pursue. Ernst Abbe, while recognizing the legitimacy of Grashof's demands, warned that the Reichsanstalt's resources were limited and that, legally speaking, it was to advance precision technology, not engineering per se. Helmholtz agreed with Abbe, although he added that the Reichsanstalt would do some engineering work. Grashof mildly rejoined that he hoped the Reichsanstalt would conduct more engineering research and he conceded that it could not now do so. [66]

Despite his seeming concession, Grashof and others on the board were still dissatisfied with the Technical Section's program. They now attempted to use an indirect approach to influence its program: the appointment of personnel. The board easily agreed to appoint Leopold Loewenherz, the choice of the Deutsche Gesellschaft für Mechanik und Optik, as the director of the Technical Section. But this and a related appointment were approved too quickly for Abbe's taste. Abbe requested that a committee choose the candidates for the remaining positions in the Technical Section. It would be the committee's responsibility, he said, to consider "how, [in light of] the limited number of positions still available for co-workers, the justified interests of the different branches of technology should be supported." Several board members agreed, and the board further charged the committee with defining the section's purposes as a whole. By the end of the first board meeting, the technologists had won from Siemens and his allies a say in determining the Technical Section's program and personnel. [67]

Grashof, perhaps emboldened by this development, again petitioned for more engineering research in the Technical Section. He wanted the board's assent to the principle that engineering research belonged within the Technical Section's domain. The board, now softened by three days of his demands, agreed to a compromise: namely, that "in the future" engineering research might be conducted in either section. "The Reichsanstalt leadership," it added, "will seek to meet, so far as means allow, the demands made of it by the engineering circles." Grashof, the Verein Deutscher Ingenieure, and the German engineering community at large had in principle gained an important concession from Siemens and his allies. [68]

Abbe, the industrial physicist, had his own doubts about the Technical Section's mission. He thought that Siemens and his allies envisaged the Technical Section as "merely . . . an extensive standards institute." He urged, instead, that it undertake "experiments and scientific investigations for the improvement of the *foundations* of scientific industry in its different fields of work." Abbe wanted to see a research-oriented Technical Section, one that would address the future, long-term needs of science-based industry in Germany. The board, however, opposed Abbe's recommendation, accepting instead Helmholtz's claim that Abbe's concerns had already been addressed by Siemens and his Scientific Section. [69]

Abbe and other critics among the technologists were not mollified, however. When Helmholtz proposed a list of topics for the Reichsanstalt to investigate during its first six months of operation, they found a new opportunity to question the Technical Section's purposes. [70] Helmholtz claimed that his list included examples of the type of work – studies of materials – recommended by Abbe, and he dangled the hope that perhaps in the future the Technical Section could undertake more such work. Grashof and the instrumentmaker Adolf Steinheil were satisfied with this. But Abbe was not. Helmholtz asked Abbe to name some specific examples of the type of work he wanted the section to do. Abbe countered that his view of the Technical Section's purposes, along with Bamberg and Loewenherz's, differed from that presented by Siemens. He vaguely asserted "that in the circles of the interested parties one expects from the Reichsanstalt's organization something other than that which was resolved yesterday." Those parties, he assured the board, would be "greatly disappointed" by the proposed work. Siemens and Helmholtz, he added, were not organizing the Reichsanstalt along the lines announced before the government and legislature. Fuess agreed and noted Abbe's demands were precisely those made by the Deutsche Gesellschaft für Mechanik und Optik in its petitioning letter to Bismarck in 1886. In the midst of this heated dispute, the board's president begged that "no sharp differences of opinion of a principled nature be allowed to arise [among the members]," which was precisely what was happening. Indeed, to Siemens the

discussion seemed to have the character of a struggle over the independence of Section II. Unlimited independence for this

Section . . . does not lie in the interests of technology. On the contrary, the latter requires that the Second Section, like the entire Reichsanstalt, be subordinated to a higher leadership through an eminent scientific power.

That "eminent scientific power" was, of course, Siemens's ally Helmholtz. Both Foerster and the government meteorologist Georg Neumayer agreed with Siemens that the mechanics' petition was an attempt to achieve independence for the Technical Section from the Scientific. They added that "the personage of the current Reichsanstalt President [i.e., Helmholtz] offers a sufficient guarantee that the essential degree of independence desired for the Second Section will not be infringed upon." They thought that precision technology should be subordinate to fundamental science and, thus, that research in it should be done in the Scientific, not the Technical, Section.[71]

Abbe remained dissatisfied. The physicists Paalzow and Kohlrausch, like Helmholtz before them, asked Abbe to name the particular studies he wanted the Technical Section to undertake. Abbe gave an example – the board's secretary did not record it – and Kohlrausch judged it to be "a pure physical, highly scientific [one], which the Second Section's workers would not be able to solve." Kohlrausch's judgment suggests, as Siemens had earlier claimed, that the artisans and instrumentmakers sought independence for the Technical Section from the Scientific, rather than merely a broadening of the Technical Section's purposes. To mollify Abbe and the Gesellschaft that he represented, the board agreed to another compromise resolution: namely, that, "so far as the extant equipment and manpower allow," the Technical Section should conduct studies on the properties of materials and on the creation or improvement of new ones.[72] It relegated final determination of the section's precise, formal purposes to a committee.

With some minor modifications, the committee's report became the Reichsanstalt's official standing business orders.[73] The purposes of the Technical Section were now declared to be fourfold: First, it was to undertake studies relevant to advancing precision mechanics "and, so far as possible, all other branches of German technology." This included studies of the properties of materials, the development of new methods for producing materials, and experimental work aimed at producing new measuring apparatus useful to both physics and industry. Second, it was to test and certify all measuring apparatus and control instruments, except for those involving weights and measures (which were already the responsibility of the Normal Eichungskommission). Third, it was to produce instruments and "other mechanical devices" for German state institutions, "insofar as their procurement encounters difficulties of supply from domestic, private workshops." Fourth, it was to produce "instrument parts for German manufacturers, insofar as the production in private workshops requires extraordinary aids."[74] As the second, third, and fourth points make clear, the section was to aid other government agencies and German industry, not compete with them. Moreover, as the first point indicates, the section's formal purposes constituted a compromise between Siemens's allies and his adversaries on the board. The various interests of the parties concerned with the Reichsanstalt had been attended to. The rift in the foundations had been bridged – at least on paper.

In principle, the Reichsanstalt's purpose was to conduct research on all physical and technological problems as well as to create and test scientific standards and instruments. However, Reichstag representatives of several of Germany's political parties had successfully fought to limit the range of the Scientific Section's activities to those problems or areas that did not compete with scientific and technical work previously undertaken by other governmental agencies, the universities and Technische Hochschulen, and private industry. Powerful interest groups had strived, rather less successfully, to force the Technical Section to concentrate more of its resources on research and to stress engineering studies in general and materials science in particular. As these industrial and engineering interests sought to influence the direction of the Institute, Siemens and his allies sought to assure them that their concerns about the amount and type of the Institute's technological research had in fact already been addressed and that, resources permitting, the Institute would (in principle) do engineering work, too. In short, Siemens and his allies had sought to make compromises with their adversaries. The result was that among the fields pursued by all German scientific and technological institutions only one was left unequivocally to the Reichsanstalt: standardization and testing. That field alone could never be cultivated by private industry or the academic centers of higher learning or technology maintained by the individual German states.

# 3

## Between Charisma and Bureaucracy

When, in 1887, Hermann von Helmholtz assumed his duties as president of the Reichsanstalt, he was sixty-six years of age. At a time in life when most men are in or nearing retirement, Helmholtz undertook a new, demanding project, that of developing and directing the Reichsanstalt. Old age and weakened health proved to be liabilities in fulfilling his new duties. Yet he more than compensated for them by bringing to his new office an unmatched set of assets: a profound understanding of science and its uses; an acute ability to recognize promising younger scientists; and years of experience in building and directing the scientific institutions that scientists needed for their work to flourish.

The fashionable portraitist Franz von Lenbach bestowed on Helmholtz the pertinent, if unofficial, title of "Imperial Chancellor of German Science."[1] The educated German public viewed Helmholtz as the first man of German science, just as they viewed Bismarck as the first man of German politics and Siemens the first man of German industry. As one of the three demigods of Bismarckian Germany and as the senior member of the German scientific establishment, Helmholtz was the ideal scientist. For many in both the German and international scientific communities, he was the most charismatic scientist of the late nineteenth century.

His charisma had its natural limits, however. The obvious problem, as Siemens frankly reminded Helmholtz two years before the Reichsanstalt opened its doors, was that Helmholtz was a mere, if genial, mortal:

> I am indeed also convinced, [Siemens wrote to Helmholtz,] that, so long as you direct the Institute with full energy and health, it would be best for scientific progress to place only diligent laborers at your side. However, you are, unfortunately,

not immortal. With an institute [i.e., the Reichsanstalt] that is meant to exist for centuries, it is certainly also necessary to consider the possibility that a less universally gifted man may direct it! At all events, it must therefore be organizationally provided for that a number of thoroughly gifted physicists obtain positions under the supervision of the Institute's director.[2]

Helmholtz's advanced age when he assumed leadership of the Reichsanstalt and the aim of building an institute that would last "for centuries" indicated the desirability of a structure and organization that would enable the Institute to flourish when the charismatic leader was gone. As this chapter argues, Helmholtz's actions during his (relatively) short presidency (1887-1894) gradually gave institutional reality to Siemens's admonition of 1885: He established the structure of a scientific bureaucracy. That is, Helmholtz established an institute that combined two elements: the actual pursuit of research and an accompanying research spirit (*Forschergeist*), on the one hand, and an efficient administrative structure, on the other. As we shall see, this structure initially served to enhance scientific work at the Institute but within a decade or two had become obstructive and burdensome.

To understand how and why Helmholtz shaped the Reichsanstalt as he did requires an appreciation of three points: first, his scientific achievements and outlook, including his vision of science – in particular, the central role of the process of measurement – and the nature of his administrative abilities and style; second, the nature of the "elements" he assembled toward a scientific bureaucracy so as to form a highly structured, centralized institution in which authority was delegated according to prescribed rules and in which scientists and technicians accustomed themselves to teamwork aimed at meeting the institution's overall goals; and third, the nature of the pressures on Helmholtz, particularly the political pressure to produce some immediately useful results for German industry in order to allay lingering doubts among the Reichsanstalt's adversaries, and the conflicting institutional pressure to initiate a long-term research program aimed at establishing physical measuring scales of the highest order for thermal, electrical, and optical phenomena. This chapter argues that, in the attempt to respond to precisely these two pressures, a causal relationship developed between the buildup of the Reichsanstalt's physical workspaces under Helmholtz, on the one hand, and the nature and timing of the Reichsanstalt's scientific and technological work under him, on the other. During the final seven years of his life, Helmholtz deployed his great knowledge, talents, and power to transform a paper institution into a scientific bureaucracy staffed by young, "thoroughly gifted physicists" and technicians working in unsurpassed physical workspaces. He did what his friend Siemens had hoped he would do.

## *"The Imperial Chancellor of German Science"*

When Helmholtz assumed his new duties, he left behind a forty-year career as one of the most creative and productive physiologists and physicists of the nineteenth century.[3] That career had its origins in a Gymnasium in the Prussian town of Potsdam, where Helmholtz was born in 1821. As a youth Helmholtz had wanted to become a physicist, but his father's insufficient means led Hermann to accept a government stipend that enabled him to study physics as part of his medical education at the Königlich Medizinisch-chirurgische Friedrich-Wilhelms Institut in Berlin, where he received his M.D. degree in 1842. In return, he agreed to serve eight years as an army doctor.

While at the medical institute in Berlin, Helmholtz also attended lectures at the University of Berlin. There he came under the influence of Johannes Müller and began a research program for attaining a physico-chemical basis for all physiological phenomena, for explaining life in terms of mechanical forces instead of vital principles. From Müller and his students (in particular Brücke, DuBois-Reymond, and Ludwig), as well as the physicist Gustav Magnus and the circle around him at the University of Berlin, Helmholtz gained a deeper appreciation of the importance of observation and experimentation in science. Indeed, although Helmholtz had excellent mathematical knowledge, most of which he acquired autodidactically and which he used throughout his scientific work, and although he always consciously conducted his scientific investigations within some explicit conceptual framework, he considered empirical facts or observed reality to be the basis of science. He held metaphysical principles, above all those of the generation of romantic scientists before him, to be irrelevant to, if not outright pernicious for, science. Following the methodological paths of men like Gauss and Weber, Helmholtz and his generation of German scientists sought to replace metaphysics in science with measurement.

His methodological commitment to measurement and to the reduction of all life phenomena to physical processes; his wide-ranging knowledge of mathematics, physics, and physiology; his passion to understand the convertibility of forces in nature; and his position among medical men – these were the major elements that at the very start of his career led Helmholtz in 1847 to his epoch-making proof of the conservation of force. Others before him had proposed one or another form of the idea of the conservation of force (or energy).[4] But Helmholtz, to the astonishment of even his closest colleagues in the newly formed Berlin Physikalische Gesellschaft, conceived the mathematical principles of force conservation, principles that encompassed and surpassed the individual findings and ideas of his codiscoverers. He provided the most general form of the conservation principle and showed its applicability to a number of previously unexplained electrical phenomena. In a single, bold publication the twenty-six-year-old medical doctor had made one of the major discoveries of nineteenth-century physics (and physiology).

Despite the widely recognized importance of his achievement for physics, Helmholtz devoted most of the first twenty-odd years of his career, from the early 1840s to the mid-1860s, to physiological optics and acoustics. After serving five years as an army surgeon in Potsdam (1843-1848), he was allowed to resign his commission early in order to teach anatomy at the Berlin Kunstakademie (Art Academy). Then, while serving in a series of university positions in anatomy and physiology – at Königsberg (1849-1855), Bonn (1855-1858), and Heidelberg (1858-1871) – he showed himself to be one of Europe's foremost physiologists, above all through the publication of his two magisterial treatises in sensory physiology: the three-volume *Handbuch der physiologischen Optik* (Treatise on physiological optics), published between 1856 and 1867, and *Die Lehre von den Tonempfindungen als physiologische Grundlage für die Theorie der Musik* (On the sensations of tone as a physiological basis for music), published in 1863. For Helmholtz, the eye and the ear were instruments, and he investigated their anatomical, physiological, and physical structures and functions. These volumes, as well as his individual papers in physiology, recorded his rich theoretical explanations and his unsurpassed mathematical conceptualization of physiological processes. They reported, furthermore, methodological and experimental results obtained by employing or creating a number of measuring devices: For example, Helmholtz investigated microscopically the nervous system of invertebrates; he performed experiments on measuring temperature changes during muscular contraction; he created a myograph for tracing muscular motion and, thereby, measured the velocity of nerve impulses; he invented an ophthalmometer for measuring the changing curvature of the cornea and other parts of the eye; he investigated elastic vibration by means of a so-called vibration microscope. And, of course, he invented (in 1850) that piece of scientific apparatus that brought him everlasting fame throughout the medical and lay worlds: the ophthalmoscope. The creation and use of these and numerous other pieces of scientific apparatus taught him to appreciate the difficulties faced by mechanical instrumentmakers and the importance of instrumentation and measurement in science.

Helmholtz's physiological work always rested on physical foundations, and during the 1860s he increasingly returned to his first love, physics. That shift in interest had several sources. With the completion of his multivolume studies in physiology during the 1860s, he felt that he had made his essential contributions to the field and believed that physiology was now a mature professional discipline, containing enough skilled practitioners to assume the tasks that he had once assumed for himself alone and that he now recognized to be beyond the power of any single individual. Moreover, unlike some of his intellectually complacent colleagues in physics during the 1860s, Helmholtz realized that there were fundamental, challenging problems – above all in electrodynamics – for the ambitious physicist to pursue.

A final, if perhaps indirect, reason for Helmholtz's return to physics was the death of his wife, Olga von Velten, in 1859, about a year after he had moved from Bonn to Heidelberg. The loss of his beloved partner of ten years and the heavy responsibility of raising their two small children left him deeply depressed. To recover, he sought new scientific interests and new social companionship. Two years later, in 1861, he married Anna von Mohl, the cultivated, gregarious, and socially ambitious daughter of the Heidelberg political theorist and statesman Robert von Mohl. Anna rekindled Helmholtz's spirit and made their Heidelberg home an intellectual salon, introducing her husband to a broader social and political world.[5] With his increasingly senior position among Germany's and Europe's scientific statesmen, Helmholtz entered the upper social echelons. His Heidelberg years brought important social changes to his life, just as they brought important changes in his scientific interests.

In addition to his 1847 memoir on the principle of the conservation of force, Helmholtz had published a number of other physical papers before full conversion to physics in the late 1860s. In 1849, for example, he laid out the principles for constructing a tangent galvanometer. During the 1850s, as part of his work in physiological optics, he published several theoretical and experimental studies on color; as part of his work in physiological acoustics, he outlined a theory of combination tones. In 1858 he published an important study on hydrodynamics. To the paper's mathematical rigor and its conceptual reasoning, Helmholtz (through the supervision of a young colleague) added an experimental dimension to this study of fluid flow. This combination of mathematical, conceptual, and experimental analysis became the distinctive mark of his work.

With his appointment in 1871 as professor of physics at the University of Berlin, Helmholtz entered a new phase of his career: Physics replaced physiology as the principal object of his scientific concerns. His transformation into a full-time physicist and his establishment of the new Berlin Physics Institute in the 1870s symbolized the new disciplinary maturity of physics. He spent much of the 1870s and beyond clarifying the conflicting electrodynamic laws and theories of Weber, Franz Neumann, and Maxwell. In particular, he raised doubts about the approaches and findings of his German colleagues Weber and Neumann, and he helped bring Maxwell's field-theoretical approach and Maxwell's view of light as an electromagnetic phenomenon to the attention of Continental physicists.

More important for our purposes, however, Helmholtz showed his command of measurement in experimental science. Although his abstract work in non-Euclidean geometry has often been noted, his more concrete work in physical measurement has not. For example, he analyzed the theoretical limits of microscopic resolution, and he did major work on the galvanic cell, in electrolysis, and in chemical thermodynamics. In 1881 he joined the German delegation to the International Electrical Congress in Paris in order to help establish internationally valid electrical units and standards. By the 1880s Helmholtz had shown himself to be one of the masters of the theory and practice of physical measurement, including the use of measuring apparatus.

By the time Helmholtz assumed the presidency of the Reichsanstalt, he was the consummate physicist. He had exercised his intellect, to a greater or lesser extent, in virtually every type and area of physics. His fellow physicists, especially the younger ones, "stood in awe," and

Figure 3.1. Hermann von
Helmholtz in 1889.
Courtesy of Siemens
Museum, Munich.

usually at a distance: Henry Rowland, making the rounds of European physics institutes in 1876, referred to him as "the greatest physicist on the Continent." [6] Michael Pupin, another young American physicist who went to study under Helmholtz, reported that "the whole scientific world of Germany, nay, the whole intellectual world of Germany, stood in awe when the name of Excellenz von Helmholtz was pronounced. Next to Bismarck and the old Emperor he was at that time [1885] the most illustrious man in the German Empire."[7] For his fellow German Max Planck, soon to become Helmholtz's junior colleague at Berlin, Helmholtz "embodied the truth and the value of his science." [8] Younger, less established physicists were reportedly so awed by his mere presence that they feared to speak without first being spoken to.[9] Even his personal features and bearing seemed to be more than human – Ostwald referred to them as "almost monumental." Ostwald, like Rowland and most of their contemporaries, thought Helmholtz "the greatest German physicist" (Figure 3.1).[10]

This doyen of the German physics discipline also brought to his presidency broad experience in the establishment and management of scientific institutes. When he accepted the post of professor of physiology at Heidelberg in 1858, he did so on condition that the Baden government build him a new physiological institute. Thirteen years later, he made similar demands on the Prussian government before accepting the post of professor of physics at Berlin. As noted in Chapter 1, he received from Prussia enough funding (1.5 million marks) to construct the largest physics institute in Germany during the last quarter of the nineteenth century.

Beyond his scientific abilities and stature, Helmholtz brought another major asset to the Reichsanstalt presidency: a vision of science and society that addressed the needs of a modernizing Germany. During the second half of the nineteenth century he became one of the foremost popularizers of science and Germany's spokesman of science. As his scientific stature grew and as his social and political consciousness matured, Helmholtz felt called upon to address the educated German public on the relations of science and society.

In his 1862 inaugural address as the prorector of Heidelberg University, Helmholtz stated unequivocally his views on the relations of science, the economy, and the nation. If he repeated the Baconian banality that "knowledge is power," he nonetheless infused it with life by noting Germany's rapid industrial development and its unprecedented use of science for technology. He believed that nature's inorganic forces were being harnessed as never before for human purposes.

His views on the relations of knowledge and political power were broad and thoughtful. He recognized that national power was based not only on material goods like machinery, cannons, and ships, goods that had become so essential to the modern state "that even the proudest and most unyielding of today's absolute governments have had to take them into account [in their plans], to unchain industry, and to concede a legitimate voice in their councils to the political interests of the working middle classes." National power, he said, was also based on the "political and legal organization of the state and on the individual's moral discipline," which "leads to the superiority of the educated nations over the uneducated." The state had to develop a diversified culture if it was to survive and prosper. It needed to support not only the natural sciences and technology, but also the social sciences and the humanities; it had to produce intangible political, moral, and cultural benefits as well as tangible material products. And to do so it had to support all of the *Wissenschaften*.

Selecting an apt metaphor for Germany of the 1860s, Helmholtz claimed that scientists formed "a sort of organized army." They supposedly worked "for the good of the entire nation and almost always on its order and at its costs." Their object was "to produce knowledge" so that industry and wealth could grow, political organization could improve and the individual citizen could develop morally, and life could be beautified. These views on the relations of knowledge and political power were quintessentially those of the educated liberal *Bürgertum* of post-1848 Germany.

Finally, he cautioned his fellow citizens that, if the sciences were to attain their social goals, then scientists must be left free of social pressure to do so. Scientists' fellow citizens "should certainly not ask about the immediate uses, as the uninformed individual so often does. Everything that gives us knowledge of nature's forces or of the mind's powers is valuable and can in due course be useful, usually where one least expects it [to be so]." He thus advocated strong support for pure science as the ultimate basis for technological development and for a more humane social and political order. "Whoever in the prosecution of the sciences," he warned, "hunts after immediate practical uses can be rather certain that he will hunt in vain. Science can only strive after [the] complete knowledge and understanding of the operations of the natural and mental phenomena." The scientist, Helmholtz argued, accumulated "intellectual capital" (*Kapital des Wissens*) that society eventually used to control nature's "hostile forces."[11] His views on the relations of science and society, enunciated a full quarter of a century before he assumed the Reichsanstalt presidency, were to find their embodiment in the Reichsanstalt.

For several reasons the older Helmholtz welcomed the opportunity to build and lead the Reichsanstalt. His scientific work had led him, at times almost daily, to use or invent scientific instruments; an institute devoted to the advancement of such instrumentation naturally found his sympathy. So, too, did one dedicated to measurement, in particular to the extensive measuring work that the limited research facilities and personnel of the universities were incapable of conducting. Moreover, he felt burdened by his teaching responsibilities at the university.[12] Both he and Siemens expected that as Reichsanstalt president he would be free of these responsibilities and thus able to pursue his advanced studies without disruption.[13] Finally, Helmholtz's sense of patriotic duty also disposed him to accept the burdens of building the Reichsanstalt: As its leader he could make unprecedented contributions to German science and society as well as pursue his own more abstract scientific interests. His sense of patriotism in building and leading the Reichsanstalt paralleled Siemens's in providing for its establishment. He once counseled his star pupil Heinrich Hertz that, when one nears the end of life and has relatively little time left for undertaking a great research program, as was Helmholtz's own case in the late 1880s, one does well to try to shape the younger generation of scientists and to devote oneself to public administration. [14] In addition to Hertz, Helmholtz nurtured a group of young physicists into scientific maturity, including Ludwig Boltzmann, Heinrich Kayser, Ferdinand Kurlbaum, Otto Lummer, Albert Michelson, Planck, Ernst Pringsheim, Pupin, Rowland, and Willy Wien.

Helmholtz's willingness to become Reichsanstalt president complemented the authorities' unanimous desire to appoint him. As suggested in Chapter 2, Siemens had in part established the Reichsanstalt for Helmholtz. Moreover, Helmholtz's close ties to the court, his military background and connections, his mastery of science, his experience in directing a physical institute, his sympathetic understanding of the importance of science for an industrializing Germany, and his unmatched prestige abroad made him the best and only candidate for the Reichsanstalt presidency.

Nonetheless, one delicate issue required resolution before Helmholtz would formally accept the presidency: his salary. As a member of the Reichsanstalt planning committee, Helmholtz himself had made the original salary proposal for the Reichsanstalt president: 24,000 marks per year plus the use of a presidential residence.[15] Confronted with governmental skepticism about the need for such a high salary as well as the use of a residence, Helmholtz vigorously defended his proposal.[16] Though another planning committee opposed part of his proposal, Siemens and Boetticher, the minister of the interior, supported it in full.[17] The minister of treasury, for his part, thought that even 15,000 marks was excessive for an Interior official, even for a Reichsanstalt president.[18] His dissent was not unreasoned: The other top German physicists, except for Helmholtz himself, earned far less than 24,000 marks per year.[19] Moreover, the proposed salary surpassed even that of an undersecretary of state.[20] The minister of treasury thus opposed the suggested salary and threatened, should the president be granted a salary of 24,000 marks per year, to deduct 9,000 marks from the Reichsanstalt's budget. He trenchantly added that "what the proposals to date do not reveal is that in the person of Privy Counselor Helmholtz the first president has already been selected."[21]

Boetticher rejoined that 15,000 marks was simply not enough to attract the best man to the post. Given the salary levels of the top physicists and "the meaning of the position of president of the Physikalisch-Technische Reichsanstalt," he proposed a compromise: Only the first Reichsanstalt president should receive an extra 9,000 marks, though this part of his salary should not apply to his pension. He added disingenuously that salary negotiations with Helmholtz had not yet begun.[22]

Fourteen months later Helmholtz entered into direct negotiations with the government. He noted that his current annual income was 32,712 marks and declared that he would accept the presidency only on condition that he receive a salary of 15,000 marks, an annual bonus of 9,000 marks, and the right to retain his position at and salary (6,900 marks) from the Prussian Akademie der Wissenschaften.[23] Boetticher's reasons for supporting Helmholtz's demands are revealing:

> The Reich administration cannot avoid a concession here. For from the very beginning and in all the scientific circles interested [in the Reichsanstalt], Dr. von Helmholtz alone has been regarded as the future director for the Physikalisch-Technische Reichsanstalt. Moreover, there is unanimous agreement that no one – not even among the most noteworthy of the other German physicists – is his equal. The calling of another scientist to that position would thus doubtless create an unfavorable impression in general and, in particular, would serve from the very outset to weaken the stature and the efficacy of this momentous institute.[24]

Boetticher added that Helmholtz's salary demands still meant a loss to the latter of about 2,000 marks per year, that in all the earlier Prussian negotiations Helmholtz had had his wishes granted, and that, therefore, he should be granted this wish too! Finally, he warned his fellow minister that "the establishment of the Physikalisch-Technische Reichsanstalt, which is excitedly awaited in many quarters," was dependent upon concluding negotiations with Helmholtz. One week later the minister of the treasury capitulated to Helmholtz's demands.[25] Two

months later Bismarck also agreed, but added the proviso that Helmholtz should receive the extra 6,900 marks only if he held weekly lectures of from one to three hours on theoretical physics at the university.[26] Bismarck probably wanted Helmholtz associated with the
university for political reasons, and Helmholtz, for his part, doubtless
wanted to maintain some academic ties to the university. He now accepted the Reich's offer.

Even before he had accepted, however, he had been busy overseeing the Reichsanstalt's organization and planning. As we saw in Chapter 2, from 1882 to 1887 he had been a major figure in the struggle to
establish the Reichsanstalt. By the latter year his involvement had become a large part of his daily business. In the first week of August, for
example, he assembled his "physical parliament," as he called it, for a
dinner. Almost all the guests were soon-to-be-named Reichsanstalt
board members: Neumayer, Kundt, Kohlrausch, Clausius, Foerster,
Paalzow, Siemens, Abbe, et al.; they doubtless spent part of their *dîner*
discussing the new Reichsanstalt. Later that week Helmholtz and Foerster devoted two days to a series of "tremendous committee meetings"
with board president Weymann to resolve more Reichsanstalt matters,
out of which still more meetings were expected. In addition, Helmholtz
and Siemens held a working breakfast in the latter's garden, after which
they visited the future quarters of the Technical Section at the Technische Hochschule Charlottenburg and inspected the progress of the
construction of the main scientific building and the president's residence.[27]

With the opening of the Reichsanstalt in October and throughout
the early years of his presidency, Helmholtz devoted much of his time
to directing the Reichsanstalt. Between 1887 and 1891 he spent a good
deal of time supervising construction work. He visited the plant site
regularly and oversaw the outfitting of the Technical Section's quarters
at the Technische Hochschule. In late 1891 he directed the transfer of
the Scientific Section into the Reichsanstalt's new *Observatorium,* its
scientific building, and provided for its equipping as well. During these
early years, moreover, he often decided about the purchase or installa-
tion of such basic items as thermometers, standard meter rods, and
boiler-security apparatus,[28] and also about such mundane matters as a
Reichsanstalt subscription to a scientific journal and a waiver of testing
fees for a scientific society.[29] And all the while he supervised the
Reichsanstalt's work and personnel, the latter of which had, by 1890,

reached some fifty in number.[30] Ironically, his new administrative burdens, as DuBois-Reymond and others reported, "far surpassed" his previous professorial duties at the university.[31] He conducted all this directorial work while simultaneously pursuing his own abstract studies on the principle of least action and other physical topics, teaching a seminar at the university, and attending numerous public functions.

The secret to his ability to fulfill his Reichsanstalt duties and still follow his other interests lay in his establishment of a hierarchically ordered institute and his willingness to delegate authority to the able young men whom he attracted to the Reichsanstalt as his subordinates. His scientific stature and authority inspired the group of young workers with whom he surrounded himself. He made them feel that they were at the forefront of their field, and so they stood ready to share his burdens. He reportedly "imparted the guiding ideas . . . and united the many heterogeneous individuals – so many of whom were not only abundantly qualified but also already accustomed to going [their] various ways – toward common goals."[32] "In no other institute has there ever been such joy and diligence in work," asserted Johannes Pernet. "Everyone worked as part of a unified program, and each labored so as to contribute to the success of the whole."[33] Helmholtz was the charismatic leader who gave scientific vision to his subjects and fused their individual creative abilities into a concerted effort. When his other interests and duties took him away from the Reichsanstalt, they administered it for him.

He supervised the Reichsanstalt in the same way that he had supervised the Berlin Physics Institute: Like a doctor in a clinic, he made "the rounds" to see how his young interns were doing, to guide and encourage them in their work.[34] He imparted a *Forschergeist*, or research spirit, to them: Thanks to his leadership they viewed the Scientific Section as an institute devoted to advanced physical research. With the aid of his assistants, Helmholtz ran the Reichsanstalt "like a physics institute."[35]

Helmholtz's obligation to teach a seminar at the university furthered this institutional *Geist,* for it meant a formal tie between the Reichsanstalt and academic science in Berlin. So, too, did the location (until 1891 and 1893, respectively) of the Scientific and Technical Sections' quarters at the Technische Hochschule Charlottenburg. From the outset, the Reichsanstalt's staff came into regular contact with the staffs of the other two most important centers of physics in Berlin: those at the university and at the Technische Hochschule. Together they formed a triangle of physics institutes that helped maintain Berlin as one of the leading centers of German and world physics.

Helmholtz's charismatic authority and importance were felt by his subordinates in still another way. The socially elite and politically powerful visited him in his Reichsanstalt laboratory and socialized with him in his "magnificent residence."[36] From the royal family on down, the Helmholtzes frequently entertained the intellectual, artistic, financial, and political elite of Germany and Europe.[37] Indeed, Helmholtz was part of that elite. On the occasion of his seventieth birthday (1891), for example, he was publicly feted by his friends and colleagues, honored by royalty and politicians – the emperor, the kings of Italy and Sweden, the grand duke of Baden, the president of the French Republic – and congratulated and awarded honors by innumerable universities, academies, and scientific societies throughout the world. In a hierarchically ordered society like imperial Germany, visits by its elite members and acts of public tribute to Helmholtz could not help but inspire dedication in his Reichsanstalt subordinates.

## The Formation of a Scientific Bureaucracy

The Reichsanstalt's gradual transformation into a scientific bureaucracy began on the day of its official enactment: The legislature made it an imperial governmental agency, relegating formal control to the Ministry of the Interior (Reichsamt des Innern). Although the Reichsanstalt continued to receive the moral and political support of Germany's monarchs, chancellors, and the legislature, that support was passive. According to one member of Parliament in 1912, the Reichsanstalt had never been a subject of parliamentary discussion.[38]

The Ministry of the Interior, whose responsibilities were as numerous as they were diverse,[39] oversaw more than it controlled the Reichsanstalt. The minister held ultimate responsibility for the Reichsanstalt's welfare, but, apart from Boetticher at the time of the Reichsanstalt's founding, he played almost no role in its affairs. Instead, he delegated authority to an undersecretary of state (*Staatssekretär*) who, among his myriad duties, served as the ex officio president of the Reichsanstalt's board. That office formally required the undersecretary to conduct the board's business, call and lead its annual meetings, report to the government on its decisions, represent it to all "foreign" parties, supervise its financial affairs, and, in accord with the Reichsanstalt president, approve all guests invited by the latter to work at the Reichsanstalt.[40] Except for three or four major decisions concerning the choice of Helmholtz's successors as president and the approval of new funds for plant construction, the board president's duties were in reality minor, supervisory ones that were largely begun and completed at the Reichsanstalt's annual board meetings. His office thus served mainly as a formal link between the Reichsanstalt and the ministry. In principle, the board president assumed responsibility for all of the Reichsanstalt's administrative and legal matters, leaving day-to- day scientific and technical matters to the Reichsanstalt president (i.e., Helmholtz and his successors).[41] In fact, however, the board president had the barest minimum of administrative duties and legal responsibilities, because the Reichsanstalt employed a staff of office workers to handle its administrative business and because it had few legal problems. Moreover, at Siemens's instigation, the board curtailed the board president's power by resolving that control of 25 percent of the Reichsanstalt's annual unmarked budgetary funds was to be "effective *solely* through the signature of the Reichsanstalt president."[42] With minimal administrative duties to perform and limited financial authority to exercise, the Interior Ministry's plenipotentiary became largely a figurehead. That role was filled by Privy Counselor H. Weymann from 1887 to 1902 and by his successor Theodor Lewald from 1903 to 1920. Their long tenures suggest a desire for continuity and a conservative attitude in the ministry's overseeing of the Reichsanstalt.

The board itself was a body of experts responsible for overseeing all aspects of the Reichsanstalt. Consisting of a president and twenty-four or twenty-five members, it held annual meetings of three to four days in March. Official contact between the board president and board members during the intervening period was infrequent. Meeting only once per year, the board could exercise only limited, general control over the Reichsanstalt. Its official duties were to review the Reichsanstalt president's report on the Reichsanstalt's work during the past fiscal year (1 April to 31 March) along with his proposed agenda and budget for the forthcoming year; to recommend a budget for two years hence; and to approve the appointment of all Reichsanstalt employees and guests.[43] To fulfill these duties in the course of three or four days meant that the board had little time for any other matters, including the setting of long-term goals. The Reichsanstalt president dominated the meetings for the simple reason that no one else was nearly so competent to discuss the Reichsanstalt's affairs; for the most part, the board rubber-stamped the president's reports and plans. As board president Weymann himself conceded, the board was "very distant from the Reichsanstalt's administration," and it was "almost impossible" for the board to make judgments on the details of the work done at the Reichsanstalt.[44]

The board represented a combination of governmental, academic, and industrial interests. Members were nominally appointed by the emperor through the minister of the interior and the board president. In fact, however, it was largely the Reichsanstalt president (above all, Helmholtz) working in cooperation with the board president (above all, Weymann) who chose the members. The emperor appointed the original twenty-four board members for a five-year period. At the end of the first period (1892), he reappointed all surviving members, and at the end of the second period (1897) he agreed to convert all current and future appointments into lifetime ones (Table 3.1).[45] These decisions further ensured continuity and conservatism at the Reichsanstalt.

Professional background, political association, and socioeconomic status were the three criteria used in selecting board members. The original board planned by Helmholtz in consultation with Weymann consisted of five ex officio members (the board president, the Reichsanstalt president, and one representative each from the army's Topographical Survey, the navy's Hydrography Section, and the government's telegraphic office); one representative each from chemistry, weights and measures, and geodesy; two each from meteorology and astronomy; and four each from physics, precision mechanics and optics, and the engineering sciences. These areas of professional interest were not strictly interpreted, however. Most board members had some training or expertise in physics, and most could have represented another area; Abbe, Siemens, and Wilhelm von Bezold, for example, could have been included under the category of physics, but were not. The common characteristic of all board members was an interest in and knowledge of physics and precision technology. Their overriding desideratum was the improvement of precision-measuring standards and instrumentation – a goal shared by academic physics, government, and industry alike.

The number of representatives in each category suggests the relative importance of the Reichsanstalt for the government, universities,

Table 3.1. *Reichsanstalt's Board of Directors, 1887–1918*

| Name | Period on board | Place of work when appointed | Area or agency represented | Type of affiliation |
|---|---|---|---|---|
| 1. Weymann, H. | 1887–1902 | Berlin | Ministry of the Interior | Government |
| 2. Schreiber, O. | 1887–92 | Berlin | Army (Survey) | Government |
| 3. Mensing, A. | 1887–92 | Berlin | Navy (Hydrography) | Government |
| 4. Foerster, W. | 1887–1921 | Berlin | Weights and measures | Government and university |
| 5. Massmann, H. | 1887–91 | Berlin | Telegraphy | Government |
| 6. Helmholtz, H. | 1887–94 | Berlin | PTR (president) | Government and university |
| 7. Landolt, H. | 1887–1910 | Berlin | Chemistry | Technische Hochschule |
| 8. Siemens, W. | 1887–92 | Berlin | Applied science | Industry |
| 9. Bezold, W. | 1887–1907 | Berlin | Meteorology | University |
| 10. Paalzow, A. | 1887–1908 | Berlin | Physics | Technische Hochschule |
| 11. Helmert, F. | 1887–1917 | Berlin | Geodesy and astronomy | Government |
| 12. Fuess, R. | 1887–1917 | Berlin | Precision technology | Industry |
| 13. Bamberg, C. | 1887–92 | Berlin | Precision technology | Industry |
| 14. Clausius, R. | 1887–88 | Bonn | Physics | University |
| 15. Kohlrauach, F. | 1887–1910 | Strasbourg | Physics/PTR (president)/physics | University/government/self |
| 16. Seeliger, H. | 1887–1924 | Munich | Astronomy | University |
| 17. Steinheil, H. | 1887–94 | Munich | Precision technology | Industry |
| 18. Zeuner, G. | 1887–97 | Dresden | Engineering sciences | Technische Hochschule |
| 19. Dietrich, W. | 1887–1914 | Stuttgart | Engineering sciences | Technische Hochschule |
| 20. Grashof, F. | 1887–92 | Karlsruhe | Engineering sciences | Technische Hochschule |
| 21. Abbe, E. | 1887–1905 | Jena | Applied physics | Industry and university |

Table 3.1 (*cont.*)

| Name | Period on board | Place of work when appointed | Area or agency represented | Type of affiliation |
|---|---|---|---|---|
| 22. Repsold, J. | 1887–1919 | Hamburg | Precision technology | Industry |
| 23. Kundt, A. | 1887–94 | Berlin | Physics | University |
| 24. Neumayer, G. | 1887–1906 | Hamburg | Meteorology | Government |
| 25. Wiedemann, G. | 1889–99 | Leipzig | Physics | University |
| 26. Quincke, G. | 1892–1914 | Heidelberg | Physics and precision technology | University |
| 27. Bach, C. | 1892–1927 | Stuttgart | Engineering sciences | Technische Hochschule |
| 28. Schmidt, A. | 1892–98 | Berlin | Army (Survey) | Government |
| 29. Scheffler, A. | 1892–97 | Berlin | Telegraphy | Government |
| 30. Siemens, A. | 1893–1918 | Berlin | Engineering sciences | Industry |
| 31. Hildebrand, M. | 1894–1910 | Freiburg i. B. | Precision technology | Industry |
| 32. Dietrich, H. | 1894–98 | Berlin | Navy (Hydrography) | Government |
| 33. Linde, K. | 1895–1921 | Munich | Engineering sciences | Industry and Technische Hochschule |
| 34. Warburg, E. | 1895–1930 | Berlin | Physics/PTR (president) | University/government |
| 35. Hagen, E. | 1895–1918 | Berlin | PTR (Technical Section) | Government |
| 36. Röntgen, W. | 1897–1920 | Würzburg | Physics | University |
| 37. Strecker, K. | 1897–1934 | Berlin | Telegraphy | Government |
| 38. Rottok, H. | 1899–1915 | Berlin | Navy (Hydrography) | Government |
| 39. Kohlrausch, W. | 1899–1927 | Hanover | Electrical engineering | Technische Hochschule |
| 40. Lewald, T. | 1903–20 | Berlin | Ministry of the Interior | Government |
| 41. Nernst, W. | 1905–35 | Berlin | Physics | University |

| | | | | |
|---|---|---|---|---|
| 42. | Schott, O. | 1906–24 | Jena | Precision technology | Industry |
| 43. | Krüss, H. | 1906–25 | Hamburg | Precision technology | Industry |
| 44. | Görges, J. | 1908–35 | Dresden | Electrical engineering | Technische Hochschule |
| 45. | Planck, M. | 1908–35 | Berlin | Physics | University |
| 46. | Rubens, H. | 1910–22 | Berlin | Physics | University |
| 47. | Dorn, E. | 1910–16 | Halle | Physics | University |
| 48. | Wien, W. | 1912–28 | Würzburg | Physics | University |
| 49. | Jaup, H. | 1913–14 | ? | ? | ? |
| 50. | Schwarzschild, K. | 1913–16 | Potsdam | Astronomy and physics | Government |
| 51. | Wien, M. | 1914–35 | Jena | Physics | University |
| 52. | Holborn, L. | 1915–26 | Berlin | PTR (Section III, Heat) | Government |
| 53. | Maurer, H. | 1915–35 | Berlin | Navy (Hydrography) | Government |
| 54. | Einstein, A. | 1917–35 | Berlin | Physics | University (Akademie der Wissenschaften) |
| 55. | Miller, O. | 1916–35 | Munich | Electrical engineering | Industry |
| 56. | Raps, A. | 1917–20 | Berlin | Applied science | Industry |
| 57. | Siemens, W. | 1918–19 | Berlin | Applied science | Industry |
| 58. | Weidert, F. | 1918–35 | Berlin | Precision technology | Industry and Technische Hochschule |

*Note:* Board members are listed in the order of their appointment. The first twenty-four members constituted the original board. *Sources:* List of original board members, copy in BAK, R43 F/2365, 4/8; "Das Kuratorium der physikalisch-technischen Reichsanstalt," 1 April 1887 to 31 December 1893, ZStAP, RI, Nr. 13146/1; "Das Kuratorium der Physikalisch-Technischen Reichsanstalt," 1 January 1894 to 1902, ibid., Nr. 13147/2; "Das Kuratorium der physikalisch-technischen Reichsanstalt," January 1903 to December 1916, ibid., Nr. 13148/3; and listings given as part of the annual meetings, copies of which are in SAA, 61/Lc 973.

Technische Hochschulen, and industry. Judged by this criterion, the board's orientation was decidedly toward the practical, technological side: As Table 3.1 shows, those representing governmental agencies, industry, and the Technische Hochschulen on the original board held eighteen of twenty-four votes (75 percent). Helmholtz himself recognized this state of affairs. In a letter to Weymann, he articulated his conception of the appropriate types and functions of board members, as well as his concerns about the board's professional orientation. Members certainly should not be involved in the Reichsanstalt's actual scientific work, he maintained, nor need they even be outstanding men in their specialties, since the latter often failed to see the more general problems. He continued:

> The special practical interests that are to be pursued by the Reichsanstalt seem to me to be sufficiently covered through the 17 representatives of the different individual specialties. Now, with the inclusion of the president there are only four representatives for pure physics – which forms the common scientific foundation of all four remaining branches represented [on the board] and which unites a large number of very diverse fields of study. It thus seems to me absolutely essential to get men for these positions whom one can trust to recognize the [Reichsanstalt's] broad and great purposes.[46]

From the very first, then, a majority of the board was concerned primarily with "practical interests"; in numerical terms, pure physics was poorly represented. The replacement of board members who died or resigned left the subsequent distribution of seats largely unchanged, at least until after 1905.

Furthermore, those whom Helmholtz chose to represent pure physics – Clausius, Kohlrausch, Kundt, Paalzow, Georg Quincke, and Wiedemann – were experimentalists and scientific traditionalists. (The sole exception here, the theoretician Clausius, died in 1888.) Except for Kundt, who died in 1894, none proved sympathetic to the exciting developments in late-nineteenth-century physics. In the late 1880s, Kundt and Kohlrausch were two of Germany's best experimental physicists. Apart from his scientific authority, Kundt was probably also selected because he was a cofounder of and advisor to the *Zeitschrift für Instrumentenkunde,* the journal in which many Reichsanstalt scientists and technologists would regularly publish.[47] As for Kohlrausch,

Helmholtz recommended him on the grounds that he was "the most distinguished" physicist in determining electrical units and setting electrical standards and was "in this area doubtless the most experienced German physicist."[48] Paalzow, to quote Planck, was "a pure experimentalist."[49] His student Heinrich Rubens confessed that for Paalzow "sobriety, even with respect to the most brilliant and most seductive theories, was the physicist's highest virtue. That he sometimes went too far in this and, especially, that he held an unjustified mistrust toward the modern views, must be conceded." [50] So, too, did Quincke, who had scarcely adapted his ideas about physics to developments after midcentury. He admired Faraday's work but failed to understand Maxwell's mathematical transformation of Faraday's electromagnetic discoveries. His talent and passion for making measurements and collecting data, in addition to his support of precision technology, however, led to his appointment to the board.[51] Finally, Wiedemann, though admittedly not quite as experimentally oriented as the other board physicists, was known particularly for his work in the precision measurement of electrical phenomena and physical standards.[52] This fact and his extensive knowledge of the German physics literature – he was the editor of the *Annalen der Physik*, the other major journal in which Reichsanstalt workers were to publish – were decisive in his appointment to the board.[53] By the time of Helmholtz's death in 1894, there was not a single physicist on the board who was either theoretically oriented or would be particularly receptive to the rapid and radical developments in physics after 1895. Helmholtz's choices of (conservative) experimental and measuring physicists helped orient the Scientific Section's work in these directions and set a pattern that proved essentially impossible to break.

In contrast to the physicists, the technologists, as noted in Chapter 2, aggressively sought to increase their representation on the board and to direct the Reichsanstalt's scientific and technical work toward their needs. Not only did the Verein Deutscher Ingenieure gain two additional seats on the board; it also sought to redirect the Technical Section's work away from problems in precision mechanics and optics toward those in mechanical engineering. The Deutsche Gesellschaft für Mechanik und Optik also successfully strove after an additional board seat, raising the number of representatives for precision mechanics and optics from three to four. Moreover, under the leadership of Fuess the instrumentmakers sought to gain yet another seat, this one for the director of the Technical Section.[54] Though they were unsuccessful in 1886 in gaining that seat, in 1895 the board approved the appointment of the director as an ex officio member.

Political association was the second of the three criteria used in selecting board members. As Table 3.1 shows, more than half of the original board members worked in Berlin, where they held high posts in major governmental or academic institutions. To show concern for the interests of the individual German states and so to ensure continued nationwide support for the Reichsanstalt, it was also necessary to select members from the non-Prussian states within the Reich: hence the representatives from Dresden and Leipzig (Saxony); Munich and Würzburg (Bavaria); Stuttgart (Württemberg); Freiburg i. B., Heidelberg, and Karlsruhe (Baden); Jena (Thuringia); and Hamburg.[55]

A certain minimum level of social respectability and income was the third criterion for honorary appointment to board membership.[56] The industrial representatives were all highly successful entrepreneurs: Siemens, Fuess, Bamberg, Steinheil, Abbe, Repsold, et al. Instrumentmakers and other technological workers obviously had more obstacles to overcome here than did government and university officials. Weymann, in reviewing the proposed nomination of two instrumentmakers to the board, once revealed the necessary and desirable socioeconomic characteristics of potential board members. In rejecting the proposed nominations, he granted that the nominees had the proper "intelligence and skill" to represent precision mechanics on the board. "Yet socially and economically," he argued, "both [men] live in such limited circumstances that it appears doubtful whether they sufficiently command the more general interests of their class." Then, too, he doubted whether they wanted "to be in a condition of maintaining the necessary [social and economic] position as members of the board."[57] The "right" social position, political association, and scientific background – these were the criteria for being a board member.

The Reichsanstalt president was the chief scientific and administrative officer. His extensive powers encompassed every aspect of the Institute's structure and the organization of its work. First, though the board charged him with ensuring the execution of those investigations approved at its annual meetings, he could change those planned investigations and reappropriate any required funds.[58] Second, he determined what work would be done, in which section, by which methods, and by whom. He functioned simultaneously as the director of the

Scientific Section and as the immediate supervisor of the director of the Technical Section. In addition, he, like the director of the Technical Section, conducted his own research. Third, the president held responsibility for all personnel matters, and virtually all his personnel recommendations to the board were ratified without debate. He not only assigned routine work to all Reichsanstalt personnel, but also could allow them to pursue their own private research at the Reichsanstalt during their free time. Fourth, he oversaw publication of the Reichsanstalt's work, determining what was to be published, where, and under whose name. Only in cases of "independent scientific investigations" were Reichsanstalt workers allowed to differ in print with the directors. Nor were officials allowed to acquire patent rights for results achieved through means provided by the Reichsanstalt. This provision sought to prohibit Reichsanstalt officials from competing with private industry and to maintain the Reichsanstalt's neutral, independent position in scientific and technological matters.[59] Finally, in accord with a long-standing desire of Siemens and Helmholtz, the president invited scientific guests, men who were "established researchers," to the Reichsanstalt, where they enjoyed the facilities and financial support for short- and long-term stays in order to conduct their own work (Figure 3.2).[60]

This rule-governed, bureaucratic system of scientific organization had a hierarchical work structure that consisted largely of small teams, not individuals. The Reichsanstalt's scientific and technical personnel broke down into six formal categories: members (*Mitglieder*), permanent co-workers (*ständige Mitarbeiter*), scientific assistants (*wissenschaftliche Hilfsarbeiter*), temporary workers (*Hülfsarbeiter*), voluntary co-workers (*freiwillige Mitarbeiter*), and guests. Members and permanent co-workers, aided by their assistants and temporary workers, undertook the assignments given and used the methods suggested to them by the directors. In turn, members and permanent co-workers directed the temporary workers and volunteers under their charge. Authority and the organization of work were hierarchical. In contrast to academic physics institutes, where research was done largely by individuals working alone, the Reichsanstalt operated as a group of teams whose scientific and technical personnel were dedicated to a series of measuring problems whose resolution required a long, and in some cases unending, period of time. To be sure, the teams were small – usually consisting of two to four workers. Yet the work of even the small-

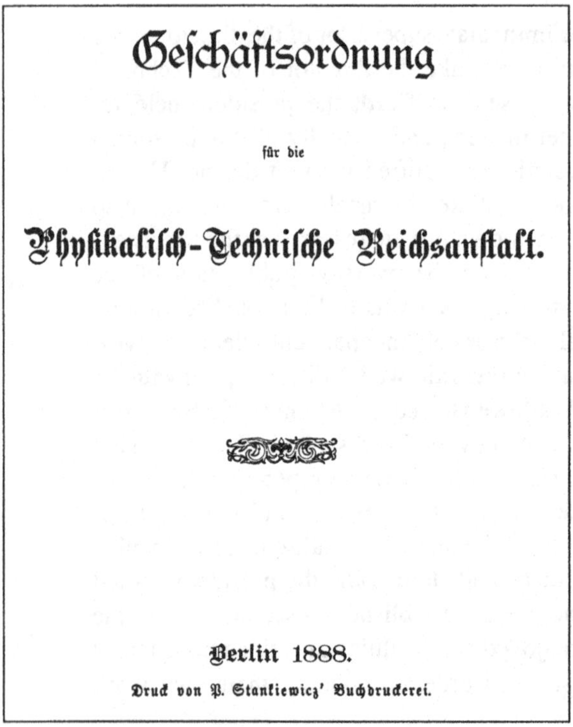

# Geſchäftsordnung

für die

## Phyſikaliſch-Techniſche Reichsanſtalt.

## Berlin 1888.

Druck von P. Stankiewicz' Buchdruckerei.

Figure 3.2. Title page of the Reichsanstalt's standing business orders. Reprinted from *Geschäftsordnung für die Physikalisch-Technische Reichsanstalt* (Berlin, 1888), title page.

est team was commonly part of some larger, overall goal. Long-term precision-measuring physics and technology *was* largely teamwork.

Helmholtz's buildup of the Reichsanstalt into a scientific bureaucracy also meant securing loyal organs of publication in which the Reichsanstalt could quickly report its results and communicate its official announcements. His achievement in this regard reflects his administrative acumen and skill, on the one hand, and the Reichsanstalt's importance to the German physical and technological communities, on the other.

The Scientific Section's long-term, extensive program of measuring physical phenomena led Helmholtz into his sole independent publishing venture for the Reichsanstalt: the *Wissenschaftliche Abhandlungen.* He modeled the Reichsanstalt's *Abhandlungen* on that of the Berlin Akademie's *Abhandlungen,* although the former appeared only as occasion warranted; for example, the first volume did not appear until 1894. In addition to containing the Scientific Section's own extensive measuring results, the *Abhandlungen* contained those of the Tech-

WISSENSCHAFTLICHE ABHANDLUNGEN

DER

# PHYSIKALISCH-TECHNISCHEN

# REICHSANSTALT

BAND I

Figure 3.3. Title page of the
first volume (1894) of the
Reichsanstalt's *Wissen-
schaftliche Abhandlungen.*
Reprinted from *WAPTR* 1
(1894): title page.

BERLIN
VERLAG VON JULIUS SPRINGER
1894.

nical Section "so far as the latter concern methods of lasting importance."[61] The volumes consisted largely of observational data and experimental results, not conceptual matters (Figure 3.3).[62]

Helmholtz found a loyal organ of publication in the *Annalen der
Physik,* Germany's premier physics journal. His task here was an easy
one, largely because he appointed his old friend Wiedemann, the editor
of the *Annalen* from 1877 to 1899, to the Reichsanstalt's board.
Wiedemann had long (since 1877) been indebted to Helmholtz, the
Physikalische Gesellschaft's advisor to the *Annalen,* for his review of
theoretical papers sent to the *Annalen.* He repaid that debt by assuring
the board that he would immediately publish all Reichsanstalt manuscripts in pure physics.[63]

Moreover, the Reichsanstalt's intimate relationship with the *An-*

*nalen* was as much an institutional as a personal one, though the point takes us beyond the Helmholtz years. When Wiedemann died in 1899, a new editorial structure was established for the *Annalen*. Supporting the new editor, Paul Drude, and Planck, Helmholtz's sucessor (in 1895) as the Gesellschaft's advisor to the *Annalen* and as Drude's advisor for manuscripts in theoretical physics, was a board consisting of Reichsanstalt president Kohlrausch, Reichsanstalt board members Quincke, Röntgen, Warburg, and (as of 1908) Planck. After Drude committed suicide in 1906, Planck became coeditor of the *Annalen* with Wien, late of the Reichsanstalt and (as of 1912) a Reichsanstalt board member. Apart from Drude, between 1887 and 1929 the only member of the *Annalen*'s board not associated with the Reichsanstalt was (from 1910 to 1920) Woldemar Voigt.[64]

The *Annalen*'s editors and its board were partial, if not (positively) prejudicial, to manuscripts issuing from the Reichsanstalt. They prized the contributions of Reichsanstalt workers to its pages. For example, in 1907 Planck and Wien had occasion to discuss whether a seemingly unfinished manuscript by Ernst Gehrcke and Otto Reichenheim of the Reichsanstalt merited publication in the *Annalen* or in the *Verhandlungen* of the Physikalische Gesellschaft, a journal devoted mostly to preliminary findings or unfinished pieces of work. Though Planck and Wien had some reservations about publishing the manuscript, Planck informed Wien that submission of manuscripts from Reichsanstalt workers was more important to the *Annalen* than maintaining a principle of accepting only finished manuscripts.[65] A week later Warburg, Helmholtz's successor once removed, persuaded Planck that the manuscript was indeed a finished piece of work and worthy of the *Annalen*. Planck wrote to Wien again, reemphasizing

> that one of the Reichsanstalt's most valuable goods is its cooperation with the *Annalen,* a cooperation that should not be lost because of some formal principle. Proof of this is the profusion of most valuable *Annalen* articles coming from the R[eichs].A[nstalt]. A loss of the same in the future could possibly be disastrous for the *Annalen.* For this reason, too, I have therefore decided to accept the article by G[ehrcke] & R[eichenheim]. Moreover, it would be regrettable if within the R.A. the idea should arise that the *Annalen*'s editors did not place a lot of weight on the R.A.'s cooperation.[66]

Planck's concern for manuscripts emanating from the Reichsanstalt doubtless also reflected his knowledge of the long-standing personal ties between the Reichsanstalt and the *Annalen,* as well as the high quality of the Reichsanstalt's work. The institution and the journal were mutual beneficiaries.

For reporting the Technical Section's results, Helmholtz selected a journal that communicated rapidly to the German technological community: the *Zeitschrift für Instrumentenkunde.*[67] The relationship between the Reichsanstalt and the *Zeitschrift* was even more intimate than that between the Reichsanstalt and the *Annalen,* for the Reichsanstalt became the latter's savior.

Founded in 1881, the *Zeitschrift für Instrumentenkunde* was the official organ of the Deutsche Gesellschaft für Mechanik und Optik. Although devoted to reporting results in the field of scientific technology, its principal concerns were with precision mechanics and optics, with establishing closer relations between scientists and precision technologists, and with furthering scientific instrumentation and measuring methods.[68] Within three years of its founding, the *Zeitschrift,* for reasons unknown, faced financial ruin. To prevent its immediate collapse, the Prussian government provided the *Zeitschrift* with a one-time subsidy of 5,000 marks and an annual subsidy of 300 marks.[69] But that support brought only a reprieve for the *Zeitschrift.* In 1886, as plans for the Reichsanstalt were underway, the Gesellschaft called for a close connection between the Reichsanstalt's Technical Section and the *Zeitschrift,* a connection that they hoped would rescue the latter from possible extinction.[70] In 1888, Fuess, Abbe, and Foerster – all three of whom were board members of both the Gesellschaft and the Reichsanstalt – argued that "only through association with the Reichsanstalt would the *Zeitschrift für Instrumentenkunde* experience an increase in circulation" and thereby become self-supporting. Until that increase occurred, they added, continued survival of the *Zeitschrift* would require an annual subvention of at least 3,000 marks.[71]

As was the case with the Reichsanstalt and the *Annalen,* the Reichsanstalt's and Gesellschaft's boards were like interlocking directorships. Of the twenty members of the Gesellschaft's editorial board of 1887, eight also belonged to the Reichsanstalt's board, and one (Loewenherz) was the director of the Technical Section. One of those eight, the physical chemist Hans Landolt, petitioned the Reichsanstalt's board at one of its first meetings for a subsidy of 5,000 marks for the *Zeitschrift.*

Abbe naturally supported Landolt's petition and added that if the *Zeitschrift* ceased publication, then the Reichsanstalt would have to find or create a new journal. Helmholtz suggested that a 3,000-mark subvention would provide sufficient proof of the *Zeitschrift*'s importance to the Reichsanstalt. The board followed Helmholtz's recommendation.[72] However, a year later Fuess, Abbe, and Foerster again requested that the Reichsanstalt's board turn the one-time subvention of 3,000 marks into an annual one, which the board, despite the ministry's reservations about the legality and propriety of such support, unanimously voted to do.[73] As a consequence, the *Zeitschrift* was now partially financed by, and henceforth edited in cooperation with, the Reichsanstalt. The Reichsanstalt now formally announced that the Technical Section would publish its results in the *Zeitschrift,* though it occasionally used other outlets as well.[74] A few years later the Technical Section took even firmer control over the journal: Between 1895 and 1911 Stephan Lindeck of the Reichsanstalt was the editor; his successor after 1912 was another Reichsanstalt employee, Friedrich Göpel. In this way, the Reichsanstalt became, as the *Zeitschrift*'s editorial board once put it, "the protector of precision technology."[75] For its protection it got what Helmholtz wanted: a guarantee of speedy publication of the Technical Section's results (Figure 3.4).[76]

Helmholtz's third concern in publishing matters – to find an outlet for its offical announcements – was a trivial one. All such announcements, for example, those stating the legally valid electrical standards, were published in the *Centralblatt für das deutsche Reich.*[77] In addition, the Reichsanstalt often published official announcements (especially those concerned with testing and certification standards) in various scientific journals, such as the *Zeitschrift für Instrumentenkunde,* where it also published the annual review of all Reichsanstalt activities.

In sum, the Reichsanstalt could quickly communicate its findings to the German physical and precision technological communities. Helmholtz found loyal organs of publication and thereby integrated the Reichsanstalt into the extant network of scientific communication in Germany; his successors Kohlrausch and Warburg maintained that integration. And where that network was found wanting, Helmholtz (and his successors) issued the *Wissenschaftliche Abhandlungen,* their own organ of publication.

# Zeitschrift für Instrumentenkunde.

*Redaktions-Kuratorium:*

Geh. Reg.-R. Prof. Dr. H. Landolt, Prof. Dr. Abbe und H. Haensch.

Redaktion: Prof. Dr. A. Westphal in Berlin.

XIV. Jahrgang.      **August 1894.**      Achtes Heft.

Charlottenburg, den 19. April 1894.

## 5ter Bericht über die Thätigkeit der Physikalisch-Technischen Reichsanstalt.
### (Dezember 1892 bis Februar 1894).*)

(Mittheilung aus der Physikalisch-Technischen Reichsanstalt.)

*Personal.*

Das Personal der Physikalisch-Technischen Reichsanstalt setzt sich folgendermaassen zusammen:

1) Der Präsident der Phys.-Techn. Reichsanstalt;

| 2) Abtheilung I | Abtheilung II |
|---|---|
| **) | 1 Direktor, |
| 3 Mitglieder, ***) | 6 Mitglieder,***) |
| 2 technische Hilfsarbeiter, | 2 technische Hilfsarbeiter, |
| 4 Assistenten, | 5 Assistenten, |
| 2 wissenschaftliche Hilfsarbeiter, | 6 wissenschaftliche Hilfsarbeiter, |
| 1 freiwilliger Hilfsarbeiter, | 2 technische Assistenten, |
| | 3 technische Gehilfen, |
| 3 Mechanikergehilfen, | 8 Mechanikergehilfen, |
| 1 Maschinist, | 1 Tischler, |
| 1 Heizer; | 1 Klempner; |

ausserdem:

| | |
|---|---|
| 1 expedirender Sekretär, | 2 expedirende Sekretäre, |
| 1 Kanzlei-Sekretär, | 2 Kanzlei-Sekretäre, |
| 3 Unterbeamte. | 5 Unterbeamte. |
| Zusammen: 22 Personen. | Zusammen: 44 Personen. |

### A. Erste (Physikalische) Abtheilung.

*I. Thermische Arbeiten.*
*1. Bestimmung der Ausdehnung von Wasser und Quecksilber.*

Die Bestimmung der relativen Ausdehnung von Wasser und Quecksilber gegen drei verschiedene Glassorten, deren absolute Ausdehnung untersucht war, ist abgeschlossen. Bei den Versuchen hat sich herausgestellt, dass aus derselben Glasröhre verfertigte Gefässe ganz dieselbe Ausdehnung haben; dagegen wurden sowohl für das Jenaer Glas XVI¹¹¹ als auch indirekt für das französische *verre*

*) Die vier bisherigen Thätigkeitsberichte datiren:
1) Vom 13. Dezember 1890 über die Zeit vom Oktober 1887 bis Dezember 1890. (Z.f.I. 1891. S.149.)
2) „ 27. Februar 1891 „ „ „ „ März 1890 „ Februar 1891.
3) „ 29. Februar 1892 „ „ Februar 1891 „ Februar 1892.
4) „ 29. November 1892 über die Jahre 1891 und 1892. (Z.f.I. 1893. S.113.)
  **) Als Direktor der Abtheilung I fungirt der Präsident der Reichsanstalt.
  ***) Davon 1 Mitglied beiden Abtheilungen gemeinsam.      21

Figure 3.4. First page of the Reichsanstalt's annual report for 1893. Reprinted from *ZI* 14 (1894): 261.

## New Quarters, Slow Science

The processes of establishing a governing board of directors, setting rules concerning internal scientific authority and organizational structure, and securing loyal organs of publication were naturally accompanied by those of finding an appropriate site for the Institute and constructing a physical plant in which the institution could function. The acquisition of land and buildings was thus still another way in which Helmholtz set the basis for transforming the Reichsanstalt into a scientific bureaucracy. To be sure, until 1887 Siemens, as we saw in Chapter 2, played the central role in providing land for the Reichsanstalt as well as a major one in designing the *Observatorium*. Moreover, Helmholtz's death in 1894 meant that his successor, Kohlrausch, had to spend much of his first two years in office overseeing the completion of the new buildings for the Technical Section. But this assistance aside, between 1887 and 1894 it was Helmholtz who planned and oversaw the construction of the Reichsanstalt.

The Reichsanstalt's home, the city of Charlottenburg, was until 1920 a self-administered political entity. The eastern city limits were about three kilometers west of the Brandenburg Gate and adjoined Berlin's western border. Until about 1880 Charlottenburg remained primarily a *Residenzstadt:* Great financiers and industrialists like Gerson von Bleichröder, Robert Warschauer, and Werner von Siemens, eminent academics like Gustav Schmoller and Theodor Mommsen, leading artists like Julius Wolff and Christian Rauch, and high government officials had villas there. Its country-like atmosphere provided an antidote to the daily environmental and social problems they encountered in Berlin.[78] The noise from Berlin was inaudible in Charlottenburg, the visitors few, and the air, as Helmholtz's student Hertz, who went to Charlottenburg's beautiful park to read books on physics and philosophy, noted, was "filled with the scent of lime trees and oranges."[79]

In the 1880s, however, the scent began to change and the beauty to fade. The growth of new industries in Berlin and Charlottenburg slowly transformed the city's character. The middle classes began settling in Charlottenburg, and the city experienced a population explosion: from about 31,000 residents in 1880 to about 306,000 in 1910.[80] During the imperial period Charlottenburg changed from a sparsely populated country-like town into a densely packed industrial city.[81] Still, in the late 1880s Charlottenburg offered enough undisturbed space for an institute devoted to exact, disturbance-free physical research and testing.

Moreover, the city housed the Technische Hochschule Charlottenburg, whose buildings were completed in 1884 and where rooms were promised for the Reichsanstalt's Technical Section. With the Reichsanstalt's establishment in 1887, Charlottenburg was on the way to becoming Germany's technopolis (Figure 3.5).

Charlottenburg was thus a natural, and the only, choice for the Reichsanstalt's location. Behind his estate and only a few hundred meters away from the Technische Hochschule, Siemens owned a large tract of land on the March Strasse. As we know, in 1885 he gave this tract – a total of 19,800 square meters valued at 566,157 marks – for the Reichsanstalt (the rectangular area *ABCD* of Figure 3.6).[82] More particularly, he offered it as the site for the Scientific Section. Both he and Helmholtz expected that this large tract would provide enough space for the Scientific Section's present and future needs. (They foresaw the Reichsanstalt's expansion even before it received governmental and legislative approval.)[83] Moreover, they hoped that either Siemens's gift of land by itself or in combination with some future acquisition of additional land would provide enough space eventually to locate the Reichsanstalt's Technical Section next to the Scientific. That acquisition occurred in 1892, shortly before Siemens's death. In order to build new facilities for the Technical Section, the Reich purchased from Siemens an additional 14,389 square meters costing 373,106 marks (the rectangular area encompassed by line *AB* and *Strassen* 4, 5, and 7 in Figure 3.6).[84] Siemens's desires to aid pure physics in general and his friend Helmholtz in particular had led him to offer the Reich land for the proposed Reichsanstalt if the Reich agreed to construct and equip a *scientific* building along with a residence for the Reichsanstalt's president. He *gave* the land for the Scientific Section; he *sold* that for the Technical. By 1893, the Reichsanstalt occupied 34,189 square meters (or about 8.45 acres) of land valued at around 939,263 marks. Before World War I, at least, it occupied the largest site and had the largest plant for physical research in the world.

Until the legislature's final approval of the Reichsanstalt in March 1887, it was Siemens who initiated and sustained the efforts to construct it. Although he offered (in 1884) his inheritance from the estate of his late brother William in order to begin construction, he warned the Reichsanstalt's (first) architect, Paul Spieker, that because the state might reject his project to establish a Reichsanstalt, the buildings "had to be so laid that in this case they will become useful for private appli-

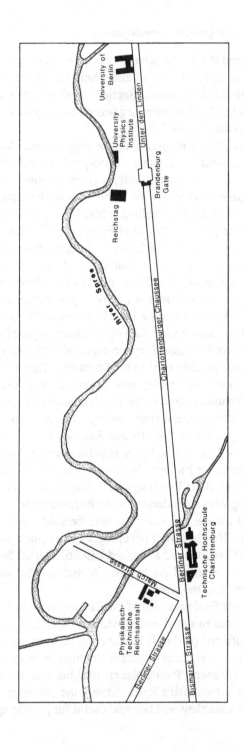

Figure 3.5. The Reichsanstalt's location in relation to the Technische Hochschule Charlottenburg, the University of Berlin's Physics Institute, and other Berlin landmarks.

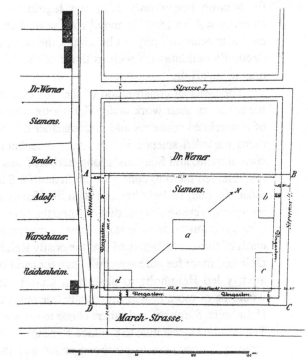

Figure 3.6. Original site plan of the proposed Reichsanstalt. Reprinted from "Denkschrift betreffend die Begründung eines Instituts für die experimentelle Förderung der exakten Naturforschung und der Präzisionstechnik." (Physikalisch-mechanisches Institut.) Vom 16. Juni 1883, *SBVR. B. Sammlung sämtlicher Drucksachen des Reichstages* (Berlin, 1887), Drucksachen Nr. IV, Beilage B, 51-58, Anlage 2, p. 67.

cation."[85] In the end, however, Siemens spent only 8,800 marks in cash on the Reichsanstalt.[86]

Spieker, Helmholtz, and, to a lesser extent, Siemens designed the scientific building (*Observatorium*) and the president's residence.[87] Spieker supervised the architectural planning of the Reichsanstalt until July 1885;[88] thereafter, Theodor Astfalck, another experienced architect of scientific and technological buildings, supervised the planning.[89] For the Scientific Section, Astfalck filled in the architectural details not provided by Spieker, Helmholtz, and Siemens. Both before and during construction Astfalck and a member of the Reichsanstalt visited physical, chemical, astronomical, and meteorological laboratories in Vienna, Munich, Paris, Strasbourg, Würzburg, and Dresden in order to examine firsthand the "essential advances in the equipment of

scientific laboratories, especially in the field of electrotechnology."
The trip led, Helmholtz stressed, to "a series of changes in the originally planned interior equipment"; if these changes had not been made,
"right from the very start the Reichsanstalt" would have been "behind
other [institutes] pursuing similar goals."[90] Construction of the Scientific Section began only after the legislature finally approved the
Reichsanstalt in 1887. It ended four years later, in 1891. As was the
case with academic physics buildings, the construction of the Scientific
Section's buildings (as well as those of the Technical Section) was a
drawn-out affair.[91]

The order of construction of the Reichsanstalt's two sections and
the nature of their work under Helmholtz were the result of a number
of interrelated concerns and circumstances. First, Helmholtz and Siemens regarded science as the basis of technology. That ideological
commitment, and Siemens's long-harbored desire to build a scientific
institute, led them to construct the Scientific Section before the Technical. Second, Helmholtz, and with him Siemens and the other board
members, were anxious to demonstrate the new institute's usefulness to
German industry and the state, thereby overcoming the lingering concerns of those who believed that the costly Reichsanstalt was being established more for pure scientific than for technological purposes. That
anxiety led Helmholtz to push both sections to produce results that
would be of more or less immediate relevance to industry. Moreover,
Helmholtz, Siemens, and others chose to use the available rooms at the
Technische Hochschule to produce some immediate results for German
industry, rather than to postpone or hamper the Technical Section's
work until new buildings for it were constructed.[92] Third, as early as
1884 an Interior official had warned Siemens and other founders of the
Reichsanstalt "that, in the interests of the Reich's present financial situation, the Reich's sacrifice not be made too large and that the new
construction be limited to the physical institute [i.e., the Scientific Sec-
tion]."[93] The inability or unwillingness of the German government to
finance the full Reichsanstalt thus also influenced Helmholtz's decision to postpone the construction of new buildings for the Technical
Section until those for the Scientific Section had been completed. The
results of these concerns and circumstances were twofold: New quarters were built first for the Scientific Section, thereby slowing the
development of science there, while old quarters were used (during the
first years) for the Technical Section, thereby allowing it to produce

some immediate technological results. The decisions taken under Helmholtz with regard to the timing of construction of the Reichsanstalt's buildings were to have a profound impact on the future course of the institution.

As an experienced builder of scientific institutes, Helmholtz knew that it would take a long time to build new facilities for the Reichsanstalt. In the meantime, provisional quarters had to be found. For the first six months, from its formal opening on 1 October 1887 until 1 April 1888, the Scientific Section operated out of rooms in Pernet's private laboratory, rented for forty marks per month.[94] Then it moved into the (already overcrowded) facilities of the Technical Section at the Technische Hochschule, where it remained until the fall of 1891 (Figure 3.7). During the first four years of the Scientific Section's existence (1887- 1891), the nature and extent of the work it would undertake was limited by the cramped, grossly inadequate workspace it occupied within the quarters set aside for the Technical Section. In particular, between 1887 and 1891 its work was restricted largely to preliminary thermometric studies needed to establish scientific foundations for the Technical Section's own thermometric work.[95] As a consequence of these logistical limitations, the Scientific Section during Helmholtz's reign produced only a minor amount of purely scientific results.

Helmholtz planned or helped plan ten buildings, five for each section, set within a finely landscaped garden that served esthetic and physical ends. The buildings were kept as distant from one another as possible so as to minimize any disturbances emanating from, and to maximize the amount of sunlight to, each building. Except for the establishment of a small thermometer-testing station in Ilmenau in 1889 and a disturbance-free magnetic laboratory near Potsdam in 1913, all of the Reichsanstalt's buildings were located in Charlottenburg (Figures 3.8 and 3.9).

The Scientific Section's main building, known as the *Observatorium,* was the Reichsanstalt's central and premier structure (Figures 3.10 and 3.11). The location and architectural style of all the other Reichsanstalt buildings were dependent on the decisions made about the location and architecture of the *Observatorium.* The building's planners sought to achieve two principal structural characteristics: maximum freedom from external disturbances, in particular alien mechanical and electromagnetic phenomena; and maintenance of temperature regularity and control.[96]

*Skizze*
*Der im Gebäude der technischen Hochschule*
*für die zweite Abtheilung der Physikalisch-technischen*
*Reichsanstalt verfügbaren Räume.*

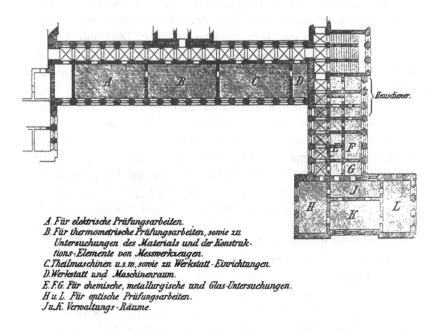

*A. Für elektrische Prüfungsarbeiten.*
*B. Für thermometrische Prüfungsarbeiten, sowie zu*
   *Untersuchungen des Materials und der Konstruk-*
   *tions-Elemente von Messwerkzeugen.*
*C.Theilmaschinen u.s.w., sowie zu Werkstatt-Einrichtungen.*
*D.Werkstatt und Maschinenraum.*
*E.F.G. Für chemische, metallurgische und Glas-Untersuchungen.*
*H u.L. Für optische Prüfungsarbeiten.*
*J u.K. Verwaltungs-Räume.*

Figure 3.7. Sketch of the rooms of the Technical Section, housed in the Technische Hochschule Charlottenburg, 1887-96. Reprinted from "Denkschrift betreffend die Begründung eines Instituts für die experimentelle Förderung der exakten Naturforschung und der Präzisionstechnik." (Physikalisch-mechanisches Institut.) Vom 16. Juni 1883, *SBVR. B. Sammlung sämtlicher Drucksachen des Reichstages* (Berlin, 1887), Drucksachen Nr. IV, Beilage B, 51-58, Anlage 4, p. 71.

To achieve a disturbance-free structure, the planners made a number of key decisions and spent a good deal of money. First, they located the *Observatorium* toward the center of the grounds, that is, as far away

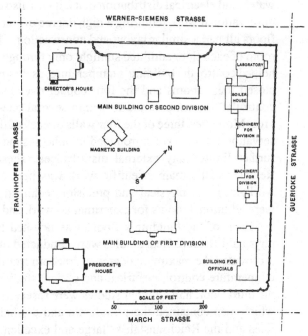

Figure 3.8. Layout of the Reichsanstalt (circa 1900). Reprinted from Henry S. Carhart, "The Imperial Physico-Technical Institution in Charlottenburg," *TAIEE 17* (1900): 565.

from street traffic and the other buildings as conditions would permit.[97] Second, they built the entire structure on a two-meter-thick concrete slab of approximately 1,000 square meters. This gave solidity to the building and helped maintain dryness.[98] Solidity as well as greater temperature control was increased through a number of other features: an especially strong double outer wall on the ground floor; an inner wall and vaulted ceilings on the main floors; a half-meter-high, barreled-arch crown covering; reinforced curved girths; and terrazzo floors (Figure 3.12).[99]

To achieve temperature regularity or maximize temperature control, especially for the interior workrooms, the planners oriented the axis of the nearly square building due north, with its front along March Strasse. The east and west sides of the building were shielded against sunlight to help maintain internal temperature control.[100] Steam heating, electric lighting, and a built-in air-ventilation system also helped keep the building's temperature at 20°C.

The *Observatorium* consisted of a basement, three main floors, and a dome.[101] The basement served principally to insulate the building's heat against ground moisture and to house the building's heating, gas, water, and electrical distribution units; it was also occasionally used for observational work requiring a high degree of stability. The three main floors all had a similar layout and construction. The center and largest room of each floor admitted sunlight only through a skylight. This, too, helped control the building's temperature, as did the thick walls that enclosed these rooms and the four copper ovens with thermostats that controlled the temperature to within several one-hundredths of a degree. Moreover, three of the four walls of each of these three central observation rooms were surrounded by a horseshoe-shaped corridor that further limited any external disturbances. These rooms, where the Reichsanstalt's main scientific work was performed, were excellent workplaces for physical and precision-technical work. Around them were additional rooms for experimental work and administration.

Each of the building's floors was devoted to a specific type of physical research: Heat studies were conducted on the first floor since it provided the maximum degree of freedom from shaking and the best temperature control; electrical and optical studies were performed on the third floor; and all other studies were reserved for the second floor, which also contained the president's and the Scientific Section's offices and the Reichsanstalt's "large and excellent library of works on pure and applied science," the nucleus of which came from the purchase of the libraries of Kirchhoff and Helmholtz after their deaths.[102]

Like much else in wealthy Charlottenburg, the *Observatorium* presented an elegant architectural façade to the world. With its yellow brick and stylized cornices, pillars, and crown, the *Observatorium* was an imposing, villa-like structure of the classical Renaissance style.[103] Altogether, it cost 387,000 marks to construct.[104] Though it cost less than a number of academic institutes, the latter's facilities were devoted principally to pedagogy, whereas the *Observatorium*'s were devoted solely to research. The Reichsanstalt got more research space per mark.

Moreover, the costs of an academic physics institute included a director's residence. The Reichsanstalt, by contrast, provided its director with his own separate residence, a large two-story structure (Figure 3.13).[105] The Helmholtzes – Hermann and Anna – helped plan it, knowing that they were planning for themselves and their family as well as for all future Reichsanstalt presidents. The building's "villa-like" exterior made it a fitting residence for old Charlottenburg.[106] Otto Warburg, Emil's son, thought the president's residence was "like a palace."[107] It was, in effect if not intent, a gift from Siemens, who lived down the street, to Helmholtz, the doyen of German physicists, Siemens's dear friend, and his daughter's father-in-law. That gift cost the Reich nearly 100,000 marks.[108] It was an elegant setting for Helmholtz and his successors to receive and entertain many of the world's leading scientists and many members of Germany's social, industrial, and political elite.

Whereas academic physics institutes provided at most a room or two for administrative purposes, the Reichsanstalt had an entire building for such business. The very existence of an administration building – the first of its kind for a physical institution – itself suggests the bureaucratic character of the Reichsanstalt. The two-story building, which cost 100,000 marks to construct, housed a business office for the Scientific Section, the board's conference room, and the board president's office and contained residences for the section's steward, its machinist, a married attendant, and two unmarried assistants.[109]

To further protect the *Observatorium* against external disturbances and varying temperature, the Reichsanstalt's planners built a separate, small, iron-free magnetic house and a separate machine house. The latter contained the Reichsanstalt's boiler plant, a Deutzer twin gas motor for operating storage batteries, and a direct-current dynamo. In addition to housing observation rooms and a chemical laboratory, the machine house also contained a room for low-temperature studies. Supplied with a Lindean "ice machine" for producing liquid air and a refrigerator for producing ice and for cooling the observation room, the Reichsanstalt's workers could easily reach and maintain the low temperatures needed for their scientific work. The machine house cost 50,000 marks to construct; the small magnetic house 8,500 marks. The availability of such funds and the care and foresight with which these secondary buildings were planned and constructed made it possible to supply the *Observatorium* with a sufficient amount of electrical, magnetic, and thermal power without thereby adversely affecting the building's goals of freedom from external disturbances and temperature control.[110]

Construction of the Scientific Section's five buildings – the *Observatorium*, the president's residence, the administration building, the magnetic house, and the machine house – cost a total of 644,754 marks. A further 232,000 marks were allowed for paving, landscaping, a water and drainage system, heating, water, gas, and electrical connections, furniture, and so on, as well as 82,310 marks for (initial) equipment and machinery for the Scientific Section (Table 3.2).[111] Thus the Scientific

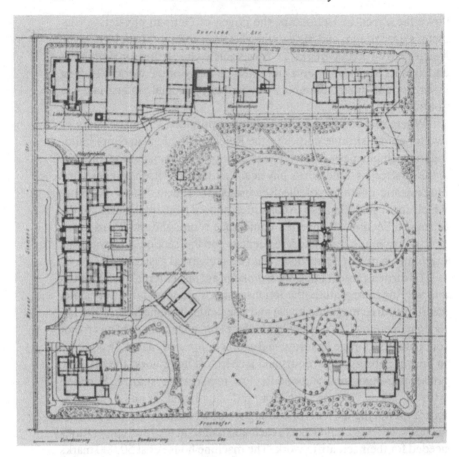

Figure 3.9 (*above*) Layout of the Reichsanstalt (circa 1906). Reprinted from Ernst Hagen and Karl Scheel, "Die Physikalisch-Technische Reichsanstalt," *Ingenieurwerke in und bei Berlin: Festschrift zum 50 Jährigen Bestehen des VDI*, ed. Verein Deutscher Ingenieure (Berlin, 1906), following 67.

Figure 3.10 (*facing page, top*) Frontal view of the *Observatorium*. Reprinted from Henry S. Carhart, "The Imperial Physico-Technical Institution in Charlottenburg," *TAIEE 17* (1900): 560.

Figure 3.11 (*facing page, bottom*) Frontal view of the *Observatorium*. Courtesy of Siemens Museum, Munich.

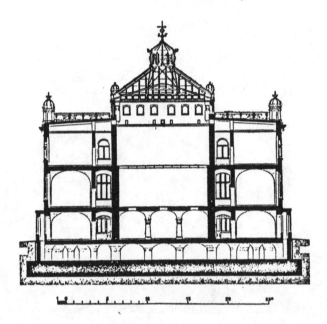

Figure 3.12. Cross-sectional view of the *Observatorium*. Reprinted from "IV. Gebäude für die Verwaltungsbehörden des Deutschen Reiches. 11. Die Physikalisch-Technische Reichsanstalt in Charlottenburg," *Berlin und seine Bauten,* ed. Architekten-Verein zu Berlin und Vereinigung Berliner Architekten, 3 vols. in 2 (Berlin, 1896), *2,* 83.

Section cost the Reich 959,064 marks to build and equip. Except for the Berlin and Leipzig Physical Institutes, the Reichsanstalt's Scientific Section was itself the most costly physical institute in imperial Germany. It provided more floor space and probably better facilities for research than any academic physics institute of the day.

The purpose of constructing new quarters for the Reichsanstalt was, of course, to provide it with appropriate facilities in which to do its work. Beyond construction of the physical plant, Helmholtz's major preoccupation after 1887 was the hiring of staff to conduct that work along with the planning of a work agenda for the Reichsanstalt. What he accomplished during his presidency can best be seen by examining the Reichsanstalt's staff and work for 1893. By then, some six years after it had begun operation and some two years after the Scientific Section's buildings had been completed, the Reichsanstalt was as fully functional as it was to become during Helmholtz's presidency.

Figure 3.13. Residence of the Reichsanstalt president. Reprinted from Henry S. Carhart, "The Imperial Physico-Technical Institution in Charlottenburg," *TAIEE 17* (1900): 558.

By 1893 the Scientific Section consisted of three laboratories – Heat, Electricity, and Optics – plus a business office.[112] It employed twenty-two individuals, of whom twelve conducted scientific or technical work, five maintained the plant and laboratories, and five manned the office. It had more professional physicists on its staff than any other scientific institution of the day.

In principle, the entire section was operated under Helmholtz's direction; in fact, Helmholtz looked to the heads of the three laboratories – the members – to supervise research in their respective domains. The members were Helmholtz's closest associates at the Reichsanstalt and his handpicked men. All three members of the Scientific Section of 1893 had studied physics at Berlin, where they received their doctoral degrees under Helmholtz. Their role was similar to that of his former university assistants, except that a Reichsanstalt member naturally had more independence and authority than a student assistant had.[113] Their salaries ranged from 4,000 to 6,000 marks per year plus a housing allowance, a level of compensation scarcely attained by a university assistant.[114] Below each of them, as well as below their counterparts in the Technical Section, stood a team of men: permanent co-workers, scientific assistants, temporary workers, voluntary co-workers, and guests (Table 3.3).

In 1893 Max Thiesen headed the Heat Laboratory. After receiving his doctoral degree at Berlin in 1878 under Helmholtz and Kirchhoff, and after lecturing there and working at the Normal Eichungskommission until 1883, he went to work at the Bureau international des poids et mesures in Paris. Helmholtz called Thiesen to the Reichsanstalt in 1890 largely for his work in extending Regnault's gas-thermodynamic studies as well as for his firsthand knowledge of French work in thermometry. He was, as Helmholtz noted, well versed in the art of taking physical measurements.[115] In addition to Thiesen, the laboratory had three others on its staff: Ernst Gumlich, Karl Scheel, and Louis Sell.

The Heat Laboratory was the largest of the section's three laboratories. During Helmholtz's tenure and beyond, its aims were threefold: to find better materials for thermometers, to make increasingly accurate temperature determinations at increasingly higher temperatures, and to understand the influence of temperature, pressure, and other parameters on the workings of heat engines. These aims coincided largely with the laboratory's overall, long-term goal: to establish a trustworthy, absolute thermodynamic temperature scale for all measurements of heat.

In keeping with these aims, in 1893 the laboratory undertook a number of problems in thermometry and related fields.[116] It investigated the relative expansion of water and mercury in different types of glass and the use of a new type of glass produced at the Zeiss and Schott factories in Jena for thermometers. It attempted (unsuccessfully) to employ mercury thermometers for achieving more precise temperature measurements. It also determined the expansion coefficients of a number of types of solids – among them glass and porcelain – used in temperature-measuring devices. The crucial instrument here was a Fizeau-Abbe dilatometer used to indicate the transition points between the liquid and solid states. (Abbe himself supplied the dilatometer.)[117] Finally, to further understanding of the influence of "hardening temperature" on the magnetization of steel and to aid in achieving a unit of luminous intensity (*Lichteinheit*), the laboratory undertook pyrometric experiments that measured temperatures up to the melting

Table 3.2. *Costs of establishing the Reichsanstalt, 1887–97*

| Scientific section | Marks |
|---|---|
| *Observatorium* | 387,000 |
| Machine house | 50,000 |
| Administration building | 100,000 |
| President's residence | 99,254 |
| Small magnetic house | 8,500 |
| Paving, landscaping, water and drainage system, supervision of construction, etc. | 174,000 |
| Heating, water, gas, and electrical connections, furniture, etc. | 58,000 |
| Initial equipment and machinery | 82,310 |
| Subtotal | 959,064 |

| Technical section | Marks |
|---|---|
| Acquisition of land | 373,106 |
| Main building | 922,000 |
| Laboratory building | 218,000 |
| Machine (and boiler) house | 180,000 |
| Director's residence | 140,000 |
| Additional plant costs | 348,000 |
| Fittings and furniture | 108,300 |
| Initial equipment and machinery | 471,390 |
| (Less reduction for the year 1895–96) | (47,500) |
| Subtotal | 2,713,296 |
| Total | 3,672,360 |

*Sources:* See the Reich budgets for the Reichsanstalt for the years 1887/88, 1891/92, 1892/93, 1893/94, 1895/96, 1896/97, 1897/98, and 1898 as given in *HEDR* (1887–98) (for exact page references see notes 103, 107–10, 193, 197, 199, 200, 202, 204–5 of this chapter); "IV. Gebäude für die Verwaltungs-behörden des Deutschen Reiches. 11. Die Physikalisch-Technische Reichsan-stalt in Charlottenburg," in *Berlin und seine Bauten,* ed. Architekten-Verein zu Berlin und Vereinigung Berliner Architekten, 3 vols. in 2 (Berlin, 1896), *2,* 80–84, on 81–82; and Henry S. Carhart, "The Imperial Physico-Technical Institution in Charlottenburg," *TAIEE 17* (1900): 555–83, on 561–62.

Table 3.3. *Number of Reichsanstalt employees and range of their salary/wages, 1893*

| Position | No. of employees | Range of salary/wages (marks) |
|---|---|---|
| President (= director of Scientific Section) | 1 | 24,000 |
| Director of Technical Section | 1 | 7,500 |
| Member | 8 | 4,000–6,000 |
| Permanent co-worker | 4 | 2,100–4,200 |
| Scientific assistant | 16 | 1,800–2,640 |
| Office worker | 3 | 2,100–4,200 |
| Legal secretary | 3 | 1,800–2,700 |
| Machinist | 1 | 1,800 |
| Mechanic/artisan | 19 | 1,200–1,800 |
| Minor official | 9 | 1,000–1,500 |
| Total personnel | 65 | |

*Note:* The figures for salary/wages do not include housing allowances.
*Source:* K. Strecker, "Die Personalverhältnisse der wissenschaftlichen Beamten der Physikalisch-Technischen Reichsanstalt" (29 August 1911), Anlage 1, BA 265, PTB-IB.

point of platinum.[118] The purpose of both the pyrometric and the luminous-intensity work was the accurate determination of high temperatures, and both were undertaken in response to requests by German industry. By the early 1890s, the Reichsanstalt, along with the Jena works of Zeiss and Schott, had become the leading center in German thermometry and had greatly helped to emancipate the German industry from French domination.

The Electrical Laboratory was Wilhelm Jaeger's fief. After receiving his doctoral degree under Helmholtz in 1887, Jaeger entered into the new Reichsanstalt's service and soon emerged as one of the foremost specialists in electrical measurement.[119] Assisting him on a part-time basis was Ludwig Holborn. Together they sought to determine the values of the fundamental electrical units (current, resistance, and voltage) and constructed the necessary measuring equipment for such work. Their work represented the core of the Reichsanstalt's efforts to fend off the French and to dominate international electrical metrology.[120] As we saw in Chapter 1, the electrical industry needed legal electrical measuring units and standards, as well as the testing and certification of commercial electrical products, for the future commercial development of telegraphy, railroads, illumination technology, time regulation, metallurgy, mechanical engineering, and so on.[121] Helmholtz was acutely aware of the large amount of capital involved here and, hence, the financial importance to Germany of quickly achieving trustworthy electrical standards.[122] The laboratory marked its first major success in establishing electrical standards (for the ampere and the ohm) with the resolutions reached at the International Electrical Congress in Chicago in 1893. However, the laboratory (in 1893) sought principally to establish commercial or legal standards, not purely scientific ones.[123] Still, its work was at once scientific, technical, and political.

The Electrical Laboratory also conducted magnetic investigations on iron and steel. Its work on the magnetic side effects and on the process of induction in different types of iron and steel was crucial to both the electrical and the steel industries – for instance, in constructing improved dynamos.[124] Moreover, the German navy petitioned the Reichsanstalt for help with its magnetic problems: The presence of a large amount of iron in its ships and torpedo boats led to deviations, sometimes by as much as 50 percent, in the directions of its compass needles. The navy asked the Reichsanstalt to determine which types of iron were free from remnant magnetism and which types of compasses best met their needs. It was a matter, they stressed, "of great importance for the imperial navy."[125]

Perhaps more than any of its counterparts, the Optics Laboratory embodied the hopes of the Reichsanstalt's founders for the mutually beneficial interaction of science and technology. Otto Lummer led both this laboratory and its counterpart in the Technical Section; both laboratories worked on different aspects of essentially the same scientific-technological problems. Like his colleagues Thiesen and Jaeger, Lummer had received his doctoral degree from Berlin, in his case with a dissertation in 1884 under Helmholtz. He stayed on at the university as Helmholtz's assistant until 1887; he then moved with him to the Reichsanstalt. His connection to Helmholtz was as much personal as scientific: He had befriended Helmholtz's sickly son Robert – himself a physicist who had been hired in 1889 to work under Lummer at the Reichsanstalt but who died before he could do so – and he visited the family often: " 'Papa' Helmholtz," Lummer later wrote, "was a second father to me and Robert a very intimate friend."[126] Both as a pure and as an applied scientist, and as a close associate of Helmholtz's, Lummer was a central figure at the Reichsanstalt. His assistants were themselves former Helmholtz students: Ernst Gumlich, who studied with Helmholtz at Berlin before receiving his degree from Jena in 1885[127] and who divided his time between Heat and Optics; Willy Wien, who took his doctoral degree with Helmholtz at Berlin in 1886;[128] and Ferdinand Kurlbaum, who took his degree under Helmholtz in 1887.[129]

The Optics Laboratory concentrated most of its efforts on photometry, the measurement of various quantities describing light.[130] It did so at the request of the Deutscher Verein für Gas- und Wasserfachmänner (German Association of Gas and Water Specialists). In 1888 the Verein had asked the Reichsanstalt, through the Ministry of the Interior, to help evaluate the currently accepted light standard and, in its place, to establish an internationally acceptable unit of luminous intensity.[131] The Verein stressed that both "scientific and technical circles" in England, France, and America, as well as in Germany, sought a standard for measuring the brightness of light. Photometric measuring standards, they argued, were urgently needed for both scientific and industrial purposes.[132] Helmholtz enthusiastically supported the Verein's request.[133] Moreover, the German navy was keenly interested in improving its photometric devices and in overcoming problems of brightness loss under adverse weather conditions.[134]

In response, Lummer and his team initially sought to develop an improved yet practical photometer, that is, an instrument for comparing light intensities from two different sources. Since 1844 photometric measurements had been made using a so-called Bunsen grease-spot photometer. In the early 1890s Lummer and his subordinate, Eugen Brodhun of the Technical Section, first improved Bunsen's device fourfold and then developed their own so-called contrast photometer (with its smooth-surfaced optical cube), which was twice as sensitive as their improved Bunsen device. By 1892 the German firm of Schmidt & Haensch, following the specifications supplied by Lummer and Brodhun, provided a better, less expensive, more durable, and more easily reproduced photometer, one that even an inexperienced worker could use. The Lummer-Brodhun contrast photometer became widely used in the gas and electrotechnical industries as well as at German and foreign physics institutes. It was awarded a prize at the 1893 International Exposition in Chicago, which gave Lummer occasion to reproach his academic colleagues for having treated photometry "rather slightingly." They had neglected the field, he claimed, until the needs of the illumination industry and the public had shown them its importance.[135]

Despite the practical success of the Lummer-Brodhun photometer, the Reichsanstalt had yet to provide the Verein and others with a unit of luminous intensity. In 1893 Helmholtz reported that the Hefner lamp – eponymously named for Friedrich Hefner-Alteneck, Siemens's top engineer and a consultant with the Reichsanstalt on this problem – fulfilled the Verein's request.[136] The Hefner lamp was supposedly superior to both Siemens's own and that of the French (Violle) and the English (Pentan lamp). Although the Hefner lamp provided a usable unit of luminous intensity for German industry, it did not satisfy international demands and it did not constitute a solution to the fundamental photometric problem of finding such a unit.

To do so, the Optics Laboratory now sought a physical or scientific standard, that is, one that was well defined, of the highest accuracy possible, and reproducible; this became its "most important and most difficult task."[137] The illuminating power of such a standard, Lummer argued, would serve as the measure of the illuminating power of all light sources, and thus as the basis for measuring the temperature of all radiating bodies.[138] In 1894 Lummer was absolved of his responsibilities for the Technical Section's Optics Laboratory. He and Kurlbaum now joined forces on a full-time basis in order to construct a bolometer sensitive enough to provide the physical standard. Their purpose in undertaking the bolometric investigations, they said, was twofold: to conduct "general experiments on the radiation of flames and radiation measurements in the visible and invisible parts of the sun's spectrum" and to bring the bolometer "into the service of photometry."[139]

The basis of their investigations was the bolometric principle, first announced by A. F. Svanberg in 1851 and greatly refined by S. P. Langley in 1881, which stated that radiation intensity could be measured in terms of the change in resistance undergone by a thin metal conductor.[140] Lummer and Kurlbaum had previously found that for their purposes pure platinum served as the best conductor and, given a constant temperature along the length of a platinum strip, always radiated the same amount of light.[141] Langley's principle and their experimental results led them to define the unit of luminous intensity in terms of the amount of light radiated by one square centimeter of glowing platinum at a definite temperature, the latter being defined in terms of the relation between two bolometrically measured radiations, the so-called whole and partial radiations. In 1894 they triumphantly reported having achieved a luminous standard reliable to within 1 percent.[142] Nonetheless, this bolometric method was too new and too undeveloped to warrant its introduction into technology.[143] Nor, as we shall see in Chapter 4, did it sufficiently meet the problems raised in measuring blackbody radiation.

In addition to its photometric studies, the Optics Laboratory also undertook polarimetric investigations. Both the German sugar industry and the German customs bureau placed special importance on the testing and certification of polarization instruments, which measured so-called optical activity, that is, the degree of rotation of a substance's plane of polarization. The Deutscher Verein für Rübenzucker-Industrie (German Association of the Beet Sugar Industry), a group with which the physical chemist and Reichsanstalt board member Landolt was particularly closely associated, had originally requested the Reichsanstalt's aid in improving their polarimeters.[144] Starting in the 1870s, Landolt had regularly investigated the optical rotating ability of organic substances, and he conducted much work on optical polarization apparatus, including the saccharimeter, a type of polarimeter designed to measure sugar content, which made it of obvious value to the sugar industry.[145] The industry hoped that the Reichsanstalt could develop a standard quartz material that would aid them in determining sugar content and, hence, prices. Finally, polarization instruments could measure the optical rotating ability of numerous organic substances (e.g., camphor and alcohol), thereby determining their content as well.[146] As was the case in photometry, polarimetric work required a standard of comparison. The Optics Laboratory studied the optical rotating ability of standard quartz plates in order to gain some comparative data. In 1893 it measured the dependency of rotation of plates at temperatures between 0° and 30°C and found that, within this interval, left- and right-rotating quartz plates had the same dependency of rotation.[147] As with its radiation work, the Optics Laboratory would continue its polarization measurements throughout the imperial period.

During Helmholtz's tenure the Scientific Section's work arose out of a mixture of scientific and industrial interests. Nonetheless, all of its laboratories conducted standards research that was of interest primarily to industry and only secondarily to pure science. In addition, the Scientific Section constructed or improved a number of measuring instruments at the request of industry or the state. Although Helmholtz imparted to the section, and to the entire Reichsanstalt, a research spirit, under his administration the section spent most of its time creating physical standards, instruments, and measuring methods. Until its new facilities were ready in late 1891, its personnel was completely devoted to helping solve the thermometric problems of the Technical Section. Thereafter, the Scientific Section began its own work. Above all, its laboratories sought to establish an absolute thermodynamic temperature scale for heat measurement, to determine a fundamental set of electrical standards, and to establish a unit of luminous intensity. It was, in short, less a section for creating new science per se than one for advancing science-based technology and technology-based science.

Helmholtz and his colleagues were aware of these trends. Reviewing the section's achievements through 1892, Helmholtz reported that the section "had in fact been kept so completely busy with the problems given to it by technology that all of its members' and assistants' time had been thereby taken up."[148] Three years later his successor Kohlrausch reported that the section's "scientific work to date primarily concerns fields whose research has been occasioned by the interests of public life."[149] Helmholtz had built new quarters for the Scientific Section, but as a consequence its scientific work developed only slowly under his administration.

## Old Quarters, Fast Technology

In contrast to the Scientific Section, the Technical Section had its own quarters – eleven rented rooms on the ground floor of the Technische Hochschule Charlottenburg – from the very start of its operations (17 October 1887) (Figure 3.7).[150] That fact allowed the section to begin its work at once and so to quickly produce results of use to German industry. The quarters were old, but the technological work came fast.

It came fast in spite of the inadequacy of the quarters for the section's needs, both for the short and the long term. The rooms were too small, improperly structured, and insufficiently equipped.[151] There was not enough space for the section's precision-mechanical, optical, and chemical work, forcing some work to be performed in the corridors; and there was inadequate heating and ventilation equipment.[152]

A remark by Helmholtz perhaps best illustrates the inadequacies: "The presence of iron rafters on the ceilings and iron bars on the windows has a very disturbing influence on electrical studies."[153] To this injury the Technische Hochschule authorities, who failed to provide all the space promised, added insult: In 1892 they asked the Reichsanstalt to vacate the rooms as soon as possible.[154] Yet despite the inadequacies and the notice to vacate, the Reichsanstalt's leaders had no choice but to keep the Technical Section in the old, provisional quarters until space became available (late in 1891) for several of its laboratories in the Scientific Section's new building and then in the Technical Section's own new buildings, which were completed in 1896 and ready for occupation the following year.

The Technical Section's high productivity under far from optimal conditions was due in no small part to the leadership of Leopold Loewenherz. As the section's first director, Loewenherz naturally had to help establish the section itself, which meant in the first instance arranging its quarters and hiring its staff. Beyond that, his (and his successors') chief duties, as Loewenherz outlined them to Helmholtz, were threefold: to purchase or build the instrumentation and to develop the measuring methods needed for the section's testing and certification work; to develop and maintain close contacts with industry, particularly through regular visits to various firms; and to publish or assist his subordinates in publishing the section's scientific/technical results, thereby making them quickly available to industry.[155] Moreover, Loewenherz had to administer *both* sections of the Reichsanstalt, including the Technical Section's extensive business correspondence and legal documentation generated by its testing and certification work. He fulfilled these duties well because he brought proven organizational and scientific experience to his new post. Unlike Helmholtz, he was a man of two worlds: science and industry.

Although an artisan's son, Loewenherz had studied mathematics and physics at the University of Berlin, where he received a doctorate in 1870 with a dissertation on synthetic geometry. At Foerster's Normal Eichungskommission, where he went to work after graduation, he developed the unitary screw thread and showed excellent skills at administering the legal and formal details accompanying testing and certification forms. During his seventeen years at the Eichungskommission, he established numerous contacts within the German gas, water, thermometer, mechanical, and optical industries. Moreover, he cofounded and became a major figure in the Deutsche Gesellschaft für Mechanik und Optik and its *Zeitschrift für Instrumentenkunde.*[156] For these reasons the contentious technologists considered him industry's man at the Reichsanstalt, their best guarantee that the Technical Section would respond to industry's needs.[157] On the other hand, his excellent scientific training and broad knowledge of Germany's "high-technology" industries made him more than acceptable to Helmholtz and the entire board, which appointed him director – he had not even submitted an application for the position – after quickly dismissing the only two applicants.[158]

Loewenherz soon became Helmholtz's man, too. Helmholtz watched with admiration as he built up the Technical Section and administered it as well as the Scientific Section. He found him "a man of distinguished [and] expert knowledge in technology," in particular one who understood the scientific foundations of technology. He prized his "great skill in administrative matters and in directing subordinate personnel," as well as his "most commendable zeal . . . for the interests of the Reichsanstalt," not least "through [his] widespread *personal* and written communications" throughout Germany.[159] In short, Loewenherz did well precisely that which Helmholtz liked least to do or could not do at all: administer the Reichsanstalt and cultivate contacts among technologists. He won Helmholtz's deep respect and trust. For his labors the director of the Technical Section earned 7,500 marks per year plus a housing allowance.[160] With a salary less than one-third that of Helmholtz's and, until 1893, no seat on the Reichsanstalt's board, the director was unmistakably the president's subordinate (Figure 3.14).

Loewenherz never came to occupy that board seat, because he died in 1892. Nor did his successor, Franz Stenger, who died within two months of assuming office.[161] The first to do so was the third director of the Technical Section, Ernst Hagen.

Although Hagen (initially) lacked Loewenherz's connections with industry, he brought his own special assets to the office. Most important, he had Helmholtz's immediate and full confidence, something that the seventy-three-year-old president desperately needed in his second-in-command. In bringing Hagen to the Reichsanstalt, Helmholtz was once again bringing in his own man. Hagen had worked under Helmholtz's old colleagues Robert Bunsen and Kirchhoff in Heidelberg, where he received his doctorate in 1875. Two years later he became

Helmholtz's assistant (1877-1883) at Berlin, habilitating under him in 1883. Following a study trip sponsored by the Berlin city government to inspect North American electrical illumination equipment, Hagen spent three years as an extraordinary (i.e., associate) professor at the Dresden Polytechnic, where he taught applied physics and established an electrical engineering program. From Dresden he moved to Kiel, where for six years he worked as a physicist for the German navy.[162] His intimate connections to both military and scientific men – he was the grandson of the Königsberg astronomer Friedrich Wilhelm Bessel and he married the daughter of Helmholtz's old friend Bezold – and his gregarious nature heavily favored his appointment.[163] As Weymann put it, Hagen was "very well qualified" for the position because he was a "man of science and technology and because of his amazing business skills."[164] Weymann's judgment was sound: Hagen held office from 1893 until his retirement in 1918.

Under Helmholtz's supervision, Loewenherz (and his successor Hagen) built an organizational structure for the Technical Section that was quite similar to that of the Scientific Section. The director of the Technical Section oversaw four laboratories: Precision-Mechanics, Heat and Pressure, Electricity, and Optics. Moreover, he also supervised the Chemical Laboratory and the workshop, both of which were established to serve the Reichsanstalt's other laboratories.

By 1893 the director commanded a staff of forty-three, of whom twenty-four conducted the actual technical work, ten provided mechanical support for the workshop, and nine manned the office. The director's principal subordinates were the four members who headed their respective laboratories as well two other members, Franz Mylius, who headed the Chemical Laboratory, and Friedrich Franc von Liechtenstein, who headed the workshop. In general, the members developed, improved, and standardized the instruments and methods used to conduct the routine testing and certification work, leaving the latter to their respective team of subordinates.[165]

Arnold Leman directed the Precision-Mechanics Laboratory from its founding in 1887 until his death in 1914. Before coming to the Reichsanstalt, Leman had conducted precision-mechanical work at the Normal Eichungskommission; his appointment to the Reichsanstalt largely represented the transference of this work from the Eichungskommission to the Reichsanstalt.[166] While at the Eichungskommis-

Figure 3.14. Leopold
Loewenherz, first director of the
Technical Section. Reprinted
from R. Fuess, "Gedenkfeier für
Dr. Leopold Loewenherz," *ZI*
*13* (1894): 177.

sion, too, he was the coeditor (1883-1887) of the *Zeitschrift für Instrumentenkunde.*[167]

In 1893 Leman's laboratory undertook six major precision-mechanical problems.[168] First, it provided the Scientific Section with precise measurements of the thickness of the quartz plates used in
polarimeters. Second, it experimented on and tested gyrometers for determining rotation speed, a task that represented a large portion of the
laboratory's time and that was justified by its "great importance" for
German motor technology.[169] Third, after industrial representatives
had approved (in 1892) the Reichsanstalt's standards for a unitary
screw thread, the laboratory regularly tested and certified such threads.
The Reichsanstalt thread found quick and growing acceptance in Germany, in part because the military adopted it and in part because a large
part of the precision-mechanical (especially for clockmaking) and the
electrical (especially for telegraphy) industries did so, too.[170] Fourth,
the laboratory investigated, as it had in previous years, the heat expan-

sion of metals, in particular aluminum. Fifth, it studied nickel-copper and aluminum alloys for use as analytical weights. And sixth, in response to the request of the military and the Ministry of Education, it continued to set standards for and to do running tests of tuning forks. (Such standardized forks were also used to study the behavior of electrical phenomena in steel.)[171]

The leader of the Heat and Pressure Laboratory, Hermann Wiebe, had qualifications similar to those of Leman and several others at the Reichsanstalt: He, too, had previously worked at the Eichungskommission, where he became closely associated with Loewenherz, Pernet, and Thiesen. Moreover, like several others later to be associated with the Reichsanstalt, he had helped found the Deutsche Gesellschaft für Mechanik und Optik. His friends in the technological community and in German industry, not the least of whom was Otto Schott, felt that Wiebe, as head of the laboratory from 1887 until 1912, represented their interests there.[172]

With its testing work for the German thermometer industry, the Heat and Pressure Laboratory conducted by far the largest number of tests of any single Reichsanstalt laboratory. Until 1887, the Eichungskommission had investigated and officially tested thermometers. Wiebe brought the Eichungskommission's thermometric work over to the Reichsanstalt. By 1889 the Reichsanstalt's thermometer-testing work had become so extensive and routine that, following a request by the thermometer industry, it agreed to open a testing station in Ilmenau, in the heart of the German glass and thermometer industry.[173] The Ilmenau station did only routine testing: In 1893 it examined nearly 12,000 thermometers, almost all of which were medicinal thermometers.[174] In Charlottenburg, the laboratory also tested nearly 1,000 other, more specialized thermometers, undertook extensive experiments on its own control thermometers, and attempted to develop more accurate thermometers capable of measuring higher temperatures.[175]

The laboratory's heat and pressure work in 1893 also included preliminary work on technical pyrometers and temperature measurements of three annealing ovens used in Thuringian glass factories.[176] Moreover, it investigated and tested a number of other heat- and pressure-measuring devices: calorimeters, barometers, manometers (of great interest and use to the German navy[177]), oil testers, viscosiometers, and several types of oils and several thousand alloy rings used as part of the security apparatus in boilers.[178]

The Electrical Laboratory stood under the direction of Karl Feussner, a physicist who had received his doctorate from Marburg in 1882 and who headed the laboratory from its formation in 1888 until 1894.[179] In Charlottenburg, Feussner used his scientific training to help develop the Reichsanstalt's measuring methods and standards in electricity.[180] He conducted nearly all of the laboratory's research, relegating all testing matters to his principal subordinates: Stephan Lindeck and Karl Kahle.

The Electrical Laboratory's avowed purpose was to cater to the German electrical industry, in particular to improve the industry's standards and measuring equipment.[181] It helped introduce into industry the standards developed by the Scientific Section's Electrical Laboratory. It conducted extensive precision measurements of current strength and voltage by means of its so-called compensator or potentiometer, a device developed by Feussner at the Reichsanstalt and one that became essential for all future work in electrical metrology.[182] Moreover, Lindeck and Kahle made running tests of and certified numerous, in some cases hundreds, of pieces of electrical apparatus: standard cells, capacitors, galvanometers, voltmeters, accumulators, batteries, and so on. Some of this testing work was done for the Reichsanstalt's own laboratories, some (the vast majority) for industry, and some for state agencies, both domestic and foreign.[183] At the request of the Berlin Elektrizitätswerke (Electrical Works), for example, the laboratory controlled and certified the electrical meters used to register the amount of current consumed by its customers. Such work was obviously "as much in the interest of consumers as in that of the Elektrizitätswerke." Only the official seal of approval of an independent organization, as the Werke rightly noted, could resolve the inevitable disputes between consumers and producers.[184] Finally, in 1893 the laboratory also investigated the magnetic properties of various steel, iron, and nickel alloys sent to the Reichsanstalt.[185] Not the least of the laboratory's accomplishments during the early 1890s was the development by Feussner and Lindeck, building on a discovery by Edward Weston in America, of manganin as an alloy for use in resistors. Because it minimized changes in resistance, manganin proved particularly useful in the establishment of a standard unit of resistance.

Lummer led the Technical Section's Optics Laboratory during Helmholtz's presidency, just as he led the Scientific Section's. His two principal subordinates in the Technical Section's Optics Laboratory were Eugen Brodhun and Emil Liebenthal. Brodhun, "the true Nestor of German scientific illumination technology," was another former assistant of Helmholtz's at Berlin, under whom he received his doctoral degree in 1887 before moving with him to the new Reichsanstalt.[186] Between 1891 and 1923 Brodhun supervised all of the laboratory's running tests. Liebenthal, after studying at Berlin and taking his doctorate at Greifswald, worked for a decade (1880-1890) at Neumayer's Kaiserliche Sternwarte (Imperial Observatory) and the Physical Institute in Hamburg.[187] He then joined the Optics Laboratory, where he remained until 1924.

The Optics Laboratory responded to the technical problems of the German illuminating industry.[188] In 1893, following the success of the Scientific Section's Optics Laboratory in developing new photometric devices, the Technical Section's Optics Laboratory began testing and certifying the Hefner lamp. It did so, as noted above, at the request of the Deutscher Verein für Gas- und Wasserfachmänner and in close coordination with them and Hefner-Alteneck of Siemens.[189] As a result of this interaction between science and industry, the Reichsanstalt (correctly) believed that "in the field of photometry Germany has attained a considerable lead over other nations."[190]

The laboratory's other photometric work for 1893 included constructing, again at the request of the Verein, a portable yet trustworthy photometric device for use in gas-illumination technology; testing, at the request of the Berlin city government, arc street lighting; running tests of electrical, gas, and kerosene lamps; investigating a ship's position light; and photometrically analyzing different types of petroleum and comparing colored light sources.[191]

By 1893 the work of the Technical Section's individual laboratories had assumed a routine character. Although the laboratories conducted a mixture of both technological research and testing, the bulk of their work was the routine testing and certification of materials, apparatus, and instruments for German industry. (In the fiscal year 1893-1894, it earned 15,279 marks for its testing work.)[192] Thanks largely to Loewenherz's leadership, the section exploited its (inadequate) old quarters, quickly establishing itself and effectively meeting many of the technological needs of German industry and the state.

Prompted by the obvious inadequacies of the old quarters, Helmholtz and the board soon began planning new facilities for the Technical Section. As early as 1889 the board called for an entirely new building for the Technical Section, one to be located next to the Scientific Section.[193] Then two years later Helmholtz appointed an architect (Astfalck) to design the new facilities, and financial provisions were made for purchasing additional land from Siemens for the section.[194] In 1892 the legislature approved the purchase of land, and in 1893 it approved 1,958,000 marks for constructing and equipping the new buildings.[195] It thus required more than four years of planning and negotiating, and then an additional three years of construction time, before the Technical Section's buildings were ready for use. Although Helmholtz died while construction was underway, he did set the train of events in motion. The new facilities were part of his legacy to the Reichsanstalt, and although we shall have to allude to a few developments that first occurred under Kohlrausch, we conclude here with a description of the Technical Section's new quarters.

Helmholtz and the Reichsanstalt leadership had five new buildings constructed for the Technical Section: the laboratory building, a machine (and boiler) house, the main or technical building, a small ventilator house, and a residence for the director of the Technical Section.[196] Astfalck, in consultation first with Helmholtz and then with Helmholtz's successor, Kohlrausch, designed and oversaw the construction of these buildings, which were architecturally similar to those of the Scientific Section.[197]

The Technical Section's buildings were located on the newly purchased plot of land, some 14,389 square meters, adjacent to the northwestern border of the Scientific Section's site. The determining factor in arranging these buildings was the need to locate the laboratory building, which housed the Chemical Laboratory on its top floor, at the extreme northern tip of the Reichsanstalt's entire site so that the noxious fumes produced in the laboratory could be blown away by the prevailing west winds.[198] The three-story laboratory building cost 218,000 marks to erect.[199] Its basement contained rooms for the High-Voltage Electrical Laboratory, a photographic darkroom, a storage and chemical room, and two rooms for experimental work in radiotelegraphy. The first floor housed rooms for studies on direct-current and electrical meters, as well as for the testing of electrical apparatus; the second, rooms for investigating self-induction and capacity (Figure 3.15).[200]

The laboratory building was connected to the Technical Section's machine (and boiler) house, which itself was attached to the Scientific Section's nearly identical machine house. The Technical Section's ma-

Figure 3.15. The laboratory building. Courtesy of Physikalisch-Technische Bundesanstalt, Braunschweig.

chine house, which cost 180,000 marks to erect, contained three large boilers for heating the section's buildings and powering the High-Voltage Laboratory's thirty-five-horsepower generator. A machine room contained machine-testing apparatus, in particular load resistances, direct- and indirect-current machines and motors, and an electrical switching panel. There was also space here for making indirect-current investigations and studying transformers. Next to the machine room were the alternating-current and high-voltage electrical rooms, equipped with an 11,000-volt battery and a 40,000-volt single-phase transformer. Finally, the machine house contained the Reichsanstalt's mechanical workshop, a plumber's workshop, a carpenter's workshop, and a forge.[201]

The Technical Section's third structure was the main or technical building, located directly opposite the Scientific Section's *Observatorium* (Figures 3.16 and 3.17). This four-story, U-shaped building cost 922,000 marks and took two years to erect.[202] Only its main walls and floors were laid down permanently; individual workrooms were easily formed by inserting or removing partitions.[203]

Figure 3.16. Main building of the Technical Section, as seen from Werner-von-Siemens Strasse. Courtesy of Physikalisch-Technische Bundesanstalt, Braunschweig.

Each of the main building's floors was devoted principally to one individual field of study: precision mechanics, optics, electricity and magnetism, or heat and pressure. The basement housed, among other rooms, residences for the steward and the machinist, a comparator room, as well as the section's Precision-Mechanical Laboratory, which was built for work demanding constancy of temperature and for studies on graduated scales (straight and circular). Moreover, below the Mechanical Laboratory was a one-meter-thick concrete plate used for studies requiring a high degree of freedom from mechanical vibrations. The ground floor was divided into two halves: one-half contained various offices (a reception room, a government lawyer's office, the main office, the cashier's office, a reference library, and the director's laboratory and office); the other housed the Optics Laboratory with its extensive area for photometric work. Heliostats were placed in the windows of the Optics and several other laboratories. The first floor was also divided into two halves: one-half contained the Low-Voltage Electrical Laboratory, the other the Magnetic Laboratory. The second floor contained the section's Heat and Pressure Laboratory along with a large conference room. Finally, a small domed area in the center of the

Figure 3.17. Staircase in
the main building of the
Technical Section.
Courtesy of Physikalisch-
Technische Bundesanstalt,
Braunschweig.

building's roof provided extra space for optical experiments and a photographic darkroom.

The Technical Section also consisted of a small ventilator house
and a home for the director. The latter housed the section's director and
contained rooms for two assistants and a porter; it cost 140,000 marks
to construct.[204] Hagen claimed that the availability of this official residence was crucial for his own scientific work. Such a residence, he
told Heinrich Kayser in 1911, "is the prerequisite for a physicist who
has to direct a laboratory and who wants to get his own [scientific]
work done too. Do you really believe," he rhetorically asked Kayser,
"that during my time here I could have done even a single one of the

numerous experiments with [Heinrich] Rubens if I had not had an official residence?"[205]

In addition to the land costs (373,106 marks) and building construction costs (1,460,000 marks) already noted, the Reich spent another 880,190 marks[206] on the Technical Section, including 348,000 marks for plant costs (paving, a sewage system, gas and heating equipment, construction design, and so on), 108,300 for fittings and furniture, and 471,390 for initial equipment and machinery. By the fiscal year 1897-1898, the Reich had provided 2,713,296 marks to purchase land for, erect, and equip the Technical Section (Table 3.2).[207]

## *"Das schwarze Jahr," 1894*

By the mid-1890s the Reichsanstalt was a flourishing scientific institution. In the fiscal year 1893-1894, for example, it employed sixty-five individuals and had an operating budget of 263,000 marks.[208] It provided scientific positions for more than a dozen German physicists. Within a few years it had come to constitute a physics institute that was complementary to those at the University of Berlin and the Technische Hochschule Charlottenburg. As the extensive publishing record of its scientists and technologists for 1893 suggests, Helmholtz had breathed a research spirit into the spanking new institution: In that year alone Reichsanstalt scientists published twenty-eight articles in scientific and technical journals, including four in the *Annalen der Physik* and ten in the *Zeitschrift für Instrumentenkunde*.[209] Moreover, they regularly reported their results to the Physikalische Gesellschaft, of which they were highly valued members. Although the Scientific Section had devoted most of its efforts to helping solve technological problems originating in industry or to problems of a largely instrumentational or metrological character, the completion of the *Observatorium* gave promise that it might soon achieve more of the substantive results that its founders had hoped for. Just as important, both sections had already effectively met a number of the standards and testing needs of German industry. The construction of new facilities for the Technical Section during the mid-1890s gave further promise that the Reichsanstalt would soon respond even more effectively in the future. To the world's physics and industrial communities, Helmholtz's Reichsanstalt was beginning to represent yet another triumph of nineteenth-century German science and technology.

To some inside the Reichsanstalt, however, the sense of triumph must have been dampened by concern about the advanced age and declining health of their charismatic leader. Especially during his Reichsanstalt years, Helmholtz suffered regularly from ill health, in particular from mental depression.[210] The death of his young son Robert in 1889 brought him still more suffering, as did, on a lesser scale, the death of his old friend Siemens in 1892.[211] To combat his depression and to recover his physical and spiritual well-being, he journeyed frequently to southern Germany, to Italy, and to Switzerland, leaving his concerns and staff in Charlottenburg far behind.

Public and professional demands, as much as private ones, also took him away from Charlottenburg. He lectured to diverse audiences throughout Germany and beyond. The longest of his absences for professional reasons became a fateful one for the Reichsanstalt. In the summer of 1893, at the age of seventy-two, he attended, at the behest and as the representative of the German government, the International Electrical Congress, held in Chicago as part of the Columbia World Exposition.[212] Earlier that year the Reichsanstalt had issued a set of recommendations for legal electrical standards in Germany.[213] In Chicago that summer, the participating scientists reached agreement both on the values of the ampere, ohm, and volt and on the experimental procedures for determining those values. After nearly two months of work and refreshing travel in the United States, Helmholtz began his return voyage to Germany. While on board ship, he had a serious accident – a head injury occasioned by a fall on some stairs – that prevented him from resuming his official duties until late November (Figures 3.18 and 3.19).[214]

Although Helmholtz had returned to work after an absence of nearly four months, he reported seeing double and needing twice as long to complete his duties as he had needed before his accident. By the spring of 1894 he had partially recovered, but he still could not make quick body motions without suffering vertigo.[215] Then in June he suffered a stroke; in early September came another, final stroke, and death.[216] The master of German science was buried on 12 September, and the Reichsanstalt, effectively leaderless during the past thirteen months, hung its flag at half-mast.[217]

Helmholtz's death followed closely upon that of his brilliant student Hertz and his valued colleague Kundt. Planck aptly called 1894

Figure 3.18. Helmholtz's last lecture, 1893. Courtesy of Bildarchiv Preussischer Kulturbesitz, Berlin 1988.

*"the black year . . . of German physics."* Helmholtz's death, he later wrote, meant the loss of "the man who stood for the best in German science, the pride of our [Physikalische] Gesellschaft, and . . . one of the last great representatives of classical physics."[218]

Like the discipline of German physics as a whole, the Reichsanstalt found it hard, indeed impossible, to replace its leader; for Helmholtz had indelibly stamped his personal mark on the Reichsanstalt. Hagen, who faithfully directed the Reichsanstalt for the twenty months (August 1893 to March 1895) during which Helmholtz was absent and ill and his successor was being sought, wrote to Kayser in November 1894:

> It is still unforeseeable how the situation here at the Reichsanstalt will develop since Helmholtz has died. Moreover, it is still undecided who his successor will be. The main problem lies in the fact that basically everything here was tailor-made for Helmholtz's *person.* Naturally nobody presumed to criticize

Figure 3.19. Helmholtz in Washington, D.C., September 1893, shortly before his return to Europe and accident aboard ship. Courtesy of Bildarchiv Preussischer Kulturbesitz, Berlin 1988.

him, and his great name covered everything, so that there could be no talk of difficulties. The latter, however, will not fail to be seen by his successor. That's why I see a rather dark future ahead. [219]

Hagen's sense of gloom, exaggerated perhaps by the burdens he had temporarily to shoulder, was mistaken. To be sure, no German physicist of the post-Helmholtzian era commanded anywhere near the authority that Helmholtz did among Germany's physicists and technologists. But Helmholtz had managed during his seven years in office to shape the Reichsanstalt into a carefully structured, well-staffed, and smoothly running scientific bureaucracy, to gain strong financial support from the government, and (nearly) to complete construction of the Reichsanstalt's plant. What the Institute needed now was not a charismatic leader who could see it through a heroic founding era. Instead, what it needed was a successor who could efficiently utilize the Reichs-anstalt's extensive human and material resources so as to meet the new challenges and opportunities in physics and technology as the nineteenth century drew to a close.

# 4

## Masters of Measurement

In the decade following Helmholtz's death, the Institute experienced prodigious institutional growth and exceptional scientific vitality. During Kohlrausch's presidency (1895-1905), the Reichsanstalt became a widely cited example of a creative, bountiful national scientific institute, whose scientific studies and technical work provided valuable contributions to academic physics, industry, and society. As we have seen, it had several domestic and foreign imitators.[1] Kohlrausch's Reichsanstalt was a success – but therein lay its problems. This chapter illustrates the dual nature of the Institute's success by examining its three most distinctive features under Kohlrausch. It also portrays the state of the Reichsanstalt in 1903 and analyzes the evolution of the Institute during Kohlrausch's entire presidency.

The first feature was the Institute's unprecedented growth during the period 1895-1905, which made it a scientific institution sui generis. By 1903, as compared with 1893, its plant had doubled in size, its staff had become nearly 70 percent larger, and its operating budget had become more than 40 percent larger. As a consequence, by the late 1890s Kohlrausch was confronting daily and long-term management problems that were of a markedly different kind and order than those faced by Helmholtz. The simple growth of the Institute's size considerably changed its nature. Helmholtz's nascent scientific bureaucracy became a full-fledged one under Kohlrausch. The Reichsanstalt's heroic era was succeeded by a bureaucratic era.

The second feature concerns the Institute's unique combination of scientific and utilitarian motives for pursuing photometric and temperature-measuring studies as part of a more encompassing research program on the measurement of light and heat radiation. Between 1893 and 1901 the physicists engaged in these studies –

Lummer, Wien, Ernst Pringsheim, Kurlbaum, and Rubens – devised theories and used the Reichsanstalt's excellent physical facilities to construct radiation-measuring instruments and to conduct experimental studies that helped foster and test a number of candidate spectral-energy distribution laws for blackbody radiation. The well-known end product, Planck's spectral-energy distribution formula and its theoretical derivation by means of a novel hypothesis about a quantum of action, constituted the beginnings of quantum physics, a subject that, in time, revolutionized physical thought about matter and radiation. This well-treated historical subject is here briefly recounted and reexamined from the viewpoint of the Reichsanstalt, the principal institutional setting for the advance of experimental research in blackbody radiation. This reexamination reveals that a more complex group of motivations was responsible for the timing and setting of the origins of quantum physics than the major interpreters of these dramatic events have allowed. In particular, it provides evidence that, in addition to the pure scientific motivation of individual physicists to discover a spectral-energy distribution law for blackbody radiation, the Reichsanstalt's institutional commitment to respond to the practical needs of the German illumination industry for better temperature measurements and a better understanding of radiation in general played an equally important part in the origins of quantum physics. At the same time, focus on the successes of Reichsanstalt physicists at blackbody-radiation work reveals an institutional problem that first seriously manifested itself toward 1900 and plagued the Institute thereafter: the loss of its best scientific manpower to the German academic systems.

The third outstanding feature of the Reichsanstalt between 1895 and 1905 was the development of a paradoxically antagonistic relationship between the Reichsanstalt and the German electrical industry. In three separate ways, the industry that the Reichsanstalt sought to advance and that benefited from its work threatened the Reichsanstalt's well-being. First, from the mid-1890s on, the Reichsanstalt's studies on electrical standards and electrical meters had matured to the point where the results merited legal enactment, yet the industry opposed (for reasons that will become clear presently), delayed, or weakened implementation of the Institute's scientific findings and recommendations. Second, the Reichsanstalt's enlarged plant and staff now allowed, and the growth of the electrical industry now demanded, more electrical testing work. Yet by consuming an immoderate proportion of the Reichsanstalt's resources, that testing work threatened the Reichsanstalt's ability to conduct research. Third, the industry, in the form of the Berlin-Charlottenburger Strassenbahngesellschaft (Berlin-Charlottenburg Streetcar Company), decided to place electrical streetcar tracks in front of the Reichsanstalt. The electromagnetic fields generated by passing electrical streetcars threatened to nullify the precision electrical and magnetic research conducted within the nearby *Observatorium*. A running battle over this problem clouded Kohlrausch's administration of the Reichsanstalt during much of his decade as president.

In one form or another, these three outstanding features of the Reichsanstalt between 1895 and 1905 – institutional growth, epoch-making scientific results, and an antagonistic relationship with the electrical industry – illustrate a deeper issue, the complex relationship between science and technology. The rise of science-based technology after 1850 meant that science and technology might be adversaries as well as allies. As he guided the Reichsanstalt into maturity, Friedrich Kohlrausch came to understand this point well.

## A New Type of President for a Changed Institute

"It is well known," wrote Planck and his fellow scientists in nominating Kohlrausch to full membership in the Akademie der Wissenschaften in 1895, "that among German physicists Friedrich Kohlrausch currently stands in the forefront." [2] What got Kohlrausch to the forefront of the German physics profession – and into the Reichsanstalt presidency – was the right combination of scientific and organizational talent, and the lucky accident of birth.

Born in 1840, Friedrich Kohlrausch was the son of Rudolf Kohlrausch, a leading German experimental physicist at midcentury. [3] The father provided his son with excellent scientific training, influenced his son's choice of style and fields of research in physics, and brought him into regular contact with Wilhelm Weber, a close colleague and a future strong influence on Friedrich. He received his doctoral degree in 1863 under Weber at Göttingen.

From his student days on, Friedrich showed a predilection for work in the precision measurement of electromagnetic, electrochemical, and elastic phenomena. He had a talent, if not an obsession, for counting and for what he called the "delightful pursuit" of measuring. He came to appreciate the historic value of that talent: "Measuring nature is one of the characteristic activities of our age," he wrote in 1900 in a rare philosophical moment. "Without [the measuring of nature], the progress made during the last century in the natural sciences and technology would not have been possible." [4] He had, moreover, a concern for good pedagogy in physics. As we saw in Chapter 1, from the late 1860s on, Kohlrausch completely reorganized and systematized laboratory instruction in physics, helping to integrate it with the traditional lecturing aspect of university physics instruction by teaching students, through his *Leitfaden der Praktischen Physik,* the art of physical measurement. Kohlrausch had a proven talent for organizing men and research, one that had manifested itself long before he came to the Reichsanstalt.

Kohlrausch also brought to his presidency extensive administrative experience: In the course of nearly thirty years of professional life before coming to the Reichsanstalt, he directed five physics institutes. His main professional way stations included appointment as codirector (with Weber) of the Göttingen Physics Institute in the late 1860s; a brief appointment (1870-1871) as ordinary professor of physics at the Zurich Polytechnic – brief because while teaching in Zurich during the midst of the Franco-Prussian War he encountered strong anti-German sentiment that offended his equally strong German nationalism; and then, after returning to the fatherland in 1871, appointments as the ordinary professor at the Technische Hochschule Darmstadt (1871-1875), the University of Würzburg (1875-1888), and the University of Strasbourg (1888-1895). In the course of fulfilling these professorial assignments, he planned and supervised the construction of one entirely new physics institute – that at Würzburg – and remodeled or reorganized three others. He also gained a reputation as a good supervisor of advanced physics students, helping nearly fifty of them to earn an advanced degree under him.

The basis of this fortunate academic career was Kohlrausch's ability as an experimental physicist – more precisely, as a measuring physicist. Like no other German physicist of his day, Kohlrausch invented, improved, and exploited precision-measuring instruments and methods, including a reflectometer, a tangent galvanometer, various types of magnetometers, variometers, and dynamometers, as well as numerous methods for measuring electrical, magnetic, and thermal phenomena. Heinrich Rubens, himself one of the foremost German experimentalists, judged that "no other physicist has surpassed Kohlrausch in the skill and care with which he used instruments and methods." Though Kohlrausch understood the physical theory behind his work and the potential implications of his experimental results for theory, he showed no interest in theoretical matters. Exact measurement was his forte, his aim, and his pleasure.[5]

In addition to developing measuring instruments and methods, Kohlrausch worked mainly on two substantive scientific topics: the determination of the absolute values for current strength and electrical resistance, and the electrolytic conductivity of dilute solutions.[6] Inspired by his father and Weber, Kohlrausch's work on determining absolute electrical values helped extend the Gauss-Weberian absolute system of measure to include the electrical and magnetic units of charge. All told, he devoted a good part of forty years of research to determining the values of fundamental electrical and magnetic quantities. Moreover, his quest for universal, trustworthy electrical standards and his representation of German interests at several international electrical standards meetings made him Germany's foremost student of electrical metrology during the last third of the nineteenth century. With the Reich's and the German electrical industry's ardent desire to lead if not dominate international electrical metrology, and with the industry's rapid expansion during the 1890s, Kohlrausch's great expertise in electrical metrology constituted yet another asset that he would bring to the presidency.

Kohlrausch was, in short, "the master of measuring physics."[7] His chosen fields of research were narrow, but he was extremely productive and the precision of his work was virtually unsurpassed. By the time of his death in 1910, he had published more than two hundred articles and books, including eleven editions of the *Leitfaden.*[8]

Outstanding ability at measuring physical (especially electrical) phenomena; talent at organizing a physical laboratory; and broad administrative experience in directing a physics institute – these were the qualities that led Planck and his associates in the Akademie to speak of Kohlrausch as being "in the forefront" of the German physics profession. And these were the essential qualities that led the Reichsanstalt leadership to select Kohlrausch as Helmholtz's successor. The Reichsanstalt needed someone who was regarded by his colleagues as one of the leading German physicists, who had a deep appreciation of and concern for precision technology, and who had proven administrative abilities, including the experience to oversee the completion of the Reichsanstalt's unfinished physical plant and to supervise its large staff. Moreover, as time passed and Helmholtz's presence faded, and as the leaderless Institute continued to operate as best it could under the circumstances, the Reichsanstalt increasingly needed someone who already understood the Institute's complex needs before he assumed the presidency. Having served on the Reichsanstalt's board since its founding, Kohlrausch could assume the presidency with a relatively sure knowledge of the Reichsanstalt's aims and means.

The Institute needed Kohlrausch far more than he needed the Institute. The year 1894, that "black year" for German physics, brought a trio of offers to Kohlrausch in Strasbourg, where he was contentedly entering his eighth year as director of the well-equipped Physics Institute. First came an offer to succeed Hertz at Bonn. Kohlrausch promptly declined, largely because he found the Bonn facilities inadequate but also because he remained unenamored of what he, like many German academics, considered the highhanded and authoritarian ways of Friedrich Althoff, the powerful Prussian official responsible for university affairs within the Ministry of Culture. Then came an offer to succeed Kundt at Berlin.[9] Kohlrausch declined this too, in part because he considered Berlin's facilities inadequate and in part because he thought the burdens of office there were simply too great for any one individual, particularly one like himself who was seeking more time for research.[10]

Then came the offer to lead the Reichsanstalt. To be precise, it did not come from the board, which according to Kohlrausch was fully ignorant of the Ministry of the Interior's planning and decision-making process;[11] instead, it came, "in a confidential communication" and nearly six months after Helmholtz's death, from the undersecretary of state (presumably Weymann) acting with the approval of the kaiser.[12] Unlike the two previous offers emanating from Berlin, Kohlrausch felt that "the circumstances simply make it a duty for me to accept the position [of president] of the Physikalisch-Technische Reichsanstalt."[13] Svante Arrhenius, then honeymooning in Germany, visited Kohlrausch in Strasbourg and reportedly told him that physicists all over Germany looked to Kohlrausch as Helmholtz's natural successor as president.[14] It was, as Eduard Riecke later said, "a position to which . . . his entire scientific work had called him."[15]

Self-interest called too. Like Helmholtz before him, Kohlrausch naturally realized that by accepting the presidency he would gain total relief from lecturing to large numbers of elementary students; the result, he imagined, would be more time for his own research.[16] Saddled with large, introductory classes during the past three decades, he now wanted "some peace and quiet in order to do some work and in order to learn something myself."[17] He knew, of course, that he would have to devote some of that relief time to the presidency's extensive administrative obligations.[18] Still, he imagined there would be increased time for his own work.

Again as with Helmholtz, the sticking point in hiring Kohlrausch was financial compensation. The Ministry of the Interior's goal of winning a top German physicist to lead the Reichsanstalt (again) clashed with the Treasury's natural desire to limit expenses. To attract Kohlrausch to Charlottenburg, Boetticher had to match Kohlrausch's current income: 17,250 marks plus a rent-free residence.[19] However, the minister of the treasury, Posadowsky-Wehner, complicated Boetticher's task by declaring that both the 9,000-mark bonus and the rent-free residence granted to Helmholtz would not be granted to his successor; that left only an annual salary of 15,000 marks. Conceding that the president's residence should not be offered on a rent-free basis, Boetticher replied that the Bundesrat had implicitly agreed to continue the bonus and that without it he could not attract the right man.[20] He therefore recommended that Kohlrausch receive a salary of 15,000 marks plus a bonus of 5,000 marks. By charging Kohlrausch 1,100 marks rent for the presidential residence, the Reich would in effect be paying him 18,900 marks per year. Anxious to secure a new president, Boetticher warned that "the decision on Kohlrausch's nomination will not bear a long delay."[21] Posadowsky-Wehner agreed to Boetticher's compromise.[22] Moreover, to preserve his (and the Reichsanstalt's) academic ties, Kohlrausch was appointed an honorary professor at the University of Berlin, where he was to teach two to four hours per week;[23] and to enhance his income, he was appointed a full member of the Akademie der Wissenschaften. Kohlrausch, who felt more at home in the southern part of Germany than in the "sand" of the Mark Brandenburg,[24] agreed to these terms. By April 1895 the Reichsanstalt had a new leader (Figure 4.1).

Kohlrausch's administrative style differed markedly from Helmholtz's – and necessarily so, for the Reichsanstalt that operated under Kohlrausch was scarcely the same as the one that was founded in 1887.

Figure 4.1. Friedrich
Kohlrausch. Courtesy
of Amerika Gedenkbib-
liothek Berliner Zentral-
bibliothek, Berlin, and
Akademie-Verlag, Berlin.

It could be said that by 1903 the heroic era of the Reichsanstalt had
ended and the bureaucratic era had begun. This transformation can perhaps be gauged by a comparison of a few statistics about the Reichsanstalt in the years 1893 and 1903.

When Kohlrausch assumed office in 1895, the Reichsanstalt consisted of five buildings; by 1903, thanks to Kohlrausch's supervision of
the construction of the Technical Section, it consisted of ten buildings
– twice as many. All told, the Reich had spent 3,672,360 marks to establish the Reichsanstalt's physical plant.[25] That was more than double
what the United States spent to construct and equip its National Bureau
of Standards and more than sixfold what England spent to construct
and equip its National Physical Laboratory.[26] By 1897 Kohlrausch
could announce that the Reichsanstalt "had now secured the most basic
conditions for pursuing its tasks . . . in a rich and appropriate style."
"And it is fair to say," he boasted, "that the noble establishment that has
been produced is matched by no other in the world." [27] The remark of
a British visitor to the Reichsanstalt in 1898 indicates that Kohlrausch's

boast was not merely Germanic chauvinism: "The question as to whether that institution [i.e., the Reichsanstalt] is not too magnificent," said the British observer, "has in fact occurred to many of those who have seen it." [28] The financial effort expended to establish the Reichsanstalt was an unmistakable sign of imperial Germany's keen interest in advancing physics and precision technology and a symbol of its quest for imperial power.

Three further indices point to the Reichsanstalt's growth under Kohlrausch. In 1893 the Reichsanstalt had annual operating expenses of some 263,000 marks; a decade later its annual operating expenses were about 375,000 marks – a growth of nearly 43 percent.[29] An equally dramatic increase occurred in the number of personnel employed: Whereas Helmholtz's Reichsanstalt of 1893 had employed 65 individuals, Kohlrausch's of 1903 employed 110 – growth of nearly 70 percent. About one-third (41) of the personnel of 1903 conducted the scientific work; about one-half (59) maintained the plant, laboratories, and equipment; and about one-tenth (10) manned the offices as administrative personnel. (The most senior scientific officials – the members – earned salaries ranging between 3,600 and 7,500 marks; junior scientific officials – the assistants – earned between 1,920 and 2,160 marks [Table 4.1].) The Reichsanstalt of 1903 had more personnel for administration alone than some academic physics institutes had in toto. Indeed, fully 69 of the 110 staff members were devoted to some form of maintenance or administration.[30] More buildings, more expenses, and more employees meant more administrative work for Kohlrausch. As these figures suggest, Kohlrausch's administrative duties far surpassed those of Helmholtz. The Reichsanstalt of 1903 had become one of the largest scientific organizations of its day. The era of Big Science, like the era of Big Business and Big Government, had begun.

The changed Institute required a new type of president. Adolf Heydweiller, Kohlrausch's son-in-law and himself a physicist, pertinently described Kohlrausch's life and work as a "paragon of order";[31] so, too, was his Reichsanstalt. Arrhenius, who worked with Kohlrausch at Würzburg in 1886-1887, gave the following illuminating description of Kohlrausch to Ostwald:

> Kohlrausch is personally a man of great charm and goodwill.
> ... Only rarely does he state his opinion, but when he does his
> opinion is well founded. He is supposed to have always lived
> as orderly as a chronometer and is in all [social?] relations a

Table 4.1. *Number of Reichsanstalt employees and range of their salary/wages, 1903*

| Position | No. of employees | Range of salary/wages (marks) |
|---|---|---|
| President (= director of Scientific Section) | 1 | 20,000 |
| Director of Technical Section | 1 | 8,800 |
| Member | 13 | 3,600–7,500 |
| Permanent co-worker | 10 | 2,400–4,800 |
| Scientific assistant | 16 | 1,920–2,160 |
| Office worker | 6 | 2,100–4,200 |
| Legal secretary | 4 | 1,800–3,000 |
| Machinist | 2 | 2,070–2,160 |
| Mechanic/artisan | 45 | 1,200–2,200 |
| Minor official | 12 | 1,000–1,500 |
| Total personnel | 110 | |

*Note:* The figures for salary/wages do not include housing allowances.
*Sources: HEDR* (1903), 2–3, 26–29, on 26–28; Staatssekretär des Innern to Staatssekretär des Reichsschatzamts, 30 November 1904, ZStAP, RI, Personalakte Warburg Sig. 645, Bl. 3–5; and K. Strecker, "Die Personalverhältnisse der wissenschaftlichen Beamten der Physikalisch-Technischen Reichsanstalt" (29 August 1911), Anlage 1, BA 265, PTB-IB.

> strict formalist. His principal [scientific] endeavour is directed at improving [measuring] methods, so as to make the probable errors smaller. Indeed, he has an all-too-great predilection for finely rounded numbers.[32]

Kohlrausch was a man who lived by rules, regulations, and regularity. "Extraordinary orderliness and conscientiousness" were, Arrhenius said, the characteristics of his measuring work.[33] And they were the characteristics of his Reichsanstalt, which, in contrast to Helmholtz, he administered more as a governmental agency than as a physical institute, stressing the observance of official regulations like "a fixed program of work and fixed working hours."[34]

Where Helmholtz had reigned and inspired, Kohlrausch administered. Unlike Helmholtz, Kohlrausch carried out most presidential administrative duties himself, despite the fact that he found them unpleasant, rather than delegating them to assistants.[35] Ironically,

Kohlrausch disliked large institutes: *"Big* laboratories," he told Arrhenius while still in Strasbourg in 1893, "are the ruin of their directors. There are always roughly 10 people working with us [at Strasbourg], and each one of them wants this or that from the director."[36] After he had moved to Charlottenburg a mere two years later, dozens of people wanted one thing or another from him as Reichsanstalt president. For Kohlrausch, the decade from 1895 to 1905 was one of "arduous, wearing *Fronarbeit"* in which he devoted himself to administrative details.[37] As with Helmholtz, his expectation that his appointment as head of the Reichsanstalt would mean increased research time proved illusory: Unwittingly, he had substituted administrative work for teaching, leaving him, if anything, still less time for scientific work. "The teaching," he told his friend Carl Barus already in 1896, "is unfortunately replaced by lots of time-consuming office work, etc., which makes the [possibility of doing] scientific work recede still further."[38] "My slow way and limited powers of work," he told Arrhenius in 1900, "are greatly absorbed by numerous administrative business."[39] "I belong to those," he again told Arrhenius, this time in 1903, "who do not have a single day free for their own work. One often becomes peevish." [40] Kohlrausch entered into and, at the same time, helped to foster a very different institutional atmosphere than had existed under Helmholtz. In place of Helmholtz's new, collegial institute with only partially realized plans came an orderly, increasingly large and rigid, yet successful organization.

To be sure, Kohlrausch had the same vision as Helmholtz (and Siemens) of the Reichsanstalt's purpose. It was to pursue both pure and applied science as well as technology and to aid academic physics institutes, industry, and government in their respective aims.[41] Nor were there great changes in the types of substantive scientific and technological work carried out under Kohlrausch's administration. The change lay elsewhere: in Kohlrausch's different style of administration and in the simple quantitative growth of the Institute, particularly the Technical Section, between 1887 and 1905.

### High Temperatures and High Science

The Scientific Section under Kohlrausch was identical in organization to that under Helmholtz, and only slightly larger. In 1903, twelve senior scientists, along with a host of technical and administrative personnel under them, conducted or directed the work of the section's three laboratories: Heat, Electricity, and Optics. The major difference between the Scientific Section of 1893 and that of 1903 lay in the outstanding performance of the Optics Laboratory. That performance, in turn, was in part dependent on the new excellent facilities for making precision measurements that had become available late in 1891 and on the Institute's goal of providing the German illuminating industry with a unit of luminous intensity as well as more energy-efficient sources of illumination. The utilitarian motives of industry, as well as the pure scientific motives of physicists, together led the Reichsanstalt as an institution to measure blackbody radiation.

Max Thiesen still headed the Heat Laboratory, which now numbered five workers: Thiesen, Scheel, and Holborn, all of whom had been with the laboratory since Helmholtz's day; and Friedrich Henning and Louis Austin, who had arrived after 1900. As in the past, the laboratory's main efforts were directed at meeting German industry's heat-technology needs. In 1903 Thiesen conducted three separate heat studies, each aimed principally at making technological improvements. First, he completed work on the expansion of water between 50° and 100°C. He had begun this work under Helmholtz in order to achieve a definitive expansion formula for temperatures between 0° and 100°C (Figure 4.2). Second, he completed another long-term project: the investigation of a platinum-resistance thermometer.[42] His third study, conducted with Helmut von Steinwehr of the Electricity Laboratory, concerned the specific heats of gases and illustrates the complex interaction among and mutual dependence of science, technology, and industry at the Reichsanstalt.

Except for two previous brief studies on specific heats,[43] the Reichsanstalt first undertook such research in response to a request from the Verein Deutscher Ingenieure. A greater understanding of the nature and effects of the specific heats of gases, the Verein argued, would mean a better understanding of the operation of a gasoline motor's cylinders and of the quantitative relationship between cylinder alteration and volume. As gasoline motors came to replace steam engines, the precise quantitative nature of reactions in gasoline motors required investigation, just as steam engines had required such investigation a century earlier. Therefore, the Verein asked the Reichsanstalt to investigate the specific heats of gases in motors operating at temperatures and pressures up to 2000°C and 50 atmospheres and to show the quantitative, causal relationship between these two parameters.[44]

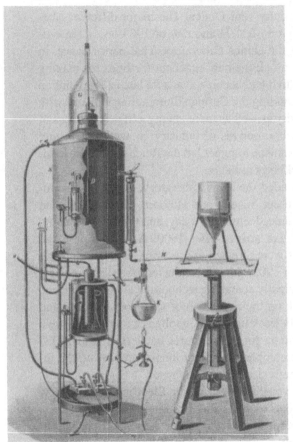

Figure 4.2. Boiling apparatus designed to maintain an internal temperature of 100°C. Reprinted from M. Thiesen, K. Scheel, and L. Sell, "Untersuchungen über die thermische Ausdehnung von festen und tropfbar flüssigen Körpern," *WAPTR* 2 (1895): 140.

The Reichsanstalt's board readily agreed to the Verein's request.[45] At the same time, it also stressed other, more purely scientific dimensions of the problem. Board member Röntgen, fresh from his recent journey to Stockholm, where he received the first (1901) Nobel Prize in physics, argued that "it is desirable that the experiments [on specific heats of gases] be conducted not only with regard to technology's needs, but also with the scientific point of view in mind." In particular, he and other board members wanted studies of the velocity of sound at high temperatures.[46] And Carl Linde, who belonged to the boards of both the Reichsanstalt and the Verein and who was one of Germany's foremost scientific engineers (see below), preached that in this matter of specific heats "experimental and theoretical work must go hand in hand." Internal-combustion engines, he noted, were twice as efficient as steam engines and therefore threatened rapidly to replace the latter, leading to the possible shutdown of large steam-generating plants. Theoretical understanding and experimental results concerning specific heats in the process of combustion were therefore "important" and required "speedy completion."[47]

Here was a scientific-technological problem tailor-made for the Reichsanstalt. In 1903 Thiesen and Steinwehr reported their first results: Small electric motors with constant and precisely measurable velocities of rotation could be used for comparing the velocity of sound in gases under various conditions. At the same time, Holborn and Austin studied the specific heats of air and carbon dioxide at high temperatures.[48]

The Reichsanstalt researched the effects of low as well as high temperatures. Indeed, it was among the first institutions to conduct systematic low-temperature studies. Its impetus for doing so came largely from Linde, who had studied at the Zurich Polytechnic, in particular thermodynamics under Clausius. Rigorously trained in science and engineering, Linde quickly became a leading figure in applied thermodynamics. In the late 1870s he developed an "ice machine" that produced unprecedentedly low temperatures and, in time, high profits for his firm: the Gesellschaft für Linde's Eismaschinen (Linde's Freezer Company). By the mid-1890s he had helped establish the field of gas liquification and helped open the new field of low- temperature physics. He (and others) provided physicists with new forms of matter like liquid air, liquid helium, and liquid hydrogen with which it later became possible to solve or investigate numerous problems in atomic physics and, of no less importance, that led to the unexpected discovery of superconductivity and to the development of solid-state physics. Linde became to low-temperature technology what Siemens had been to electrotechnology.[49]

When Linde joined the Reichsanstalt's board in 1895, he informed his new colleagues that he had obtained liquid air by using the Joule-Thomson effect. The incredulous board called for a full report. Kohlrausch first dispatched Holborn and Wien to Linde's Munich laboratory to learn from and help Linde and his group with their low-temperature measurements; then he went south to see for himself.[50] Convinced of the soundness of Linde's work, the Reichsanstalt started conducting (in 1896) its own low-temperature studies. In 1903, for example, Henning tested a formula originally designed for high tempera-tures, reporting its validity at temperatures as low as -190°C.[51] Work in low-temperature physics became a natural complement to the Reichs-anstalt's work in high-temperature physics and placed it at the forefront of all temperature-measuring physics.

The work of the Heat Laboratory's other major figure, Karl Scheel, was scientifically undistinguished and yet demands attention, for during the course of his long career at the Reichsanstalt (1888-1931) Scheel became a prominent figure within the Institute and throughout the German physics community. He owed that prominence to his im-portant administrative responsibilities: first, to his supervision of the Reichsanstalt's exchange of scientific and business communications with other institutions and individual scientists; and second, to his in-tendancy of the Reichsanstalt's large and ever growing library, which, thanks largely to Scheel, had by 1904 become "an important center for physics periodical literature, one increasingly used [by individuals] from outside [the Reichsanstalt]." [52] As to his prominence throughout the German physics community, he owed that to his extensive editorial work, which began in 1899 when he assumed the coeditorship of the *Fortschritte der Physik*. Published by the Berlin (later Deutsche) Physi-kalische Gesellschaft since its founding in 1845, the *Fortschritte* was the world's premier review journal of physics. With the death of Gustav Wiedemann in 1899, however, it faced imminent collapse. Scheel saved it in the form of the *Halbmonatliches Literaturverzeichnis der Fortschritte der Physik* (1902-1919).[53] Moreover, between 1902 and 1936 Scheel edited the Gesellschaft's *Verhandlungen,*which, until 1919, was bound with the *Halbmonatliches Literaturverzeichnis* to form the Gesellschaft's *Berichte*. Along with the monopolization of the *Annalen der Physik*'s editorial board by Reichsanstalt board members and the numerous publications by Reichsanstalt workers in the *Annalen* itself, Scheel's editorial work helped place the Reichsanstalt at the cen-ter of the German physical community's publishing efforts. His indis-pensable editorial work for the Gesellschaft represented a significant Reichsanstalt contribution to that community (Figure 4.3).

Wilhelm Jaeger headed the Scientific Section's Electrical Labora-tory during Kohlrausch's presidency, as he had during Helmholtz's. In addition to his longtime colleague Holborn, his principal aides by 1903 were Hermann Diesselhorst, Edward Grüneisen, and Steinwehr. Moreover, Kohlrausch and several members of the Heat Laboratory

Figure 4.3. Karl Scheel.
Courtesy of Physikalisch-
Technische Bundesan-
stalt, Braunschweig.

also conducted electrical studies. The laboratory's work addressed both
the immediate, technical problems in electrical and heat measurement
that confronted the Reichsanstalt and the less urgent instrumentational
needs of German physicists, chemists, and technologists.

As in the past, the establishment of trustworthy electrical standards
and the accompanying measuring instrumentation and apparatus continued to be the laboratory's overriding goal and responsibility. In 1903
Jaeger and Diesselhorst compared the standard resistances of the Scientific and Technical Sections, a perennial topic at the Reichsanstalt.[54]
(Diesselhorst soon became one of the Reichsanstalt's few experts on
mathematical and theoretical topics in electromagnetism, heat, and,
above all, the application of measuring methods to electromagnetic and
thermal phenomena.)[55] Jaeger also investigated the use of rotating galvanometers in different types of resistance movements. And Steinwehr
reinvestigated the behavior of mercury sulfate and polarization phenomena in standard cells (Figures 4.4, 4.5, and 4.6).[56]

The Reichsanstalt's studies on electrolytic conductivity began in
1896, shortly after Kohlrausch arrived in Charlottenburg. In 1903
Grüneisen and Kohlrausch continued their ongoing study of the electri-

Figure 4.4. Wilhelm
Jaeger. Courtesy of
Deutsches Museum,
Munich.

cal conductivity of hydrous solutions. Grüneisen also continued his
earlier and related work on the internal friction of salt solutions, while
Kohlrausch measured the increase in conductivity of water exposed to
a radium preparation. In other work, Holborn and Austin analyzed the
scattering of glow-current cathodes in air and hydrogen. And
Kohlrausch, Holborn, and Henning continued their attempts to develop
a disturbance-free (i.e., astatic) torsion magnetometer, the relevance of
which we shall see later in this chapter.[57]

The Electrical Laboratory also investigated particular problems requested by academic scientists. The work of Steinwehr and Jaeger in
constructing a platinum-resistance thermometer that, between 1° and
2°C, was ten times as precise as the mercury thermometer then in use
illustrates the point – and more. The new thermometer enabled them to
determine the value of water in a Berthelotian calorimeter in terms of
international electrical units.[58] Their inducement to make this thermometer and these measurements came from Kohlrausch's good
friend, the Berlin chemist Emil Fischer, who in 1903 sought a standardized Berthelotian calorimeter as the means of measuring the heats
of combustion for pure organic compounds.[59] Three years later, when
Steinwehr and Jaeger's laborious measuring work was finally

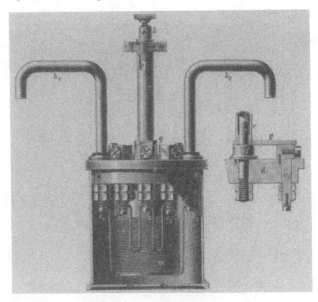

Figure 4.5. Apparatus (a so-called branching box) used to compare the values of wire resistances against that of a standard wire resistance. Reprinted from K. Feussner and St. Lindeck, "Die elektrischen Normal-Drahtwiderstände der Physikalisch-Technischen Reichsanstalt," *WAPTR* 2 (1895): 525.

completed, Fischer, speaking "in the name of thermochemistry," thanked them for having helped secure the foundations of that specialty.[60] Moreover, he noted that their work was "of importance not only for chemistry but also for heating technology, biology, and medicine." It was fundamental, demanding measuring work, the type that helped establish the experimental foundations of science and technology. Such work, Fischer said, was "little known in public" and so "often underestimated." "Precisely in this regard, however, the accomplishments of the Physikalisch-Technische Reichsanstalt are unsurpassed examples, ones to which it preeminently owes its great reputation in scientific circles at home and abroad."[61]

By contrast, the Optical Laboratory's work during Kohlrausch's tenure had no shortage of recognition among scientists, technologists, and others. The search by various Reichsanstalt physicists for a spectral-energy distribution law for blackbody radiation culminated in what soon became one of the centerpieces in the transition from "classical" to "modern" physics: Planck's law and quantum hypothesis of 1900-1901. Planck's quantum approach to a satisfactory radiation law received most of its empirical support from data supplied by the Reichsanstalt. In turn, that law helped win an international reputation for the Reichsanstalt as a center for fundamental science.

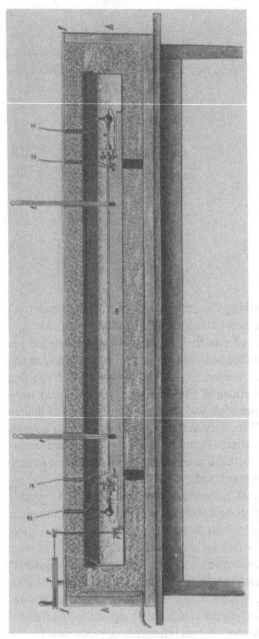

Figure 4.6. Oil-filled "bath" apparatus maintained at 0°C and used to make electrical measurements of mercury resistances. Reprinted from W. Jaeger, "Die Quecksilber-Normale der Physikalisch-Technischen Reichsanstalt für das Ohm," *WAPTR 2* (1895): 415.

The story of Planck's route to his law has been told often, in great detail, and well.[62] But it has never been told strictly from the Reichsanstalt's point of view. To do so, even if in an abbreviated fashion, is to appreciate the principal institutional resources that permitted, and the combination of scientific and practical motives that led to, the Reichsanstalt's participation in blackbody research. In particular, it becomes clear that industrial interests were indeed behind the Reichsanstalt's blackbody-radiation work. This point is at issue not only because it has either been neglected or never been proved, but also because it has been explicity denied by the late Hans Kangro, who, in his otherwise outstanding account of the background to and the experimental foundations of Planck's blackbody-radiation law, rejected the possibility that the search for that law and the associated radiation measurements was motivated by the "practical interests of industry."[63] Here, evidence of those "practical interests" at work at the Reichsanstalt is presented. To appreciate them, as well as the pure scientific interests motivating the research in blackbody radiation, is to appreciate more deeply the setting and timing of the origins of quantum physics.

Although Helmholtz had never shown much interest in the problem of heat radiation, it is a fact that all but one of the Reichsanstalt physicists who conducted blackbody-radiation research between 1893 and 1901 had studied with him at Berlin. The Reichsanstalt physicists who held permanent positions there owed their training and appointments to him: Lummer, who led the Optical Laboratory from 1887 until 1904; Wien, who worked at the Reichsanstalt between 1890 and 1896 and was its leading theoretician as well as an able experimentalist; and Kurlbaum, who worked with Lummer from 1891 until 1901. Moreover, those physicists who held guest or volunteer status owed their appointments either directly to Helmholtz or, in the case of the one individual appointed by Kohlrausch, indirectly to him in that he had created such positions in order to further the research of young, talented scientists. Included in the category of guests or volunteers were Pringsheim, who promoted (1882) and habilitated (1886) under Helmholtz and who was a volunteer at the Reichsanstalt between 1893 and 1904; and Rubens, who promoted (1889) and habilitated (1892) under Kundt at Berlin and who joined the Technische Hochschule Charlottenburg in 1896, becoming at the same time a guest worker at the Reichsanstalt.[64] Rubens needed that guest status because the Technische Hochschule's physics research facilities were virtually nonexistent.[65]

The immediate origin of the Reichsanstalt's interest in radiation research lay in its photometric studies. That work, as we saw in Chapter 3, came in response to a request from the Deutscher Verein für Gas- und Wasserfachmänner to develop better light-measuring devices and a better unit of luminous intensity. By 1893 the Reichsanstalt had (temporarily) satisfied the Verein's request. But unsolved scientific and technological problems remained: to find a scientific basis for the improved photometric devices and to develop a universally acceptable unit of luminous intensity; more generally, to furnish the German scientific and industrial communities with more accurate temperature measurements based on an absolute temperature scale.[66] From 1893 on, when the Scientific Section's new facilities were available, the Reichsanstalt investigated these broader problems in light and heat radiation.

The conceptual basis of the Reichsanstalt's radiation work lay in the older results of Kirchhoff (and, independently of him, Balfour Stewart), who in 1859 had announced a law describing a constant ratio between any body's emission and absorption of heat radiation, a ratio independent of the nature of the body and equivalent to some unknown universal function whose only parameters were the absolute temperature $T$ and the frequency $\upsilon$. Kirchhoff called a body that absorbed all incident radiation and reflected none a "blackbody"; at equilibrium, its spectral-energy distribution could be described by the unknown universal function. In 1879 and 1884, respectively, Josef Stefan and Boltzmann showed that the energy $E$ emitted by a blackbody increased as the fourth power of its temperature, a result soon known as the Stefan-Boltzmann law,

$$E = \sigma T^4,$$

where $\sigma$ is a constant. In 1893 the Reichsanstalt's Willy Wien (Figure 4.7) employed the Kirchhoff and Stefan-Boltzmann laws, along with Maxwell's equations and the second law of thermodynamics, to derive his own, so-called displacement law. Wien showed that "in the normal emission spectrum of a blackbody ... the product of the wavelength [ $\lambda$ ] and the temperature is a constant," [67] that is,

$$\lambda T = \text{constant.}$$

A year later he (more rigorously) rederived his law, showing that, as a consequence of Kirchhoff's law, blackbody radiation could be defined "as a condition of stable heat equilibrium." From this definition it fol-

Figure 4.7. Willy Wien
circa 1890. Courtesy of
Physikalisch-Technische
Bundesanstalt, Braun-
schweig.

lowed that, once the blackbody's spectral-energy distribution was known for any given temperature, the temperatures for all other energy distributions could be derived.[68]

Wien's seminal studies of 1893 and 1894, published as so-called private communications, that is, as unofficial work, led the Reichsanstalt to investigate all aspects of blackbody radiation. Wien's scientific results were seen as potentially useful to the German illuminating and heating industries, which sought to increase the range and exactness of their temperature measurements. Moreover, his results also had broad implications for the Reichsanstalt's own program to develop better temperature-measuring instruments and an absolute temperature scale. When Kohlrausch assumed the presidency in the spring of 1895, Wien apprised him of his results and urged him to allow the Optical Laboratory to test the validity of his radiation law as a regular, official part of the Reichsanstalt's program. Although skeptical of highly abstract work like Wien's, Kohlrausch reluctantly yielded to Wien's entreaties. He presumably did so in part because Wien's results fit in nicely with similar, experimentally oriented interests of other Reichsanstalt workers in light and heat radiation.[69] In particular, Lummer and Kurlbaum had recently presented results using a bolometer to achieve a more satisfactory unit of luminous intensity and had reported on the use of platinum as a standard for studying spectral-energy distribution.[70] Results like these led Kohlrausch to hope that the spectral-energy work would yield "a more direct definition of the unit of luminous intensity" and "reduce the energy distribution of various ordinary light sources to that of the platinum unit of luminous intensity."[71]

To confirm Wien's law and the Stefan-Boltzmann law and to measure light radiation photometrically, Lummer and Wien proposed constructing a new type of blackbody along with an associated method for testing the radiation laws.[72] This blackbody was essentially a hole in a furnace maintained at a certain temperature. Lummer and Wien prophesied that the results of work done with it would be "*as important for technology as for science*."[73] Kohlrausch agreed, reiterating that blackbody research might lead to a better means of defining a unit of luminous intensity.[74]

Despite the promise of Wien's displacement law and the blackbody radiator, the universal function describing the spectral-energy distribution of a blackbody remained undetermined. Using the seemingly secure experimental results of Friedrich Paschen of the Technische Hochschule Hannover and his own set of highly conjectural hypotheses, in 1896 Wien promulgated a spectral-energy distribution law,

$$E = C \lambda^{-5} e^{-c/\lambda T},$$

where $C$ and $c$ are constants.[75] During the closing years of the century Planck rederived Wien's law more rigorously; hence it became known as the Wien-Planck law.[76] In the meantime, its truth and utility remained in doubt.

One of the first believers in the ultimate utility, if not truth, of the radiation laws was the ever boastful, ever loquacious Lummer. In 1897 he delivered a lecture at the Berlin Polytechnische Gesellschaft (Polytechnical Society) entitled "Light and Its Artificial Production," boldly articulating the technological implications of the recent radiation research by physicists at the Reichsanstalt and elsewhere. The ultimate goal of research and technological development in artificial illumination, he explained, was to separate light from heat. The Reichsanstalt sought an improved, absolute basis for comparing the economic value of different light sources to replace the present, relative one of price per candlepower. To do so, it first needed to understand "the physical laws connecting the luminous intensity of a body with its temperature and nature." This meant understanding Kirchhoff's law, the Stefan-Boltzmann law, the theoretical conditions for light emission, and

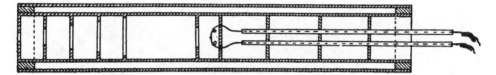

Figure 4.8. Schematic diagram of a blackbody (with a thermocouple inside). Reprinted from [Friedrich Kohlrausch], "Die Thätigkeit der Physikalisch-Technischen Reichsanstalt in der Zeit vom Februar 1899 bis Februar 1900," *ZI 20* (1900): 147.

Wien's displacement law. These laws, together with the absolute blackbody then being constructed at the Reichsanstalt for testing them and for acting as an emission standard (since it gave information on the minimum temperature at which light would become visible), should make possible the comparative measurements of the light-radiation properties and characteristics of carbon, the metals, and their oxides. "With the assistance of our knowledge of the radiating power of luminous substances and of the temperature at which they become luminous," Lummer bragged, "I can compare theoretically the intensities of different light sources. Unfortunately," he had to concede, "considerable difficulties are met with in the measurement of high temperatures." Moreover, he admitted that the Reichsanstalt faced many difficulties in making these physical laws technologically useful. Yet he had no doubt that industry would someday benefit from the Reichsanstalt's work "if the directions pointed out by science are followed."[77]

Those directions became clearer later that year, when Paschen, as well as Lummer and Pringsheim, largely confirmed the validity of the Stefan-Boltzmann law.[78] Then in 1898 Lummer and Kurlbaum announced the realization of the blackbody. Their device, an instrumentational breakthrough three years in the making, consisted of a thermocouple placed inside a completely closed platinum cylinder whose blackened interior walls could be brought by electrical means to thermal equilibrium (Figures 4.8 and 4.9). Their blackbody was superior to the devices of Violle, Siemens, and the Reichsanstalt's own previous creations (i.e., their improvements of the Hefner lamp) for achieving a satisfactory unit of luminous intensity.[79] It helped give definite experimental proof of the validity of the Stefan-Boltzmann law and, since it was an excellent temperature-measuring device, promised to be of use for the Reichsanstalt's more practical problems. Kohlrausch, in addi-

Figure 4.9. The Reichsanstalt's radiation-measuring setups circa 1900, including (in front of A) a Lummer- Kurlbaum blackbody and (near E) a Lummer-Pringsheim blackbody. Reprinted from *Müller-Pouillets Lehrbuch der Physik und Meteorologie*,10th enl. ed., 4 vols. (Braunschweig, 1909), Vol. 2, Otto Lummer, *Die Lehre von der Strahlenden Energie (Optik)*, Tafel 1. Courtesy of Physikalisch-Technische Bundesanstalt, Braunschweig.

tion to noting the new blackbody's scientific uses – temperature and spectral-energy measurements, the testing of Wien's displacement law and a recent formula of Paschen's that described various energy curves – emphasized that it "will also be of value for the experiments [for establishing] a unit of luminous intensity."[80] Lummer and Kurlbaum soon reported improvements in their blackbody that made it trustworthy up to 1900 K, confirmed the Stefan-Boltzmann law between 90 and 1,700 K, and attempted to use it to establish a unit of luminous intensity. Those results encouraged Kohlrausch's hope that "the fundamental laws of heat- and light-radiation," once determined, would help solve "[temperature-measuring] problems of illuminating and heating technology" (Figures 4.10 and 4.11).[81]

The chief unconfirmed law was that of Wien-Planck. Paschen,

Figure 4.10. Otto Lummer.

operating in part with equipment supplied by and thermometers standardized at the Reichsanstalt, believed that he had experimentally confirmed it.[82] Lummer and Pringsheim, however, obtained experimental
results that conflicted with Paschen's, and so raised doubts about the
law's validity. In the first part of their classic study of the spectral-
energy distribution of a blackbody and of polished platinum, they
showed that, at higher temperatures and greater wavelengths than had
previously been investigated, there occurred deviations from the Wien-
Planck law. Their measurements challenged theory and found it
wanting. Moreover, in the second part of their paper, a part neglected
by historians of radiation physics, they reported their efforts to determine the temperature of ordinary solid incandescent bodies by means
of the radiation laws. This was a problem, they said, "of scientific and
technical interest," and they sought "to show that the radiation laws
found for the blackbody and polished platinum are a helpful means

Figure 4.11. Ferdinand Kurlbaum. Courtesy of Tech-
nische Universität Berlin,
UB (HA, HGS).

toward solving this problem." By assuming the validity of Kirchhoff's
and Wien's displacement laws, they claimed to be able to determine the
temperatures of any illuminating body whose radiation properties lay
between the extremes of those of a blackbody and polished platinum.
Included here were carbon – "one of the most important substances in
illuminating technology" – candles, oil lamps, gas lamps, acetylene
lamps, carbon-filament electrical lights, and so forth.[83] Lummer and
Pringsheim's results, Kohlrausch hoped, would help the Reichsanstalt
"reach a decision on the economy of light sources."[84]

The Reichsanstalt's blackbody radiation studies reached a climax
in 1900. Lummer and Pringsheim now obtained and exploited
blackbody radiation for long waves that further suggested that the
Wien-Planck distribution law was invalid for higher temperatures and
longer wavelengths.[85] That work, along with some theoretical results
of Lord Rayleigh,[86] stimulated Lummer and E. Jahnke to propose an

Figure 4.12. Heinrich Rubens. Courtesy of American Institute of Physics, Niels Bohr Library, New York.

empirical formula that covered all measured temperature and wavelength values; it also stimulated Thiesen to replace Wien's formula by one of his own.[87] However, these ad hoc formulas and the relatively restricted wavelengths used in Lummer and Pringsheim's important study meant that still further testing of the Wien-Planck law at still greater wavelengths was needed, as were, perhaps, a new law and explanation.

Fresh leadership in the experimental realm now came from Rubens, working as a guest at the Reichsanstalt (Figure 4.12). Building on earlier work of E. F. Nichols, Rubens and Nichols had developed their method of residual radiation (*Reststrahlen*) that allowed them to eliminate all rays except those of a desired, narrow wavelength band.[88] In 1900 Rubens and Kurlbaum employed residual radiation to demonstrate unequivocally not only that Wien's distribution law was invalid but that, with increasingly higher temperatures and longer wavelengths, it increasingly deviated from the observed results.[89] Shortly after they had completed their study, Rubens visited Planck at the latter's home to report their results. That very day (7 October 1900) Planck devised a new formula that accounted for all experimental results:

$$E = \frac{C\lambda^{-5}}{e^{-c/\lambda T} - 1}.$$

Within a fortnight (on 19 October) he presented it to a session of the Physikalische Gesellschaft.[90] Immediately after the session Rubens compared his experimentally determined data with Planck's new formula. The next morning Rubens again visited Planck at his home, this time to report the excellent fit between Planck's formula and Rubens's data.[91] As a result, several months later Planck proposed his hypothesis of a quantum of action as the new conceptual element needed to provide the physical foundations for deriving his radiation law.[92] Physicists at the Berlin triangle of institutes – the University of Berlin, the Technische Hochschule Charlottenburg, and the Reichsanstalt – had led the way in establishing a blackbody-radiation law. As a consequence, a new view of radiation (and, in time, matter) began to emerge, one that revolutionized physical thought.

The Reichsanstalt's Optics Laboratory had proved its mettle vis-à-vis academic physics institutes and had earned high praise from its academic brethren. Yet as quantum physics now (slowly) developed, the Reichsanstalt did no research in this emerging field; instead, it continued to test Planck's and other radiation laws, to determine the values of the constants in Planck's law, and to develop a temperature scale based on the radiation laws. It did so because, unlike the new quantum physics, the blackbody-radiation laws encouraged hopes of utility for industry as well as for the Reichsanstalt's own work in light and heat measurement. The newly confirmed radiation laws showed applied physicists and technologists the way to build, among other items, improved optical pyrometers that permitted unprecedently accurate and high temperature measurements of illuminating bodies. They showed, too, how best to recognize, and hence exploit, the light and heat emission properties of illuminating substances. As the new century opened, Lummer and Pringsheim felt authorized to declare that "the radiation laws seem to be well enough understood and advanced that we believe it is again time to try . . . to make them serviceable for temperature measurements."[93] (The Nobel Committee in Physics also pointed [in 1908] to the practical use of Planck's radiation law.)[94] The radiation laws were now to form the basis of a new temperature scale, one based not on the thermodynamics of gases but rather on the radiation of a blackbody.[95]

In 1903, for example, Lummer and Pringsheim reconfirmed and extended the range of experimental validity of Planck's law. Moreover, they determined the temperature of a number of illuminating substances and processes. They expected that their work on measuring the differences between blackbody and polished-platinum radiation would be important for optical pyrometry. Furthermore, Lummer and Ernst Gehrcke showed the effectiveness of plane-parallel plates in resolving very fine spectral lines and argued that the plates warranted use in the construction of an interference spectroscope. The latter, originally conceived by Lummer and later developed by himself and Gehrcke, provided high resolving power and proved to be an essential spectroscopic tool for atomic research. Lummer and Gehrcke expected that their spectroscopic work would lead to improved measurement (in wavelengths) of the meter, of the rotation of the plane of polarization, and of the Zeeman effect. The two men also improved an apparatus for related investigations involving plane-parallel plates, and they began preparations for a series of investigations on capillary tubes and gas illumination. In addition, Lummer and Pringsheim studied the anomalous dispersion of gases, a topic of considerable importance for the theories of light refraction and emission in general and for understanding solar phenomena in particular.[96]

Much of this high-quality and pure scientific work was to be of importance for modern physics. Nonetheless, after the turn of the century the laboratory devoted most of its efforts to using the radiation laws to develop an absolute temperature scale, one of use for illumination and heating technology, and to determining the temperatures of everyday light and heat sources. The Reichsanstalt's workers believed that the blackbody, scientifically anchored to the radiation laws, now provided the means for achieving an absolute temperature scale.[97] By 1903 the Reichsanstalt's Optics Laboratory had succeeded in establishing an exact, absolute, radiation-based thermometric scale valid in the range from -200 to +2,000 K. But by then the laboratory had essentially exhausted its earlier line of scientific research.

To conclude this brief account of the Reichsanstalt's radiation work, there were two major reasons for the Reichsanstalt's leading institutional role in the investigation of blackbody radiation. First, be-

cause the heart of its mission coincided with the essential observational
and experimental desiderata of such studies – the time-consuming precision measurement of thermal and optical phenomena and the construction of the relevant precision instruments for making such
measurements – and because it had outstanding facilities for undertaking such studies, the Reichsanstalt was fully prepared and willing to do
so. Precision measurement was central to understanding the behavior
of blackbody phenomena; and to perform the highest quality of precision measurement meant having able measurers, adequate workspace,
and good instruments. The completion of the Reichsanstalt's *Observatorium* in late 1891 provided the Institute with unmatched material conditions for conducting radiation-measuring work. By contrast, the
gifted experimentalist Paschen, the Reichsanstalt's principal competi-
tor at testing the various heat-radiation laws, had pathetically little institutional support at the Technische Hochschule Hannover.[98] Still
worse was the situation of Rubens at the Technische Hochschule
Charlottenburg, whose research facilities for physics were virtually
nonexistent. Many of his outstanding experimental results at the turn of
the century arose from his opportunity to work at the nearby Reichsanstalt.[99]

Second, the Reichsanstalt and its physicists were motivated by a
combination of pure scientific and utilitarian considerations. The Institute's physicists who conducted blackbody research were motivated, it
would seem, largely by a desire to achieve a theoretically well justified
and empirically well confirmed law of blackbody radiation. In addition, however, there existed utilitarian motives for pursuing this radiation research: Such research would eventually advance the temperature-measuring needs of and contribute to the development of more
energy-efficient lighting and heating sources for the German illuminating and heating industries. Lummer, the leader of the Optics Laboratory, clearly recognized the technological possibilities that might issue
from the Reichsanstalt's radiation work. And even a pure scientist like
Wien acknowledged that "from the laws of heat radiation" one could
determine "the economy of light emission of an illuminating body." If
Wien's (and his fellow physicists') goals were largely to establish those
laws, he (and they) nonetheless recognized their potential implications
for industry.[100] Understanding the distribution of spectral energy
raised hopes of generating greater profits for industry and greater
savings for consumers. To be sure, the Reichsanstalt, in the person of

Kohlrausch, had to be pushed into radiation work by industry and by its own knowledge-hungry physicists. Yet even the so-called private communications of Reichsanstalt physicists on radiation research – which in fact resulted from the use of Reichsanstalt materials, equipment, and time – quickly became part of the Institute's official research program.

By 1901 the heady race to confirm the radiation laws was largely over. The Reichsanstalt had played an essential role in confirming those laws. Both the Institute as a whole and the individual physicists who had engaged in blackbody-radiation research received international recognition. Helmholtz had brought most of them to the Reichsanstalt as young, promising physicists. By 1901 their demigod-mentor had been dead for seven years, and they had turned promise into performance. As mature scientists they now clamored for new professional roles and new scientific problems to work on. In 1901 Kurlbaum, desirous of heading his own laboratory, became the leader of the Technical Section's High-Voltage Laboratory.[101] In 1904 Lummer became ordinary professor of experimental physics at the University of Breslau.[102] Some six months later, in 1905, his good friend and long-time collaborator Pringsheim joined him in Breslau as professor of theoretical physics.[103] And in 1906 Rubens became professor of experimental physics at the University of Berlin, where he at last had adequate research facilities.[104] Ironically, the very successes of its individual physicists led to new problems for the Institute.

## Measuring for Marks

The construction of the Technical Section's new buildings in the mid-1890s was the principal source of the Reichsanstalt's institutional growth under Kohlrausch. During that period, the employees of the section devoted much of their time to constructing new equipment and apparatus (mostly of the electrical variety) for the new quarters instead of conducting scientific work. Between early 1896 and the spring of 1897 they lost still more time because they twice had to move the section's operation: once from the inadequate quarters at the Technische Hochschule Charlottenburg into temporary quarters in the Scientific Section's *Observatorium* and once from the latter into the Technical Section's new, fully equipped buildings. With its expanded facilities fully in place by 1897, however, the section undertook an increased amount of work, especially research and testing work for the electrical industry. Serving the German electrical industry became the outstanding feature of the Technical Section's work under Kohlrausch.[105] Yet as its link to the industry tightened, a clash of interests and antagonism arose.

By 1903 the section employed approximately twenty-five senior scientific officials, or about twice as many as the Scientific Section, as well as a team of mechanics, minor officials, and office workers. As in Helmholtz's day, there were four laboratories, along with a workshop and Chemical Laboratory, all of which had remained since 1894 under Hagen's direction (Figure 4.13). From the outset the Technical Section required a far larger number of technical assistants and administrative personnel than did the Scientific Section because of its need to conduct tests, issue certificates, and administer its extensive business with industry and others. Starting in the mid-1890s and especially after the move into its new, large quarters in 1897, the number of office personnel required to meet the growing and burdensome administrative problems attached to testing work increased greatly.

Arnold Leman still directed Precision Mechanics. He and his chief aide since 1887, Arnold Blaschke, continued to play central roles in the Deutsche Gesellschaft für Mechanik und Optik: Blaschke was the Gesellschaft's long-time secretary, and both men belonged to the editorial board of the Gesellschaft's *Vereinsblatt* (known from 1898 on as the *Deutsche Mechaniker-Zeitung*), which appeared as a supplement to the *Zeitschrift für Instrumentenkunde*.[106] These responsibilities helped maintain close connections between the Reichsanstalt and the German mechanical and optical industries. The scientific basis of that connection was the testing and certifying of precision-mechanical devices: gauge blocks, graduators, end-measuring bolts, rings, measuring and graduated screws, screw threads, water-meter threads, tuning forks, and so forth.[107] The work was routine, mundane, and essential to the German precision-mechanical industry.

During Kohlrausch's presidency, the Electrical Laboratory became the second largest Reichsanstalt laboratory. In the late 1890s its thirteen workers were distributed into three sublaboratories: High-Voltage, Low-Voltage, and Magnetics. The Technical Section's new buildings meant a sizable expansion in the laboratory's work: Whereas previously it had lacked space for such items as accumulators, after 1896 it could conduct studies on alternating and polyphase currents, on high voltages, and on generators and turbines. These were the new, techno-

Figure 4.13. Ernst Hagen
circa 1900, from an original
drawing by George Varian.
Reprinted from Ray Stannard
Baker, *Seen in Germany*
(London, 1902), 182.

logically important areas for the electrical industry. The Electrical
Laboratory contributed to the latest industrial developments and
addressed industry's changing needs. It also expanded its work on setting electrical standards, testing parts and products, and so on. "The
great economic meaning that [the commercial development of electric-
ity] has already acquired and will acquire even more in the near future,"
Kohlrausch wrote in 1895, "makes it obviously important to treat
thoroughly and to test experimentally the many practical and scientific
questions that are here raised."[108] Even more than under Helmholtz,
the Reichsanstalt under Kohlrausch gave to the electrical industry. The
growth of the Technical Section's Electrical Laboratory paralleled that
of the German electrical industry.

From 1901 to 1904 Friedrich Kurlbaum supervised the High-Voltage Laboratory, the first of the Electrical Laboratory's three sublaboratories. As noted earlier, after a decade's worth of pioneering
scientific research as Lummer's associate in the Scientific Section's

Optical Laboratory, Kurlbaum hankered after more responsibility and independence. In 1901 he was named a member of the Reichsanstalt and, after a brief training period at Siemens & Halske, placed in charge of the Technical Section's High-Voltage Laboratory. At the same time, Kurlbaum (with Holborn) continued his radiation measurements by means of an optical pyrometer. His new testing responsibilities as well as his old scientific research apparently did not, however, satisfy his desires for career advancement: In 1904 he left for an academic appointment as professor of physics at the Technische Hochschule Charlottenburg.[109]

Of the eight men under Kurlbaum's direction, five were responsible for the direct-current and three for the alternating-current work. The laboratory routinely tested four classes of objects: measuring apparatus (including direct-, alternating-, and polyphase-alternating-current dial apparatus for measuring voltage and current strength, as well as electrical meters, capacitors, optical pyrometers, and other measuring apparatus for resistance, insulation, and induction standards; motors and transformers – all told, in 1903 nearly 600 pieces of measuring apparatus); various materials (varnishes, oils, porcelain insulators, insulator tubes), wires, and cables; and a miscellaneous class of objects ranging from safety fuses to railway rails.[110]

In addition to its routine testing work, the High-Voltage Laboratory also did some research. First, Kurlbaum and a colleague used an optical pyrometer and spectral analysis to study bodies at high temperature (ca. 1,800-2,400 K). Second, the laboratory reinvestigated the so-called Dolezalek electrometer used in alternating-current measurements. And third, it continued to evaluate the use of an optical pyrometer for testing alternating-current ammeters and constructed a new self-induction variometer for use in self-induction measurements.[111] At the turn of the century, the High-Voltage Laboratory had to increase its research, testing, and certification work on alternating-current systems, self-induction standards, and the associated measuring instrumentation in order to solve the technical problems occasioned by the rapid rise of wireless telegraphy and telephony. Moreover, its work on the optical pyrometer yielded increasingly accurate temperature measurements that were of use in the procelain and iron industries.

Stephan Lindeck was the sole member of the Low-Voltage Laboratory, the second of the Electrical Laboratory's three sublaboratories. Most of his work consisted largely of routine testing of hundreds of resistances (many of which had been sent to the Reichsanstalt from foreign countries), galvanometers, accumulators, conduction and resistance materials, as well as standard Clark and standard Weston cells.[112] Lindeck's more creative scientific work lay in his helping to establish the international standard value of the ohm and the related measuring methods.[113] All this standards and testing work was of obvious importance to the German electrical industry. In addition, Lindeck's editorial work for the *Zeitschrift für Instrumentenkunde* from 1895 to 1911, his membership on the board of the Deutsche Gesellschaft für Mechanik und Optik, his organization of Germany's exhibit of mechanical and optical apparatus at the Paris International Exhibition of 1900 and at the St. Louis Exhibition in 1904 represented further Reichsanstalt links to the German mechanical and optical industry.[114]

The Magnetic Laboratory, the third of the Electrical Laboratory's three sublaboratories, was the bailiwick of Ernst Gumlich and P. Rose. In 1903, as in other years, they tested magnetic measuring apparatus and materials: scales, coils, steel, cast iron, dynamo sheet metal, as well as nonmagnetic materials. In addition to their routine testing work, during 1903 Gumlich and Rose also conducted some research. At the urging of the German navy they studied the magnetization of steel.[115] Moreover, at the request of the Verband Deutscher Elektrotechniker (Association of German Electrical Engineers) they studied the properties of sheet metal, with their work leading to better measuring methods for wattmeters. (At the turn of the century, Gumlich measured the magnetic properties of different alloyed sheet metals. He found that the addition of silicon greatly reduced the loss of iron in sheet metal, and so raised the performance levels of transformers and dynamos. His work proved to be of great economic importance for the German electrical industry.) In three other studies, Gumlich and Rose investigated magnetization by means of both direct and alternating current; determined the electrical conductivity of rod-shaped magnetic materials in the hope of improving the construction of transformers and other electromagenetic devices; and determined the initial permeability of a large number of materials needed to construct disturbance-free galvanometers.[116]

Starting in 1901 the Electrical Laboratory acquired another responsibility: supervision of the new electrical testing offices around Germany. The Reichsanstalt's Office of Electrical Testing (Elektrisches Prüfamt) owed its existence to and derived its responsibilities from Germany's new law of 1898 on electrical units and standards. Such a law was needed, the legislators wrote, because of "the tremendous expansion of the industrial application of electricity and the large amount of money involved in paying for supplying electric current and electrical apparatus." The lawmakers noted the widespread use of electrically powered machinery in factories and workshops; the emergence of new electrical streetcars; the use of electrical illumination (arc lights and incandescent lights); as well as the use of electricity in heat production, electrochemical decomposition, magnetization, electrical clocks and other types of signal transmitters, and so forth. Above all, the legislators pointed to the need for state regulation of the electrical meters used to measure the amount of electricity consumed by the utilities' customers.[117]

The law, which became effective in 1902, defined Germany's legal electrical units as the ohm (resistance), the ampere (current), and the volt (electromotive force); all other electrical parameters were to be derived from this triumvirate. Furthermore, it empowered the Reichsanstalt to set electrical standards and to test and certify electrical apparatus.[118] The Reichsanstalt's authority in electrical matters now matched that of the Normal Eichungskommission in matters of weights and measure.

Nevertheless, the law lacked teeth: It did not force the industry to certify its electrical meters or other apparatus. It went no farther than stating, first, that all commercial electrical measures be ultimately reducible to the three legally defined electrical units and, second, that electrical apparatus give the correct data to within a certain (unspecified) range of error.[119] Moreover, the official standards were not set solely on the basis of the Reichsanstalt's scientific findings. They were, instead, the result of negotiations between the Reichsanstalt, on the one hand, and the Verband Deutscher Elektrotechniker and the Vereinigung der Elektrizitätswerke (Alliance of Electrical Works), on the other. In addition, there were, in Kohlrausch's words, "exhaustive consultations with a selected group of representatives of the electrotechnical interests" – including AEG, Siemens & Halske, the Union-Elektrizitätsgesellschaft (Union Electrical Company), and Schuckert & Co. – and a questionnaire was sent to more than 1,000 electrical firms before the Reichsanstalt drew up its new regulations for testing electrical products.[120] Regulation occurred only after the industry's advice and approval had been received. The Reichsanstalt maintained a modus vivendi with the industry it sought to regulate and promote. The law was only a first step toward regulating the commercial relations between industry and consumers.

The electrical industry was mistrustful of the Reichsanstalt's regulatory powers and authority. Supervision of the Reichsanstalt's obligations to control meters and other devices fell to Karl Feussner of the Electrical Laboratory. From 1903 on, he undertook annual visits to the Office of Electrical Testing's five regional offices in Ilmenau, Hamburg, Munich, Nürnberg, and Chemnitz. His main concern was to ensure the accuracy of the offices' own standard meters and the availability of trained personnel and good equipment. But because the law did not require manufacturers to certify their meters, the regional offices had little work to do. Hagen reported that the offices probably controlled only around 10 percent of all electrical products; the manufacturers themselves continued to control their own products. The big firms in particular had their own testing departments, and they feared the offices' rights of inspection.[121] Moreover, Feussner reported that industry was less interested in improving its electrical apparatus than in protecting them from misuse and in simplifying their operation (Figures 4.14 and 4.15).[122] Both in setting legal electrical standards and in controlling electrical meters, the industry acted as if its interests were opposed to those of the Reichsanstalt. At the same time, the Electrical Laboratory was burdened by routine testing work for the industry that it sought to promote, increasingly leaving laboratory workers less time for research.

Despite the enormous growth of the Electrical Laboratory, the Heat and Pressure Laboratory remained the largest laboratory at the Reichsanstalt. That was due to the fact that, during 1903 alone, Hermann Wiebe and his crew of four tested nearly 20,000 pieces of scientific and technical apparatus. The bulk of their work consisted of testing nearly 17,000 thermometers, more than 14,000 of which were medicinal thermometers (Figures 4.16 and 4.17). Wiebe also supervised the Reichsanstalt's thermometer control stations in Ilmenau, where in 1903 more than 46,000 medicinal and nearly 2,300 other types of thermometers were tested, and Gehlberg, where still another 800 medicinal thermometers were tested. Moreover, the laboratory tested nearly 1,000 other pieces of heat-measuring apparatus; the bulk of this work included nearly 700 LeChatelier thermocouples, the vast majority of which had been sent in by industry. Finally, it also tested nonther-

Figure 4.14. First direct-current electrical meter certified (1903) by the Reichsanstalt under Germany's new electrical standards law of 1898. Reprinted from Wilfried Hauser and Helmut Klages, "Die Entwicklung der PTR zum metrologischen Staatsinstitut," *PB 33* (1977): 461.

mometric devices, including spring manometers, barometers, oil-testing apparatus, alloy rings used in boiler-security apparatus, and indicator springs used in determining the influence of temperature on spring gauges.[123] All this testing activity, executed for only a small fee, afforded self-evident benefits to Germany's thermometer industry.

Like the section's other laboratories, Optics spent most of its energies on routine testing. After Lummer left the laboratory in 1894, Eugen Brodhun took over its management. His principal associates were Liebenthal and Schönrock. In 1903 Brodhun and Liebenthal performed more than 600 photometric tests, including tests on Hefner, Nernst, osmium, and arc lamps and on carbon-filament electric lights and incandescent gas-light apparatus. They also standardized a large number of carbon-filament electric lights sent to them from abroad. Brodhun, moreover, participated in sessions of the International Photo-

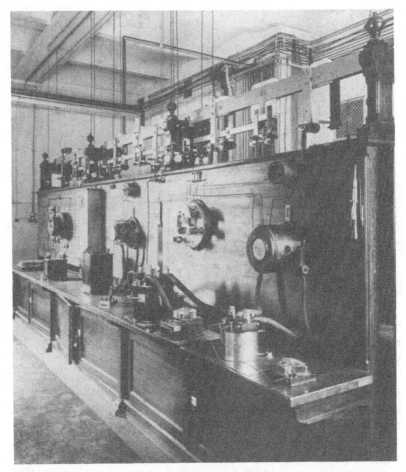

Figure 4.15. View of a meter-testing laboratory circa 1900. Courtesy of Physikalisch-Technische Bundesanstalt, Braunschweig.

metric Committee attempting to reach an international agreement on the unit of luminous intensity. As of 1903 a practical and internationally recognized standard had not been agreed upon. However, the committee did establish a (temporary) set of numerical relationships between the English, French, and German national standards (Figure 4.18).[124]

The laboratory also greatly aided the German sugar industry through Schönrock's polarimetric research and saccharimeter testing.

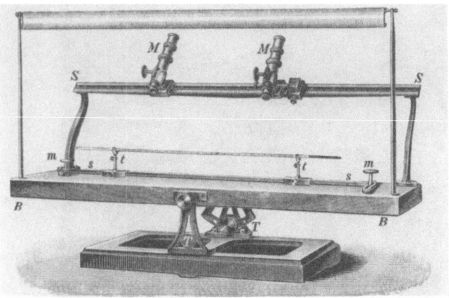

Figure 4.16. Drawing of a double-microscope apparatus used for calibrating thermometers. Reprinted from J. Pernet, W. Jaeger, and E. Gumlich, "Thermometrische Arbeiten betreffend die Herstellung und Untersuchung der Quecksilber-Normalthermometer," *WAPTR 1* (1894): 44.

By the turn of the century he had developed a trustworthy standard quartz plate and a polarization apparatus that provided an absolutely secure basis for measuring sugar content, be it on the production line or in the marketplace, and that helped stimulate international trade in sugar. Together with Brodhun, he tested for optical purity saccharimeter quartz plates sent to the Reichsanstalt. In order to achieve a still better measure of sugar content, his research in 1903 sought to determine the influence of temperature and wavelength on the temperature coefficient of sugar's angle of rotation. He also investigated how and why different mixtures of light and different absorption processes produced different values for sugar rotation.[125]

One final piece of work conducted under the Optical Laboratory's auspices in 1903 deserves mention, not least because its exceptional nonutilitarian nature contrasts with the normal type of work done in the Technical Section. Between 1902 and 1910 Hagen and Rubens investigated the relationship between the electrical conductivity of metals and their optical properties in the infrared region of the spectrum. They

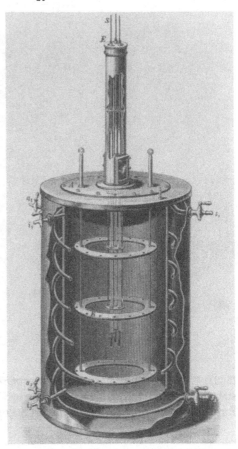

Figure 4.17. Drawing of an apparatus used to compare thermometers. Reprinted from M. Thiesen, K. Scheel and L. Sell, "Thermometrische Arbeiten, betreffend die Vergleichungen von Quecksilberthermometern unter einander," *WAPTR* 2 (1895): 18.

demonstrated empirically a clear relationship between the reflectivity and conductivity of metals, thereby further confirming Maxwell's electromagnetic theory of light.[126] Such "pure" scientific work on the electromagnetic and optical properties of metals was, to be sure, of potential use to the metallurgical and electrical industries, in particular for wireless telegraphy and telephony. But actualizing that potential was something that would occur only in the future. Research work like this, conducted by the director of the Technical Section and by a Reichsanstalt guest on a topic that had no immediate practical use, was exceptional. Most of the section's staff spent most of its time doing routine testing work. The point is perhaps best illustrated by the fact that in 1903 the Reichsanstalt earned 56,000 marks from testing fees (most of

Figure 4.18. *Measuring the Candle-Power of Electric Lamps* at the Reichsanstalt circa 1900, from an original drawing by George Varian. Reprinted from
Ray Stannard Baker, *Seen in Germany* (London, 1902), 185.

which came from industry) – more than three times what it had earned
in 1893.[127]

## Bad Streetcars, Bad Nerves

"You would not believe the multiplicity of demands placed on the
R[eichs]A[anstalt] and on me," Kohlrausch complained to Arrhenius in
1898.[128] It was not only the Institute's size and Kohlrausch's style of
administration that made leadership of the Reichsanstalt (for him at
least) a time-consuming, difficult, often unpleasant duty. In addition to
his normal burdens of office, he was confronted with an extraordinary,
taxing problem: the need to protect the Reichsanstalt against foreign
physical disturbances from electrical streetcars. Between 1895 and

1901 he passionately battled against the combined forces of the city of Charlottenburg, the firm of Siemens & Halske, and the Berlin-Charlottenburger Strassenbahngesellschaft, which sought to run electrical streetcars in front of the Reichsanstalt. The battle originated from the fact that electromagnetic fields emanate from unshielded electrical cables placed at street level, pass through the earth as so-called vagabonding currents, and render all nearby magnetic and many electrical measurements nugatory. The Strassenbahngesellschaft wanted to place such cables along its tracks, which lay only 280 meters from the Reichsanstalt.[129] Moving streetcars, moreover, sent mechanical vibrations through the ground and air that could likewise invalidate precision-measurement work. The electrical industry, which by the 1890s had become heavily dependent on the Reichsanstalt's expertise, threatened to nullify much of the Institute's work.

Already Helmholtz had anticipated that the placement of electrical streetcar tracks near the Reichsanstalt would pose a threat to the Institute's precision-measurement work.[130] To counter that threat, as well as to provide for future expansion, the Reichsanstalt bought an additional piece of land surrounding the Scientific Section's buildings. That tactic proved insufficient.

As a Reichsanstalt board member, Kohlrausch had known of the potential threat to the Institute's work before his assuming the presidency.[131] Within a month of taking office the electrical streetcar problem was causing him "a lot of work and concern" and preventing him from feeling at home in Charlottenburg. "We are supposed to be able to present ourselves with the title of world's foremost physical laboratory [*das erste physikalische Laboratorium der Welt*]," he wrote to a friend shortly after assuming the presidency, "yet in fact all finer electrical and magnetic work . . . lie abandoned. That is simply unacceptable. Hopefully, we can still find a way out."[132] The presence of electrical streetcars constituted a grave political problem involving industry and local government.

The problem was, moreover, aggravated by Kohlrausch's attitude toward technologists and by his bad public relations work. In 1895 he addressed the Elektrotechnischer Verein on the disturbances to scientific institutes caused by electrical streetcar systems and, more generally, on the relations of science and technology. Electrotechnologists, he claimed, were generally unsympathetic to physics and physicists. He noted, too, that the Reichsanstalt's limited resources did not allow it to respond to all of technology's calls for help. Quoting Siemens, he pointed out that science must not be made dependent on "material interests"; instead, it should be left alone in the hope and expectation that one day it would contribute to those very interests. Physicists, he said, "have the duty to defend themselves" against excessive, inappropriate demands on their institutes. "They would not be physicists if they did not do so." The Reichsanstalt could not allow the destruction of its scientific work by the unchecked plans of industrial technology, including those of the Berlin-Charlottenburger Strassenbahngesellschaft, he said.

As might have been expected, his audience of technologists reacted unsympathetically to his problems: "The session had finally turned into a critique, I do not want to say of physicists – although that too was partly the case – but of physical methods."[133] In private, he gave (to Barus in far-off America) even fuller expression of his views, fuming that one "cannot serve two masters well." The technologist, he said,

> sees things from the technical point of view, that is, that of the use for others and for oneself. That is then no longer a scientific point of view; it [i.e., technological work] can therefore [!] also be just as completely satisfying as a nursery garden or a shop.
>
> . . . Almost without exception one thus finds that an electrotechnologist, even if he had also been a good physicist, immediately disregards the physics (circumstances permitting, pays little attention to the latter) – indeed, even hates [it]. I can often observe this in my current conflict with the electric railway.[134]

Such a haughty, hostile attitude, combined with his lack of connections to industry before 1895, produced little love between himself and the electrical industry.

As Kohlrausch indicated in his speech to the Verein, the Reichsanstalt's battle with its local municipal and commercial interests was not unique. With the development of the dynamo in the 1870s by none other than Werner von Siemens and his firm of Siemens & Halske, electrical streetcar systems became possible; in 1879, Siemens demonstrated such a system in Berlin and, in 1881, set up an experimental system in Charlottenburg.[135] Because of its size and the presence of Siemens & Halske, Berlin and its suburbs received one of the first and most extensive systems.[136] From the mid-1890s on, the construction of electrical streetcar systems in the major cities of Western Europe and North America caused numerous battles between city residents and various types of urban scientific institutes, on the one hand, and the streetcar companies and their municipal government allies, on the other. Physicists in particular engaged in a "fierce fight" with the companies over their plans to lay uninsulated tracks near physics institutes.[137] The directors of physics institutes in Darmstadt, Halle, Hanover, Jena, Königsberg, Munich, and elsewhere suffered from, protested against, and sometimes successfully resisted the laying of tracks near their institutes.[138] As physical research became ever more precise, it became correspondingly more important to protect that research against foreign disturbances. The ability of urban physics institutes, above all the Reichsanstalt, to conduct precision measurements was at stake. "My first year, or perhaps the first few years, in office was a bad time," Kohlrausch later wrote. "Having departed Strasbourg in conflict against the earth currents [produced] by the electrical streetcars, I found the situation vis-à-vis the public in Charlottenburg even more difficult." He considered it particularly unfortunate and unpleasant that the Reichsanstalt's interests conflicted with those of Siemens & Halske as well as with the city of Charlottenburg:

> The Reichsanstalt – still in the process of growing – was still foreign to the spirit of Charlottenburg. And now this new organization [i.e., the Reichsanstalt] opposed the [city's] interests. The mayor, the city council, and the townspeople felt themselves injured and insulted by the intruder. I got a cool reception from the mayor. There were hard words at the town council meetings, especially since it turned out that the officials making the decisions took seriously the Reichsanstalt's objection to the presence of earth currents. Antipathy toward the Reichsanstalt's other interests also came in the form of limited cooperation from city officials. It was really a bad time, this struggle with one's own city as well as with the electrical streetcar company and with Siemens & Halske. One could not deny the opponents' right, and that one's own heart was actually on the opponents' side. Perhaps the unfair judgment against my point of view helped me in this uncomfortable situation, for one was also no longer bound to people after they acted with hostility in public![139]

It took six years, until 1901, before the Reichsanstalt finally settled with the city and the Strassenbahngesellschaft. The Reichsanstalt allowed the Gesellschaft the right to run its cars in front of the Reichsanstalt, but only with doubly shielded overhead electrical cables. Moreover, the Gesellschaft compensated the Reichsanstalt with 100,000 marks to be used toward the future construction of a disturbance-free magnetic laboratory.[140] Shortly thereafter, Kohlrausch and others managed to develop a disturbance-free (i.e., astatic) torsion magnetometer; they replaced the magnetometer needle by an astatic system that made the fields generated by the streetcars inconsequential.[141] With an astatic magnetometer, as well as with a so-called *Panzergalvanometer* developed by Henri Du Bois and Rubens, the Reichsanstalt could conduct its precision-electrical and -magnetic work without fear of external disturbances. By 1902 the first trains went past the *Observatorium.*

Kohlrausch's successful stand had prevented an intolerable infringement upon the Reichsanstalt's ability to do precision work – but at the cost of his own health. His was a delicate and sensitive nature, which was partly why he was such an excellent measuring physicist.[142] The battle with the Strassenbahngesellschaft, he wrote, "heavily burdened my mood and was perhaps the main element in shattering my nervous system."[143] As a result of his career change from academic life to the Reichsanstalt, he had lost – and sorely missed – the extensive (intersemester) vacation time offered by the former.[144] As early as 1900 he was referring to himself as "an old physicist."[145] A year later he confessed to Barus that he was experiencing increasing difficulty in simultaneously doing scientific and administrative work; he felt old and happily alluded to retirement.[146] His nerves had experienced too much stress for too long. In 1904 he told Fischer, not for the first time, that he could no longer remain in office "without injuring the Office [of president] and myself."[147] The battle over the streetcars had left him "with an unhealthy aversion toward every personal conflict" and led to heart, digestive, and sleep problems. "Toward the end of [my] Charlottenburg tenure, such nervous disburbances . . . had grown to an intolerable limit and had even impaired the sensory functions, particularly those of the eyes."[148] The legal settlement of 1901 brought only limited physical and psychological relief to his overstrained system.

By 1904 his earlier musings about retirement had turned into a definite, irrevocable decision. In November he wrote to Wien, telling him that seven months earlier he had informed the Ministry of the Interior of his decision and of his "passionate" wish "that a result finally be reached." He continued:

> I am still here involuntarily and with a bad conscience, the latter because I am not up to the constantly exacting intellectual and visual work. To read and consider an MS makes me useless within a short time. I have to avoid really observing (e.g., thermometers) with the eyes. That certainly is not proper for an official of the R[eichs]A[nstalt]! I absolutely will not remain here beyond Easter.[149]

Like Helmholtz before him, Kohlrausch was often physically incapacitated during his last years at the Reichsanstalt. Even extended vacations did not help him recover.[150] Nevertheless, his strict sense of duty compelled him to remain until April 1905, when his successor, Emil Warburg, relieved him. Scheel reported that Kohlrausch characteristically spent "the final hour of his administration, the midnight hour of 31 March 1905, working alone in the Reichsanstalt's administration building in order to leave everything ordered for his successor."[151] Warburg arrived the next morning, and Kohlrausch retired to the quiet university town of Marburg, where he quickly convalesced and spent the remaining five years of his life doing precision physics in his home laboratory, just as he had done as a youth.[152]

By the time of his retirement in 1905, Kohlrausch's Reichsanstalt was markedly different from Helmholtz's of 1894. Kohlrausch had inherited a partially completed institute and a team of scientists and technologists dedicated to Helmholtz personally and to his idea of building a Reichsanstalt. During its first years (at least), Helmholtz's new institute had suffered from natural start-up problems. Helmholtz had infused it with a scientific spirit, but its scientific accomplishments were minimal. Kohlrausch completed what Helmholtz had begun. By 1903 the Institute's plant had doubled in size, its staff had grown by nearly 70 percent, and its operating budget had increased more than 40 percent over that of 1893. Kohlrausch had taken over a nascent scientific bureaucracy from Helmholtz and overseen its growth into a full-fledged one. The bureaucratic aspect, however, placed enormous administrative demands on him, demands that eventually exceeded his considerable capacities.

Although Kohlrausch lacked his predecessor's broad understanding of science and his demigod status within Germany, he brought his own qualities to the Reichsanstalt: a love of order and precision that brought out the best experimental skills of his enlarged team, and an authority in measuring physics that helped turn the Reichsanstalt into the institutional master of measurement at the turn of the century. Under Kohlrausch's direction the Institute made incalculable contributions to science by establishing unprecedented degrees of accuracy and precision in setting scientific standards, especially for electrical and thermal phenomena, and in perfecting older and inventing new types of scientific instruments and measuring methods. Its experimental work in blackbody radiation, in particular, placed it at the forefront of modern physical research and showed it to be a world-class scientific institute. Even if these scientific contributions were partly motivated by their potential benefit to German industry, they nonetheless helped introduce new ideas about nature that were to have immense scientific and philosophical implications. By 1905, moreover, the Reichsanstalt had repeatedly proved its importance to German industry, above all to the "high-technology" industries of electricity, optics, and mechanics. That importance lay not only in research that led to industrial novelty, but also in the testing of established wares. Its research and testing work helped Germany to challenge Britain as the foremost economic power of the age. All told, in 1903 the Reichsanstalt published fifty-two articles, including five in the *Annalen* and eight in the *Zeitschrift für Instrumentenkunde;*[153] by contrast, in 1893 it had published only twenty-eight articles, including four in the *Annalen* and ten in the *Zeitschrift*. Yet the Institute's very successes led to the emergence under Kohlrausch of new problems: loss of some of the Institute's best scientists to the academic world; enormous testing burdens that hampered the Institute's ability to conduct research; and a paradoxically antagonistic relationship between the Institute and the electrical industry.

Kohlrausch transformed Helmholtz's plans to conduct scientific research into reality. His Reichsanstalt created new science; it applied extant science to the advancement of technology and industry; it devised and standardized precision instrumentation for the academic and industrial worlds; and it brought nation-wide electrical standards to Germany and helped bring international ones to the industrialized world. Despite the old concerns of some, it had managed to complement rather than compete with academic institutes and industry.[154] Kohlrausch

matched his predecessor's accomplishments at the Reichsanstalt by
guiding the young institution to maturity. Less than two decades after
the Reichsanstalt had been established, its founders' hopes had been
fulfilled. And yet the Institute was soon to stand in need of reform.

# 5

## The Search for Reform

When Emil Warburg assumed office in 1905 as the third Reichsanstalt president, he inherited a mature, highly successful scientific institute. As we have seen, the Reichsanstalt had made numerous important contributions to German science, technology, and industry up to this point. But Warburg and others closely associated with the Reichsanstalt knew that it now had also to confront a number of older, internal problems as well as a number of new challenges from without. By 1905, the Reichsanstalt's heroic years and its decade-long building program were long past. Now in its eighteenth year of operation, the Institute needed to make changes if it was to continue to flourish. It needed reform.

The sources of the Institute's problems were twofold: First, especially after 1901 some of the Reichsanstalt's leading scientists left the Institute for academic (as well as industrial and other governmental) positions. The excellent resources that they had enjoyed as young researchers at the Reichsanstalt were now also to be found at many German universities and, increasingly, at the Technische Hochschulen and in industry as well. With improved opportunities to do research elsewhere, the Reichsanstalt needed to revitalize its own ability to do scientific and technological research. In particular, it had to rebuild the Scientific Section's Optics Laboratory and greatly expand the Technical Section's High-Voltage Laboratory; it had to provide higher salaries and more rapid promotions to prevent its best people from leaving for academic and industrial careers; and it had to provide more money for more instrumentation and equipment.

Second, the Reichsanstalt needed a way of overcoming the enormous burden of testing work that prevented its personnel from conducting more research. As we saw in Chapters 3 and 4, long before 1905 the Technical Section had to devote most of its efforts to testing products and materials, while the Scientific Section devoted a good deal of its efforts to developing the standards and instruments used in such tests. Those trends, as a discussion of the Reichsanstalt in 1913 and an analysis of the Reichsanstalt's testing work from 1887 to 1914 will show, continued throughout Warburg's presidency. To a far greater degree than its founders had ever envisioned, the Reichsanstalt had become a standards and testing institute. German industry, through its testing demands on the Reichsanstalt, was in effect controlling many of the Institute's resources. Reform was needed to enable the Scientific Section to devote itself exclusively to scientific research and the Technical Section to increase its minimal amount of technological research.

It proved far easier to find a reformer than money to carry out reforms. Kohlrausch's weakened physical condition during the final years of his presidency, like Helmholtz's during his, dramatized the need to appoint a strong, energetic leader who could meet the challenging administrative burdens of directing a large scientific bureaucracy. An institute with an annual operating budget (for 1905) of more than 400,000 marks and employing more than 100 individuals working in ten different buildings required a leader of outstanding administrative skills as well as high scientific attainments. A new boss was needed, someone who could protect what the Reichsanstalt had already achieved and, at the same time, who recognized the new developments in physics and technology and could lead the Institute in participating in them. In Emil Warburg, the Reichsanstalt found the fresh, indefatigable, scientifically perceptive leader it needed.

During the first nine years of his presidency (1905-1914) Warburg sought to reform the Reichsanstalt. He saw to the appointment of new board members who themselves had made some of the latest, most important advances in physics. He reinvigorated the Scientific Section's Optics Laboratory. He had a new High-Voltage Laboratory built for the Technical Section and argued for more funds to build other new laboratories as well. He established new, nongovernmental sources of funding for the Reichsanstalt's research. And finally, he dissolved the Reichsanstalt's original division into two sections, replacing it with an alternative structure aimed at enhancing research. In these and other ways, Warburg renewed and reorganized the Reichsanstalt.

Yet institutional and political forces beyond Warburg's control ultimately blocked full reform of the Reichsanstalt. In part, the Reichsanstalt's own institutional inertia prevented full reform. As a mature scientific institution, the Reichsanstalt was saddled with set scientific-technical problems and work routines that were not easily, if at all, abandoned. Moreover, it suffered from a shortage of senior scientific posts, leaving precious little room for the advancement of younger men who had already performed well. To the internal structural rigidity preventing reform there were added two major political obstacles: The first was that the imperial government refused to supply the Institute with more money to create more and better-paying senior positions. The second was of quite another order: World War I. Even the most determined, reform-minded president supplied with sufficient funds from a fully supportive government could not have overcome the multiplicity of deficiencies brought on by war; it was the undoing of reform at the Reichsanstalt.

## The New Boss and the New Physics

Unlike the appointments of Helmholtz and Kohlrausch to the presidency, Warburg's appointment was (ultimately) the result of a genuine search. To be sure, the initial attempt to secure a successor for Kohlrausch was made solely by board president Lewald, who wanted Röntgen to head the Reichsanstalt. His choice was an understandable mistake.

Lewald chose Röntgen, professor of experimental physics at the University of Munich and a specialist in precision physics, mainly because of his public prominence. Röntgen's discovery of X-rays in 1895-1896 had catapulted his name into public life. Newspapers around the world reported on Röntgen's invisible rays, often with an accompanying picture of an X-rayed bone (e.g., that of a human hand or skull) that made the utilitarian implications of his discovery dramatically clear to the otherwise mystified layperson. Overnight, Röntgen became Germany's best-known scientist, a status that was reinforced when his discovery brought him the first Nobel Prize in physics (1901).[1] He had the public and professional recognition that Lewald judged to be an unbeatable combination in a Reichsanstalt president. In August 1904, Lewald asked Röntgen, who had been appointed to the Reichsanstalt's board in 1897, to succeed the ailing Kohlrausch.[2] Following a two-month delay, Röntgen declined Lewald's offer, because, to use his own words, "I was not the right man for the post."[3] He doubtless meant what Lewald had failed to see: that he had poor administrative ability, lacked tact, and was uninterested in the technological applications of physics.[4] Fortunately for himself and the Reichsanstalt, Röntgen wisely chose to remain in Munich.

Following Röntgen's refusal, Lewald asked several of Germany's leading physicists to suggest candidates for the presidency. Four names were given serious consideration.[5] Three of them – Walther Nernst and Woldemar Voigt of Göttingen and Ferdinand Braun of Strasbourg – were dismissed quickly. Nernst effectively excluded himself by deciding during the midst of the Reichsanstalt's search to leave Göttingen for Berlin, where in 1905 he succeeded Hans Landolt as professor of physical chemistry. That was just as well, since Nernst held a number of investments in illumination manufacturing firms that required or sought Reichsanstalt certification for their products, thus raising the issue of a potential conflict of interest on Nernst's part. As for Voigt, he was too much the theoretician and too little the experimentalist, more specifically too little the precision physicist, to lead the Reichsanstalt. That was presumably the main reason that, in Kohlrausch's view, German physicists and technologists, including those at the Reichsanstalt, would have been "highly critical" of Voigt's appointment.[6] The third unsuitable candidate, Braun, was indeed an experimentalist who had made several important, mainly instrumentational, contributions to physics and technology – above all the discovery of the rectifier effect, the invention of the cathode-ray oscilloscope, and a number of advancements in wireless telegraphy that would bring him a share of the Nobel Prize in physics for 1909. But Braun's candidacy had two shortcomings, the second of which was an insurmountable barrier to his nomination: First, he had relatively little experience in the field of measuring physics; and second, his contributions to the technology of wireless telegraphy led him to become intimately involved with and to acquire financial investments in that industry. In particular, in 1903 he had become a major figure and shareholder in a new firm known as the Gesellschaft für drahtlose Telegraphie (Wireless Telegraphy Company), which had arisen as a result of the merger of the firms Braun-Siemens and AEG-Slaby-Arco and which would eventually be known as Telefunken.[7] His extensive business interests were, Kohlrausch said, "absolutely incompatible" with the holding of a position at the Reichsanstalt.[8] So, too, was his holding of several crucial patents for wireless telegraphy, though Kohlrausch did not mention this. Patent holdings and business interests like those of Nernst and Braun reflected the fact that by the end of the century physics had become increasingly useful industrially. But Reichsanstalt physicists had to be, and be seen to be, completely impartial in the setting of standards and the testing of products.

The quick dismissal of Nernst, Voigt, and Braun as candidates left only one name: Warburg.[9] In recommending him, Kohlrausch cited Warburg's excellent reputation among German physicists and judged him "by far the most suitable candidate."[10] Moreover, he reported that Warburg was ready to accept the presidency if the Ministry of the Interior met his salary demand. (As with Helmholtz and Kohlrausch before him, satisfaction of that demand alone stood between Warburg and the presidency.) Warburg's current income as professor of physics at the University of Berlin and as a member of the Akademie der Wissenschaften totaled nearly 40,000 marks, a figure that was considerably greater than the Reich's (initial) salary offer.[11] Warburg told Lewald that he would not accept the presidency unless he was offered a minimum annual income of 33,000 marks. More anxious than ever to name a new president, Lewald offered Warburg a salary of 15,000 marks, an annual bonus of 10,000 marks, and a further 8,000 marks for being an honorary professor at the University of Berlin, where he was to teach two to four hours per week. (Again as with Warburg's predecessors, the honorary professorship was also offered to help maintain the Reichsanstalt's tie to academic physics, a tie that the ministry was keen to preserve.) In addition, Warburg was naturally allowed to retain his membership in the Akademie, where he continued to receive a (small) income.[12] Lewald's package deal added up to the 33,000 marks that Warburg had demanded. In April 1905 Warburg became the Reichsanstalt's third president.

Warburg brought outstanding, and appropriate, scientific qualifications to his new post. A distant relation of the famous German-Jewish banking family,[13] Warburg had enrolled at the University of Heidelberg in 1863 to study chemistry with Bunsen. Inspired by the lectures on theoretical physics by Bunsen's colleague Kirchhoff, Warburg switched to physics and, in 1865, to Berlin. There he conducted experimental work in Magnus's laboratory, became a close friend of Magnus's assistant, Kundt, and studied Helmholtz's papers on theoretical physics. There, too, he earned his doctorate (1867) and habilitated (1870).[14]

Following service in the Prussian army – he won the Iron Cross for his contributions to the Prussian effort in the Franco-Prussian War – he accepted the extraordinary professorship of physics at the new Kaiser Wilhelm University in Strasbourg, where Kundt had just been appointed to the ordinary professorship. Between 1872 and 1876 the two men experimentally confirmed two predictions of the new kinetic theory of gases and formulated another prediction. That excellent piece of experimental work helped make Warburg's reputation as one of Germany's finest experimental physicists.

It also helped him achieve academic independence. In 1876 he left Strasbourg to become the ordinary professor of experimental physics at the University of Freiburg. There he continued his older experimental work on the kinetic theory of gases and also took up new studies, discovering and explaining hysteresis cycles and conducting numerous studies on electrical discharges in gases, liquids, and solids (especially glass). Though Warburg's motivations were solely those of the pure scientist, much of his work yielded important technological applications. For example, that on the relation of inner friction and heat conduction in gases later became of practical use in constructing electrical illuminating lamps; that on magnetic hysteresis proved of profound importance for electrotechnology; and that on electric discharges in gases became crucial for the exploitation of the glow-discharge tubes used in telephone equipment.[15]

Beyond his strictly scientific credentials, Warburg also brought the "right" views and the sure-handed executive ability needed to lead the Reichsanstalt. As to the former, Warburg believed in the ideology of pure science and the subordinate role of technology to science. Speaking of electrotechnology, he innocently declared that it owed its origins to "great discoveries" that were "made without any consideration for practical results. Therefore," he concluded,

> scientific work is occasioned and conducted by excluding every practical secondary aim. Natural science is motivated solely by the desire implanted in the human mind spurring us to find connections between natural phenomena. [This state of affairs] is not only in the interests of pure science, but also in [the interests of] its practical applications as well.[16]

Like Siemens, Helmholtz, and Kohlrausch before him, Warburg believed that science and technology were separate but hierarchically related fields.

As to his executive ability, Warburg possessed sure-handed organizational talent and extensive managerial experience, the latter earned through the directorship of two physics institutes in the course of three decades. Like Helmholtz and Kohlrausch before him, Warburg had planned and overseen the construction of a physical institute – that at Freiburg from 1888 to 1891 – before being offered the Reichsanstalt presidency. In 1895 he left the small Freiburg Physics Institute to become ordinary professor of experimental physics at Berlin and director of Berlin's large Physical Institute. His new appointment was to the position held previously by Kundt, the very one that Kohlrausch had declined because he considered it too demanding for one individual. Too demanding for most, perhaps, but not for Warburg: Between 1895 and 1905 Warburg managed to meet his extensive teaching duties, conduct his own research – studies on electrical discharges in gases, ozone formation due to spark and point discharges, and the electrolysis of gases – and build the largest and probably the best school of experimental physics in Germany. One laboratory visitor reported that Warburg "has apparently no difficulty in guiding the researches of his thirty [!] students, among whom he makes a regular round of visits every morning."[17] During his decade of teaching in Berlin his sixty-two doctoral students produced 225 pieces of published work in virtually every field of physics and went on to pursue careers in academic physics, industry, and government. "One had to work thoroughly," reported one of those students, "for on this point Warburg tolerated no fooling around." "There was no limit to the working hours. It was a rare Sunday when the activity stopped. One could go to Warburg at any time, even in the late evening." [18] An American visitor saw the laboratory working conditions less positively. Warburg's assistants, he said, worked "like slaves."[19] If so, Warburg was a slave to his own system: In addition to running his university laboratory, conducting his own research, sitting (from 1895) on the Reichsanstalt's board, and participating in meetings of the Akademie der Wissenschaften, Warburg also chaired (from 1897 to 1905) the Berlin (Deutsche) Physikalische Gesellschaft. Although he resigned the chairmanship in 1905, he continued to play an active role in the Gesellschaft, remaining on the board of the *Annalen der Physik* until his death in 1931. In the fifty-eight-year-old Emil Warburg, the Reichsanstalt had found an indefatigable leader who possessed the capabilities and stamina needed to administer a large institute and to bear its enormous burdens (Figure 5.1).

Figure 5.1. Emil Warburg circa 1906. Courtesy of Bildarchiv Preussischer Kulturbesitz, Berlin 1988.

Unlike Kohlrausch, Warburg understood and welcomed the theoretical advances and the flurry of new experimental results that appeared in physics after 1890 and that have since sometimes been cumulatively referred to as the New Physics. Following Hertz's experimental demonstration of electric waves in the late 1880s, a number of physicists – above all, G. F. Fitzgerald, H. A. Lorentz, and Henri Poincaré – sought to reconcile Newtonian mechanics and Maxwellian electromagnetic theory and, in some instances, questioned the mechanical foundation of physics. After 1905, Einstein's special theory of relativity, which resolved the contradictions between mechanics and electrodynamics, served as the new foundation for future developments in mechanical, electromagnetic, and optical theory. In so doing, it made physics a more abstract and, very soon, more mathematical science. Along with the emergent quantum theory and the increased importance

of statistical mechanics, special relativity demonstrated more deci-
sively than ever the leading role theory might play within physics. For
many physicists after 1910, these new theoretical developments in the
foundations of physics constituted the most exciting work in their discipline; yet those developments lay far beyond the Reichsanstalt's interests. Helmholtz had built the Institute at a time when the Newtonian,
mechanical world view – the reduction of all physical phenomena to
the laws of mechanics – still reigned supreme, an era in which physical
measurement was something only to be practiced, not itself analyzed.
The Reichsanstalt's work was implicitly predicated on the mechanical
world view. The new, emerging ideas of quantum and relativity physics had no practical relevance for it, yet they subtly undermined its ability to keep pace – to understand and to exploit – the central developments in mechanical, electromagnetic, and optical theory.

Less irrelevant and less threatening, but no less sensational, was the
experimental dimension of the New Physics. From 1895 onwards a series of striking experimental discoveries increasingly drew the attention
of both experimentalists and theorists: X-rays, radioactivity, the
Zeeman effect, blackbody radiation, and the electron, to name only the
most important. In principle, these experimental discoveries were
much closer to the Reichsanstalt's interests than the new theoretical advances were. However, they and their closely related theoretical explanations were in one way or another intimately tied to the emerging
problem of atomic structure. And that was a problem that before 1918
scarcely touched on the Reichsanstalt's more practical interests.

The issue of the relevance of the New Physics to the Reichsanstalt
was a paradoxical one. The Institute's contributions to radiation physics between 1893 and 1901 had placed it at the forefront of modern
developments. But blackbody radiation, and Planck's new quantum
theory of blackbody radiation in particular, remained of peripheral interest to physicists before 1905.[20] By the time a quantum approach to
physical problems did indeed start to interest an increasing number of
physicists and even chemists (in particular, after the first Solvay Congress in 1911),[21] the Reichsanstalt had lost all of its former outstanding
contributors to radiation physics: Wien, Lummer, Pringsheim, and
Kurlbaum had all left for academic positions.[22] Moreover, the Reichsanstalt's institutional interest in studying radiation stemmed, as we
have seen, principally from its need to develop a better temperature
scale and more accurate temperature measurements for German in-

dustry. Although the Reichsanstalt continued its blackbody-radiation studies after 1901, those studies no longer stood at the leading edge of physical research.

Kohlrausch's retirement and Warburg's assumption of office in 1905, Einstein's *annus mirabilis,* presented an excellent opportunity to introduce some of the research topics of the New Physics into the Reichsanstalt. Warburg had often recognized and participated in the most important current developments in physics. He was, for example, among the first to recognize the significance of Röntgen's discovery of X-rays, of Bohr's model of the atom, and of Einstein's general theory of relativity.[23] Moreover, he was the second most successful nominator of Nobel Prize winners in physics between 1901 and 1915,[24] a fact that reflected his fine ability to recognize the outstanding results and men of contemporary physics. He understood contemporary physics and wanted the Reichsanstalt to participate in it through increased research work.

The challenge to Warburg and the Reichsanstalt was to find ways to increase its research work in general and its contributions to the exciting problems in contemporary physics in particular. Board members were aware of, and some sought to bridge, the growing gap between the Reichsanstalt's concerns and those of pure physics. At board meetings in 1905 and 1906, for example, Röntgen pointed to the necessity of taking up some of the new problems in physics. He noted, with much dissatisfaction, that "all of the Reichsanstalt's work lies in the fields of practical and technical physics." He urged, instead, more work in pure physics, pointing in particular to the need to test experimentally some of the new ideas concerning "the foundations of mechanics." He specifically recommended that the Reichsanstalt conduct experiments on "the movement of the ether with [respect to] bodies" and on "the ratio of mass to charge" in various (unspecified) particles, and in this way test some of the latest claims about mechanics and several newly discovered particles. He (and others) added that Siemens and Helmholtz had hoped that the Reichsanstalt would conduct precisely these types of experimental investigation.[25]

The Reichsanstalt's constitutional warrant for conducting such experimental work in pure physics was that it did *not* open up a new field for research but rather only verified and extended work that had been begun elsewhere.[26] As in the past, the board wanted to support, not compete with, academic science. In 1907, Warburg publicly restated this policy. Echoing Röntgen's remarks, he noted that the Reichsanstalt's founders intended the Institute to conduct both "extensive technical-scientific and pure scientific investigations, independently of technology's momentary needs." While he stressed the Reichsanstalt's mission to do pure science, he interpreted that mission quite narrowly as one of developing fields of physics that had already been opened up by academic scientists. The Scientific Section's formal purpose, he said, did not include the making of "new discoveries." That was (largely) the realm of academic science. Though he did not want to make a "sharp distinction" between academic and Reichsanstalt science, he had no doubt about the existence of that distinction:

> It will . . . always be more of the academic teacher's responsibility to open up new paths into the unknown primeval forest, while the Reichsanstalt's responsibility is more to improve and smooth the known paths for general use, not to start but to complete, i.e., to refine the instrumental aids and to increase the precision of measurements.[27]

In effect, Warburg declared that the Reichsanstalt's institutional role in advancing pure physics was a subordinate one. However, in so doing he had at least defined where its expanded research program should concentrate its efforts. That was a sine qua non for doing more research.

So, too, was a board more sympathetic to increased research, one that better understood the problems of contemporary (pure) physics. With the Interior Ministry's support, Warburg selected new board members who were closely involved with the latest developments in physics. Of the eleven new members appointed between 1905 and 1914, four (Planck, Rubens, Dorn, and Willy Wien) were pure physicists, one (Nernst) a physicist and chemist who contributed to both pure and applied science, one (Karl Schwarzschild) an astronomer and astrophysicist of the "pure" variety, and two others (Johannes Görges and Max Wien) applied physicists. Warburg thus strengthened physics' representation on the board; the appointed scientists, as expected, counseled the Institute to conduct research work – of the measuring sort, to be sure – that would help resolve some of the problems in contemporary physics. As Lewald told the ministry in nominating Planck in 1907: "His appointment is especially desired in light of the pure scientific investigations that, following a suggestion by Professor Dr.

Röntgen, are soon expected to be begun at the Reichsanstalt."[28] The entire process of bringing more physicists onto the board in order to boost pure physics at the Reichsanstalt culminated in the appointment of Einstein in 1917.[29] Indeed, in 1912 Warburg had even offered Einstein a position on the Reichsanstalt's staff. Einstein, who had just accepted a position at the Eidgenössische Technische Hochschule (Federal Institute of Technology) in Zurich, declined.[30]

The main difficulty in conducting research work like that recommended by Röntgen was less one of legitimacy and proper advocacy than of finance. Within a year of Röntgen's recommendation, the Reichsanstalt had begun a study of the ratio of the charge to mass of cathode rays. But further work required additional funding. One potential source of funding was the so-called Experimental Fund, used for special financial aid to Reichsanstalt physicists and technologists. However, the fund had first to reach 70,000 marks before it could achieve its original purpose of helping both groups, Warburg stated. He noted, too, that the Institute's increased size and the growth of its testing activities had led its technologists to overtax the fund's resources, leaving the Institute unable to meet the needs of its physicists. He therefore called for a board resolution, to be presented to the ministry, on the Reichsanstalt's financial needs. The resolution declared that

> 1. due to the requests by technology, the Reichsanstalt's personnel and resources are increasingly used for the solution of problems having a predominantly technical interest; that, however,
> 2. according to the Reichsanstalt's original plans and its standing business orders (I, §2), fundamental problems of a theoretical interest should also be treated by the Reichsanstalt. [Execution of the latter problems] are not completely possible because of the demands of technology noted under point 1. Moreover, because of the increased testing activities, a considerable part of the Experimental Fund has been claimed for the remuneration of the laboratory mechanics entrusted with the testing work.

The board asked Lewald "to demand" from the ministry an increase in funds for experimental work in next year's (1907) budget.[31]

But the ministry made only a small increase. And so the expected occurred: In 1909 Lewald reported that the money needed for conducting testing activities was exhausting the fund.[32] In 1911 Warburg noted that the fund's "tight resources" provided only 15,000 marks for the Scientific Section. With a gas compressor (sorely needed for work by Holborn and Max Jakob) alone costing 6,000 marks, Warburg was forced to ask for (and received) 3,000 marks in financial aid from the Akademie der Wissenschaften.[33] The Reichsanstalt's inadequate financial resources for hiring new physicists and supplying the Scientific Section with adequate material resources was constantly forcing it to delay or abandon work of a pure scientific nature. The lack of sufficient funds for research led Wilhelm Kohlrausch – professor of electrotechnology at the Technische Hochschule Hannover, a board member since 1899, and brother of Friedrich Kohlrausch – to state the obvious:

> The scarcity of [financial] means at the Reichsanstalt leads to a situation where needed [scientific] investigations must be postponed again and again. That injures the progress of science in the German Reich and the reputation of the Reichsanstalt at home and abroad.[34]

Despite the board's vocalized concerns and its urgent petitions for increased funding to advance the Reichsanstalt's pure scientific work, the ministry remainded almost completely unresponsive.

### Scientists on the March – Out

The failure of board petitions to gain more money for research from the ministry led Warburg to try a new tack: He sought to demonstrate the Reichsanstalt's personnel problems by means of an irrefutable report. Conducted in 1911 by board member Karl Strecker, the report based its analysis on a complete statistical history of the Reichsanstalt's scientific personnel between 1887 and 1911, including, among other things, data on wage and salary ranges, the number and increase of Reichsanstalt employees per year, ages and rates of promotion, length of employment and (where applicable) reasons for leaving.[35]

Not surprisingly, Strecker's report revealed what Warburg and the rest of the board had long known, namely, that "for years" the Reichsanstalt had suffered from two major personnel problems: first, that there was a seriously insufficient number of scientists to solve the Institute's "numerous and important problems" and, second, "that far too many of the most able scientific officials leave the Institute because they get a better and more distinguished position [elsewhere], in particular through a call to a university or Technische Hochschule."[36] As a consequence of the continuous loss of "outstanding officials" to the academic world, the report concluded that "the average level of talent and scientific importance of the [Reichsanstalt's] officials is declining." In addition, it stated that "studies of *pure physics* . . . have long had to be postponed." By 1911 the situation had become a "vital question" ("*Lebensfrage*") for the Reichsanstalt.[37]

To be specific, the report showed that between 1901 and 1910 twenty-four scientists left the Reichsanstalt.[38] We know, moreover, that between 1890 and 1914 a significant number of the Reichsanstalt's best scientists left it to pursue academic careers at a university or Technische Hochschule: Included here (in their order of leaving) were Pernet, Wien, Richard Wachsmuth, Lummer, Kurlbaum, Pringsheim (a long-term Reichsanstalt guest), Rudolf Rothe, Siegfried Valentiner, Baumann, Diesselhorst, Leithäuser, Ernst Orlich, Heinrich Fassbender, and Erich Hupka. These individuals belonged to, and in good part constituted, the most creative and productive part of the Reichsanstalt's scientific staff. For these men the Reichsanstalt had become, in effect, a postdoctoral training institute for future professors of physics. It took recent graduates trained in pure physics from the universities and returned seasoned scientists to them and the Technische Hochschulen. Conversely, apart from its first three presidents, it received only one or two experienced scientists from the academic world.[39]

A good instance of the career aspirations, if not patterns, of such men is that of Willy Wien. Ten days before his new appointment to the Reichsanstalt began, the young and highly ambitious Wien wrote to Hertz:

> On 1 July [1890] I shall assume a position at the Physikalisch-Technische Reichsanstalt in Charlottenburg and simultaneously habilitate at the University of Berlin. I also hope at that time to bring my theoretical studies on the localization of energy to a preliminary conclusion. In any case, I shall direct my main attention to pursuing a university career.[40]

Six years later, in 1896, Wien left the Reichsanstalt for an extraordinary professorship at the Technische Hochschule Aachen, "although," as he later said, "the position was in no way a brilliant one."[41] But those that followed were increasingly "brilliant": In subsequent years he became

professor of physics at Giessen, Würzburg, and Munich. His professional goals and strategy were doubtless similar to those of numerous other good Reichsanstalt scientists.[42] What distinguished him from
most of them was his rapid and extraordinary success, not his ambitions. The departure of highly talented men like Wien greatly hurt the
Reichsanstalt's scientific well-being. The concerns expressed at the
time of the Reichsanstalt's founding that the Institute might harm academic science through (unfair) competition turned out to be not so
much false as ironic: It was competition with the academic institutes
that hurt the Reichsanstalt.

Not only the academic world attracted the Reichsanstalt's scientists. So, too, did industry and other governmental agencies. The report
revealed that of the twenty-four scientists at the Reichsanstalt in 1901
but gone by 1910 (apart from two unknown cases, two deaths, and one
retirement), one had become a business consultant, six had transferred
to another governmental agency, six had gone to work for industry, and
six had joined a university or Technische Hochschule. In gross numbers, losses to industry and other governmental agencies were twice as
great as those to academic institutions. However, of the four tenured officials who were gone by 1910, three had left for an academic institution and the fourth one had died.[43] Academic institutions attracted the
Reichsanstalt's best senior men, industry and government its junior
men.

There were several reasons for these attractions, two of which – the
rate of promotion and salary – can be analyzed with some precision. Already in 1888 Helmholtz had anticipated difficulties in retaining
Reichsanstalt scientists. He wrote, "Since advancement at the Reichsanstalt is difficult, . . . , [and] given a nonincrease in salary rates, the
more able personnel, after they have exploited the Institute's advantages, will try to go to a university, and in so doing would be lost to the
Institute."[44] That melancholy prediction proved, as the case of Wien
and others showed, to be all too accurate. The Reichsanstalt had always
offered, as the report of 1911 noted, a "quite handsome" salary to, and
so could attract, "a very able young physicist."[45] In 1900, for example,
the Reichsanstalt's scientific assistants earned annual salaries ranging
between 1,920 and 2,620 marks.[46] By comparison, assistants at the
Technische Hochschulen had a median annual income of 1,800 marks,
with a range between 600 and 5,000 marks, and university assistants
averaged 1,400 marks, with a range between 800 and 2,500 marks.[47]

The Reichsanstalt's starting salaries for its youngest scientists were noticeably higher than those offered by the universities and Technische Hochschulen.

The Institute's personnel problems thus did not lie in hiring new, young physicists but rather in promoting and retaining the best of them. The chances of obtaining a tenured post, that is, of becoming a permanent co-worker or member, were minimal. Between 1901 and 1910 only four positions for permanent co-worker and seven for member had become available; on average, the Reichsanstalt had only one permanent post per year to offer to its nontenured employees, the sole source of its tenured officials. The report declared:

> This situation afflicts the entire scientific corps of Reichsanstalt officials. The assistants who for years have put their energy into serving the Institute do not get ahead, or, when they are finally promoted, for the most part are already further advanced in years than the corresponding salary available to them.[48]

It was a sobering conclusion.

Comparative statistics on the salary ranges of tenured academic and Reichsanstalt scientists bear out the claim that the latter did not receive adequate financial compensation. Around 1900 extraordinary and ordinary professors earned considerably higher salaries than their respective Reichsanstalt counterparts, the permanent co-workers and the members. The annual income (salaries and fees) of extraordinary professors at the Technische Hochschulen then averaged 4,500 marks; at the universities it averaged 6,000 marks and ranged between 3,000 and 10,000 marks. At the Reichsanstalt, by contrast, permanent co-workers earned between only 2,400 and 4,800 marks. At the senior level, the annual income of ordinary professors at the Technische Hochschulen averaged 8,000 marks and ranged between 4,000 and 15,000 marks; at the universities it averaged 14,000 marks and ranged between 4,000 and 37,000 marks. At the Reichsanstalt, again by contrast, the income of a member ranged between only 3,600 and 7,500 marks.[49] In other words, the Reichsanstalt's tenured scientists were paid considerably less than their academic brethren. As the report, with biting understatement, put it, many of the Reichsanstalt's best physicists naturally preferred "the considerable incomes [i.e., salaries plus fees] of an ordinary professorship at a well-attended university . . . to the not excessively high salary at the Reichsanstalt."[50] The discontent over inadequate pay felt by many Reichsanstalt scientists was reinforced by the high cost of living in the capital, where, to compensate Berlin academics, Prussia paid them higher salaries than their colleagues in the provinces.

The seemingly obvious way to remedy the departure of Reichsanstalt scientists was to raise their salaries to a level commensurate with those of their colleagues at German (and in particular Berlin) academic institutions. Yet it was precisely this that the Reichsanstalt could not do. It could not do so because, being an official governmental agency of the Reich, the salary ranges of its officials had to be more or less equivalent to those of similar officials at other Reich agencies.[51] To raise salaries at the Reichsanstalt would require raising those of similar officials at all Reich institutions.[52] The Reichsanstalt was entrapped within the imperial bureaucratic system and so had no hope of ever offering the higher financial rewards of the academic world.

In addition to better financial compensation for middle- and senior-level men, a career at a German academic institution had two other major, if less tangible, advantages over one at the Reichsanstalt. First, an academic career offered far greater opportunities for free or undirected research, not least because it also offered considerably longer vacations.[53] An academic scientist could choose the research topics that he currently found to be most interesting or important; he was not subject to anyone else's directives to conduct a particular piece of work. A Reichsanstalt scientist, by contrast, was directed either by the president or by his laboratory leader to undertake a certain project. To be sure, he had access to the Reichsanstalt's resources to conduct "private" investigations on his own time. Moreover, in certain cases he might be able to convince the president that a given research topic fell within the scope of the Reichsanstalt's responsibilities and merited the use of its resources. But apart from these rare exceptions, his work was directed by and done for others during officially regulated business hours and within a hierarchical, bureaucratic organization. As Warburg himself publicly admitted, "It is . . . scarcely to be doubted that the academic teacher's freedom of choice in his field of work provides a more favorable milieu than that of an official working under greater constraints and with problems that are in many cases assigned to him."[54] Second, an academic position commanded greater prestige and held greater status than one at the Reichsanstalt. This intangible but widely sought social benefit was particularly useful in a class-ridden society like imperial Germany.

The obvious consequence of slow promotion, low pay, restricted scientific freedom, and lower status was that, to quote one of the Reichsanstalt's employees, "precisely the best workers at the Reichsanstalt" left it for a call to the universities and the Technische Hochschulen.[55] Yet the ministry dismissed the board's demands for more money. As a ministry official wrote in 1912 in response to a Reichsanstalt request for salary increases:

> To date, the Physikalisch-Technische Reichsanstalt has shown a completely satisfactory development. Its accomplishments enjoy general recognition and esteem. I therefore harbor doubts about granting salary increases in order to attract and maintain outstanding scientists.[56]

Complacency, not financial support, was the ministry's response.

Exit from the Reichsanstalt was predicated on more than academic ambitions and discontent with the Institute, however. Most of the Reichsanstalt's dissatisfied scientists could not have left the Institute for German universities and Technische Hochschulen had there not been a dramatic increase in the number and research needs of German academic physics institutes during the late nineteenth century.

When Siemens and Helmholtz sought to found the Reichsanstalt in the mid-1880s there were only five new physics institutes in Germany (plus two under construction), and all already suffered from heavy teaching obligations. But the 1880s and 1890s, as we saw in Chapter 1, was a period of unprecedented growth in the number, size, and research capacities of academic physics institutes. Between 1878 and 1905, nineteen new academic physics institutes were built, fourteen of them between the Reichsanstalt's formal opening in 1887 and Warburg's assumption of the presidency in 1905.

These new and expanded institutes conducted an increasingly large amount of research. If the establishment of new institutes must be ascribed largely to the pedagogical functions that they were intended to serve, this did not prevent German physicists from increasingly using their enlarged facilities and staff for research. By the 1880s and 1890s the research ethos had become the institutionalized mission of all German physicists. By 1906 a German physicist went so far as to claim that "the task of physics today consists essentially in this: to pursue research."[57] German physicists felt fully justified in employing their new and increased resources for research purposes.

When the Reichsanstalt was established in 1887, some feared that because of its extensive resources it would injure German academic institutes through unfair competition. But some ten to fifteen years after its founding, just the opposite was happening: It was the new German university physics institutes with their extensive resources, along with the old but expanding institutes at the Technische Hochschulen and the nascent industrial physics laboratories, that were injuring the Reichsanstalt by attracting its best physicists.

Moreover, when the Reichsanstalt's founders had agreed in 1887 that the Institute should not undertake investigations already being pursued by academic institutes, that agreement had relatively little consequence in that most institutes then lacked sufficient resources to undertake extensive experimental work. The Reichsanstalt was supposed to pursue only those topics that demanded more time, personnel, and material resources than were available at German academic institutes. It was to complement and supplement, not compete with, academic institutes.

But by the turn of the century, when the new or refurbished physics institutes with their increased research capacities and their enlarged staffs eager to do research dotted the academic landscape, the limitation on the scope of the Reichsanstalt's work had become far more important. By then the Reichsanstalt began to lose its unique and most precious asset: unmatched research facilities manned by highly skilled scientists supplied with the time and material resources to conduct research. By 1900 the complaints of Siemens and Helmholtz during the 1880s that German academic institutes lacked the resources to conduct research in physics had lost their validity. The Reichsanstalt, itself part and parcel of the expanded German research facilities in physics, took cognizance of these facts.[58] Its principal scientific function vis-à-vis the German physics community now came to be almost exclusively that of providing measurements and the means for making measurements and thereby extending, confirming, or disconfirming known results. Those of its physicists devoted exclusively to measuring physics had an excellent setting in which to pursue their work; those with greater ambitions had alternatives.

Having failed to win more support from the Ministry of the Interior, Warburg and the board decided to develop new and private sources of funding. Under Warburg's leadership, the Reichsanstalt established two such funds to enhance its pure scientific research. The first and most important of these was the Helmholtz Fund for scientific research, conceived in October 1912 on the occasion of the Reichsanstalt's twenty-fifth anniversary.[59] Its general goal was to solicit financial contributions so that "through the collection of private monies for its scientific studies the high reputation that the Institute enjoys at home and abroad [can be] consolidated and increased."[60] Private funds had to be marshaled to provide what the government refused to support adequately: pure science.

The schemers behind the Helmholtz Fund – virtually all board members, including Foerster, Kohlrausch, Lewald, Linde, Nernst, Planck, Röntgen, Rubens, Warburg, and Wien – explained in their confidential memorandum calling for financial support that, because the Reichsanstalt was a governmental agency, the income of its scientists was considerably less than that received by "outstanding physicists in industry or at the universities." The Reichsanstalt's relatively low salaries had led, they said, to the loss of some "outstanding employees" and the failure to obtain others. Moreover, they argued that, despite the increase in its annual budget over the years, the Reichsanstalt had not obtained enough funding to conduct scientific research at a pace and level that would keep it "in the flow of scientific development."[61] Thus the purpose of the fund was to provide supplemental, nongovernmental financial aid to hire and retain excellent scientists; to permit Reichsanstalt employees to undertake study trips in Germany and abroad; and to purchase costly instruments that the Reich failed to provide for.[62] By May 1913, the date of its official founding, the fund had accumulated assets of nearly 260,000 marks that yielded an annual income of more than 10,000 marks.[63] It attracted more than 200 donors who pledged either 1,000 marks per year for five years or simply paid a one-lump-sum of 5,000 marks. Major benefactors included such firms and individuals as Friedrich Krupp A.G. (50,000 marks); Siemens & Halske A.G. and Siemens-Schuckertwerke G.m.b.H. (50,000); Arnold, Wilhelm, and Carl Siemens (10,000 each); Karl Zeiss and Schott & Co. (10,000); Carl Linde (10,000); Bayer & Co. (5,000); BASF (5,000); Otto Reichenheim (5,000); and Dillinger Hüttenwerke (5,000).[64]

Shortly after the Helmholtz Fund was established, Emil Rathenau and the board of directors of his AEG set up, in December 1913, the Emil Rathenau Foundation (Stiftung). Its purpose was "to allow the

Physikalisch-Technische Reichsanstalt to promote investigations in the field of electricity and magnetism." It sought to do so by providing monies to further the same three concrete goals – to hire and retain scientists, to fund study trips, and to purchase instruments – as the Helmholtz Fund. The foundation's assets consisted of 100,000 marks in the form of AEG bonds; as with the fund, the annual income derived from the bonds was the source of the foundation's awards.[65] The establishment of the foundation, like that of the Helmholtz Fund, demonstrated that, a quarter of a century after Siemens had established the Reichsanstalt, German industrialists still believed that supporting science would eventually yield gain for both science and industry and that the Reichsanstalt was the appropriate institution for realizing their belief.

New presidential leadership abreast of the problems of contemporary physics; new board members who had themselves made outstanding contributions to that physics; new sources of funding for research in physics – these reformers and reformist measures sought to retain the Reichsanstalt's best scientists by increasing salaries and by enabling them to conduct more physical research. The reform efforts were obstructed, however, not only by governmental complacency and parsimony, but also by the very size and complex demands of the institution itself.

By 1913 Warburg's Reichsanstalt had an operating budget of nearly 670,000 marks, which was almost 80 percent more than its budget of 1903 and greater than 150 percent more than that of 1893.[66] It employed 139 individuals: Slightly more than one-third (50) of these constituted the major scientific and technical personnel; one-half (70) maintained the plant, laboratories, and equipment; and the remainder (19) devoted themselves full time to administering the Reichsanstalt (Table 5.1). The number of personnel had increased by approximately 26 percent since 1903 and by approximately 114 percent since 1893. Together, the scientific and technical workers published ninety-two articles, more than three-quarters as many again as the Reichsanstalt of 1903 had published and more than three times as many again as that of 1893.[67] Not even a man of Warburg's administrative talent and energy could easily reform such a large and complex scientific institute.

The Scientific Section of 1913, like that of 1903, consisted of three laboratories: Heat, Electricity, and Optics. All three operated under Warburg's immediate direction, just as they had under Helmholtz's and

Table 5.1. *Number of Reichsanstalt employees and range of their salary/wages, 1913*

| Position | No. of employees | Range of salary/wages (marks) |
|---|---|---|
| President (= director of Scientific Section) | 1 | 27,000 |
| Director | 2 | 8,000–11,000 |
| Member | 17 | 3,600–7,500 |
| Permanent co-worker | 14 | 2,700–6,600 |
| Scientific assistant | 16 | 2,400–3,200 |
| Office worker | 6 | 2,100–5,000 |
| Technical secretary | 9 | 1,800–3,300 |
| Legal secretary | 4 | 1,800–3,200 |
| Machinist | 2 | 1,800–3,200 |
| Mechanic/artisan | 59 | 1,400–3,000 |
| Minor official | 9 | 1,200–1,700 |
| Total personnel | 139 | |

*Note:* The figures for salary/wages do not include housing allowances.
*Sources:* HEDR (1913), 6–7, 38–43; and for the salary range of the scientific assistants, K. Strecker, "Die Personalverhältnisse der wissenschaftlichen Beamten der Physikalisch-Technischen Reichsanstalt" (29 August 1911), Anlage 1, BA 265, PTB-IB.

Kohlrausch's. The section consisted of seventeen scientists; it had added six new workers – nearly a 50 percent increase – since 1903, although most of these were lower, nontenured personnel. Most of the Reichsanstalt's leading physicists met regularly with other Berlin physicists by participating in the university's physics colloquia.[68]

With Thiesen's retirement in 1910, Holborn became the new leader of the five-man Heat Laboratory. His full-time collaborators included Friedrich Henning, Hugo Schultze, Wilhelm Heuse, and Jakob. In addition, Scheel, Steinwehr, Jaeger, and Walther Meissner provided further help. The laboratory was still, in 1913 as in 1893 and 1903, the largest laboratory in the Scientific Section. Most of its sixteen-odd studies conducted in 1913 were fairly closely related to similar studies done under Helmholtz or Kohlrausch.

The hiring of Meissner, who had received his doctorate under Planck in 1907 and had joined the Reichsanstalt the following year, was an important element in Warburg's plan to boost scientific work there. Meissner proved to be one of the leading students of low-temperature physics, in particular one of the leaders in superconductivity research after World War I. In 1913, Warburg entrusted him with setting up low-temperature equipment for studying the behavior of liquid hydrogen. With it Meissner measured the optical refractivity of liquid hydrogen.[69]

Like Meissner, Henning conducted precision-measurement studies at low temperatures. During 1913 he investigated the platinum-thermometer temperature scale and related it to the hydrogen scale between -193° and 0°C. He also used the platinum thermometer for determining the boiling points of oxygen and carbonic acid and the freezing points of mercury, benzol chloride, chloroform, carbon disulfide, and ether. Moreover, he related these findings to the hydrogen thermometer, which in turn could be reduced to the thermodynamic scale. Finally, he measured the resistance of different types of platinum, copper, and lead at the boiling point of hydrogen, and he developed an empirical equation that related different platinum thermometers to one another between -193° and 0°C.[70] Henning's (and Meissner's) measuring work on low temperatures had both technological and scientific goals.

The laboratory's low-temperature measurements also included studies on the specific heats of gases.[71] Scheel and Heuse investigated the polyatomic gases methane, acetylene, ethyl, and ethane, and Holborn and Jakob studied the specific heats of air at pressures up to 200 atmospheres. In order to understand more precisely the gas thermometer's deviations from the thermodynamic scale, Holborn and Schultze determined the isothermal lines of air, helium, and argon at various temperatures and pressures.[72] The study of argon, in particular, was undertaken on Planck's (and Nernst's) initiative. In 1908 Planck had asked the board to have the Reichsanstalt perform "a systematic series of experiments on the thermodynamic behavior of a carefully selected substance," suggesting that argon would be the most appropriate substance for study. Nernst seconded Planck's motion and provided the argon.[73] Such work on extending the experimental foundations of thermodynamics had been a prime Reichsanstalt topic since Helmholtz's day.

The laboratory's other long-term heat studies included Jaeger's and Steinwehr's determination of the temperature coefficient of mercury resistance used in thermometers and Steinwehr's calculations of the heat of solution of cadmium sulfate by means of a differential calorimeter. Scheel and Heuse used a Fizeauian apparatus for investigating the absolute expansion of quartz down to the temperature of liquid hydrogen. Their work had industrial goals: It contributed in part to the Reichsanstalt's ongoing studies on the behavior of different glasses for thermometric use and in part to ongoing studies on developing standards for the quartz plates used in saccharimeter testing. Jakob studied, "on request from industry," the thermal conductivity of four different insulating materials. Heuse determined the atomic weight and density of helium, and he and Warburg constructed three aneroid instruments displaying relatively limited elastic aftereffects. Both pieces of work had technological as well as scientific aims.[74] Finally, among new work, Meissner investigated thermal and electrical conductivity in metals at temperatures ranging between 20 and 373 K.[75]

The Electrical Laboratory of 1913 was the section's smallest laboratory; it was still headed, as it had been under Helmholtz, by Jaeger. His full-time senior associates were Fassbender, Hupka, and Carlo Müller, with Steinwehr working on a half- time basis. During 1913 the laboratory conducted only five projects, at least four of which were meant primarily to serve technological needs. Fassbender published a summary of his measurements on standard resistances and their replications; Steinwehr published his on the silver voltammeter and continued work on standard cells; and Fassbender and Hupka together discovered two types of vibration in the so-called Pouleian electric-arc lamp. In addition, thanks to the "not inconsiderable means" furnished by the International Solvay Institute for Physics and by four private firms, Warburg, Hupka, and Müller began constructing a high-voltage influence machine capable of reaching nearly 400,000 volts and intended for use in cathode- and radium-ray experiments.[76]

The revitalization and expansion of the Optics Laboratory constituted one of Warburg's major effective reforms at the Reichsanstalt; it lay at the heart of his efforts to increase the Institute's pure scientific work and to make it once again an important contributor to the latest developments in pure physics. His principal problem here was one of personnel: He had to replace the group of first-rate physicists – Lummer, Pringsheim, and Kurlbaum – who had recently left the laboratory for academic appointments with men of similar potential or accomplishment.

Warburg's success had two sources: first, his own solid understanding of the important problems in radiation physics and his excellent judgment about promising younger physicists; and second, the board's and ministry's support. In 1911, he got the board to petition the ministry for new funds – at least 20,000 marks – to enable the Reichsanstalt to pursue radioactivity research and testing, in particular to purchase radioactive materials and to create a new senior position for an "established researcher in the field of radioactivity."[77] What probably motivated the ministry to provide (for once) the new funds was the growing need of physicians and hospitals for radioactivity standards and dosages. During the early 1900s the diagnostic and therapeutic use of radioactive sources had spread throughout the medical world; radiological institutes were then being established at Heidelberg, London, Vienna, and elsewhere.[78] Radiation units of measurement, standards, and dosages now had to be established. Here again, nationalistic aims came into play.

With the money in hand, Warburg knew where to turn for what he wanted: the laboratory of Ernest Rutherford in Manchester, England. There worked a young – and therefore inexpensive – but experienced and gifted researcher in radioactivity: Hans Geiger (Figure 5.2). After receiving his doctorate in 1906 at the University of Erlangen, Geiger became an assistant at Manchester to Arthur Schuster and then to Schuster's successor, Rutherford. Between 1907 and 1912 Rutherford and Geiger devised a number of electrical techniques for counting $\alpha$-particles and determined the latter to be doubly charged. With Ernest Marsden, Geiger also worked on $\alpha$-scattering experiments, paving the way for and later confirming Rutherford's interpretation of their findings as evidence for the existence of a nucleus within the atom.[79]

Despite his successful research in England and Rutherford's attempt to keep him at Manchester, Geiger yearned to return to Germany.[80] Offers from Tübingen and the Reichsanstalt made the desirable possible.[81] He chose the Reichsanstalt over Tübingen presumably because Warburg convinced him that he could do more research there than he could at Tübingen and because Warburg promised to obtain the needed (and expensive) radioactive materials and apparatus. In October 1912, just a month after he had begun his new duties, he wrote to Rutherford about his expectations for radioactivity research at the Reichsanstalt and about the latter's resources:

Prof. Warburg has no objection to my working on the $\alpha$-particles and I shall get all the necessary apparatus. He told me that you had spoken with him about it in Brussels and I think he is only glad if that kind of work is being started at the Reichsanstalt. I am afraid that it will take a long time though to get the necessary apparatus together. There is nothing here at present which could be of any use for my work. All the apparatus are divided up between the different departments and there is only a very small collection of old fashioned instruments which are general property. I like the system quite well, since all the apparatus which I shall get in the course of time will then entirely be in my charge.

Prof. Warburg has tried to get Radium from Vienna at the same price at which the standards are sold; but without success, we have to pay full price. . . . There are about 15 mgr at the Anstalt already which were brought by Kohlrausch about ten years ago.[82]

In short, Geiger had to start radioactivity research in Charlottenburg from scratch, but with the assurance that he had full backing to set up a proper laboratory.

Geiger's principal duty at the Reichsanstalt was to formulate standards for and to test the radioactive preparations sent in by individuals, hospitals, commercial firms, and government agencies, in other words, to meet medicinal and commercial needs.[83] His standardizing and testing work constituted the sine qua non for allowing him to pursue, in the time remaining, his own scientific research. That arrangement satisfied Geiger. A year after he had begun working in Berlin he wrote to Niels Bohr, his former colleague in Rutherford's laboratory, that he "like[d] it very much here at the Reichsanstalt."[84] Warburg, for his part, was more than pleased with Geiger's first-year performance. He informed the ministry that Geiger's work had been so good and his radioactivity testing so extensive and important for the public that, his youth and brief tenure notwithstanding, he deserved to be named a Reichsanstalt member. The ministry complied.[85]

Geiger joined an excellent team of senior scientists in the Optics Laboratory, now headed by Ernst Gehrcke. In addition to Geiger, Warburg had brought three other major scientific workers to the laboratory: in 1907 Ludwig Janicki, in 1912 Rudolf Seeliger, and in 1913 Walter Bothe, a student of Planck's.[86] Joining these five full-time workers were three part-time assistants and two guests. The former category included Hupka, Müller, and Meissner; the latter, Otto Reichenheim and James Chadwick. Chadwick, who later became a leading student of

Figure 5.2. Hans Geiger.
Courtesy of Amerika-
Gedenkbibliothek Berliner
Zentralbibliothek, Berlin,
and Akademie-Verlag,
Berlin.

atomic and nuclear physics (not least for discovering the neutron in
1932), had been a student of Rutherford's at Manchester, where he met
Geiger. Upon winning an 1851 Exhibition Scholarship, Chadwick had
to find a new institution in which he might acquire further research experience. "The obvious thing for me to do," he later said, "was to go to
Geiger if he would have me." "There was no other place," he explained,
"where I could have continued on the radioactive side except Paris."
(And Mme. Curie's interests, he added, were mostly on the chemical
side.) As desired and expected, Chadwick spent the years 1913-1914
working with Geiger as a Reichsanstalt guest; as undesired and unexpected, he spent 1914-1918 as a prisoner of war in Germany.[87]

By 1913, then, Warburg had rebuilt the Optics Laboratory into a
world-class scientific center. The laboratory engaged in fifteen projects
that year. Janicki continued his observations on the interference of vari-
able path differences in wedge-shaped plates. He also measured the arc
spectrum of iron between 4,282 and 5,140 $\lambda$ and investigated the effect
of current strength and length of exposure on wavelength.[88] With the

rise of the wireless telegraphy industry, understanding the changes in wavelength of different materials under different conditions had become increasingly important.[89] Janicki, working with Seeliger, studied the light emission of metal vapors during glow discharge. Seeliger, working alone, also determined quantitatively the intensity of gaseous light stimulation as a function of the velocity of cathode rays. The work was "of far-reaching theoretical interest for the mechanism of light stimulation."[90]

Gehrcke supervised or participated in three projects during 1913. With Reichenheim he measured the velocity and charge-to-mass ratio of anode or positive rays, which he and Gehrcke had codiscovered in 1906. (Reichenheim also continued his previous investigation of anomalous anode fall and the origins of anode rays.) Their studies on the behavior of anode rays advanced understanding of gas discharge. Second, Gehrcke and Seeliger continued their investigation of surface charges in metallic conductors in a vacuum. And third, Gehrcke, Geiger, and Meissner investigated whether hard $\gamma$-rays emanating from radium could be bent through crystals as could X-rays and, as Rutherford had recently shown, very soft $\gamma$-rays. They reported that experimental variations failed to produce a bending of the rays.[91]

Hupka conducted three X-ray studies in 1913. First, following work on the X-ray diffraction of crystals by a number of physicists (Max von Laue, Walter Friedrich, Paul Knipping, and the father-and-son team of William Henry Bragg and William Lawrence Bragg), Hupka and August Steinhaus performed experiments on the reflection of X-rays in crystals. Second, Hupka measured X-ray intensity. And third, he irradiated thin pieces of commercial lead with X-rays, demonstrating that they yielded good pictures on photographic plates.[92]

Geiger and his associates studied and tested radioactive preparations. With Bothe, he tested radium and mesothorium preparations used for medicinal purposes. Their successful work led to a large demand for radioactive preparations, which in turn increased the testing demands placed on them. During 1913 they tested nearly 2,300 milligrams of radium sent to them.[93] (Despite – or because of? – the onset of war, during 1914 their testing work more than tripled to greater than 8,200 milligrams.)[94] Virtually all radium and mesothorium traded in Germany now came supplied with a Reichsanstalt certificate.[95] In January 1914, Geiger wrote to Rutherford: "The testing we have to do at the Reichsanstalt is still increasing. We had nearly one Gramm of

Ra[dium]-Metall [*sic*] to measure up in small fractions of about 20-30 mgr each in one month only. Medical people are getting very enthusiastic here about the good results they obtain by the application of Radium."[96] Although testing radioactive preparations slowed down Geiger's scientific research,[97] he nonetheless found time to invent his point counter – a forerunner of the well-known "Geiger counter" – which he (along with Bothe and Chadwick) then employed in counting both α- and β- particles. Geiger's counter had evolved from a detector that he and Rutherford had conceived in 1908 in Manchester; it would later evolve into the Geiger-Müller counter and prove to be an important tool in early cosmic-ray and nuclear physics.[98]

Warburg conducted most of his own scientific work in the Optics Laboratory. Like Helmholtz and Kohlrausch before him, he had expected that his new position as president of the Reichsanstalt would allow him more time for research than he had had at the University of Berlin; indeed, for him, as for them, this was probably the principal reason for accepting the post. His expectations went unfulfilled, just as theirs had. As a consequence of his time-consuming administrative duties, throughout his seventeen-year tenure at the Reichsanstalt Warburg had to restrict his scientific investigations to two major topics. First, he conducted his pioneering investigations of energy transformation in photochemical processes, in particular comparing in 1913 his experimental findings on the photochemistry of ozone deformation with Einstein's theoretical predictions from his photochemical equivalence law.[99] "I follow the experiments in your photochemical workshop," Einstein wrote Warburg in 1912, "with great attention. You are making real that which for years I had only vaguely dreamed about."[100] During his years at the Reichsanstalt Warburg created what Einstein called "quantitative photochemistry."[101] Second, with the aid of Hupka, who had trained under Planck at Berlin, and Müller, who had trained there under Warburg himself and under Rubens, Warburg further refined the values of the constants in Planck's blackbody-radiation law. Also in 1913, the three men conducted preparatory work toward a more precise determination of temperatures based on the Stefan-Boltzmann and Wien displacement laws.[102]

Although this laborious radiation work did not fit Warburg's scientific tastes, he undertook it out of a sense of duty to the Reichsanstalt.[103] Like Kohlrausch before him, he continued to hope that it might eventually form the foundation of a national and international unit and standard of luminous intensity. So, too, did German industry. In 1909, for example, the Deutscher Verein von Gas- und Wasserfachmänner and the Verband Deutscher Elektrotechniker informed the Ministry of the Interior (the former not for the first time) of the importance of an established, definitive unit and standard of luminous intensity for the worldwide illumination industry. The Germans had their Hefner lamp; the English, French, and Americans their Pentan lamp, which threatened to become the worldwide standard despite the fact that, according to the Germans, it was inferior to the Hefner lamp. The German illumination industry, the Verein and the Verband stressed, had "a tremendous export of illumination goods," and so greatly needed a single unit and standard that had universal validity. The way to a "rational unit of luminous intensity," they said, lay in the work of Lummer "on the absolute blackbody" and required the aid of the Reichsanstalt as well as that of other national metrological institutes. By 1909 that unit and standard had become so important to them that they offered to pay the costs of any German scientists who would work on establishing it.[104]

Warburg supported the industry's plans and, to a lesser extent, its hopes. "It is true," he informed the ministry, "that the introduction of a unit of luminous intensity recognized by all civilized countries is of great importance for both technology and science." But because the Reichsanstalt had labored at this problem for nearly a decade and a half, Warburg had a more sober view than did industry of the difficulties involved. He continued:

> Furthermore, when the petitioners see in the so-called blackbody, which has been frequently investigated at the Reichsanstalt and elsewhere during the past years, a means for producing a satisfactory unit of luminous intensity, this insight can be, to a certain extent, supported as a possibly practicable means [for achieving such a unit]. [However,] right now one cannot foresee whether this means will indeed achieve its goal. Only experiments, which are already partly underway at the Reichsanstalt and partly about to be begun, will be able to tell.[105]

Warburg told the board much the same thing: With the Reichsanstalt's blackbody radiator now operating at 2,000°C "it appears promising to define a unit of luminous intensity in this way."[106] In the event, however, it was not until 1941 that an international standard, one indeed based on blackbody radiation, was achieved.[107]

The Scientific Section's work during Warburg's administration, as during Helmholtz's and Kohlrausch's, consisted largely in investigating practical, technological problems of interest to industry and the public. Most of its work originated in requests from industry or government. That which did not was usually justified (to the public, at least) by the practical benefits that it might someday bring to private industry and to the consuming public.

Nonetheless, Warburg strove to expand pure scientific research at the Reichsanstalt. If he did not get more money from the ministry to halt the exit of the Reichsanstalt's best scientists to academic institutes and elsewhere, he at least managed to attract several promising younger physicists. In so doing, he rebuilt and even expanded the Optics Laboratory, which, as in the past, conducted far more pure scientific research than any other Reichsanstalt laboratory. Yet even here the scientific work of Geiger and his associates in radioactivity found its justification, and hence governmental support, in the practical needs of the German health industry and medical profession.

The setting of standards remained the Reichsanstalt's premier duty and strength, and Warburg naturally sought to maintain, if not increase, the Reichsanstalt's supremacy and international prominence in metrology. Like his predecessors, he organized and participated in international metrological conferences: In 1905, he hosted an international conference at the Reichsanstalt on electrical standards, and in 1908 he headed a group of Reichsanstalt workers participating at a similar conference in London, which aimed at revising the international agreement on electrical standards arrived at during an earlier 1893 meeting.[108] Moreover, he led the way to the international agreement of 1913 on copper standards, so important to electrotechnology, and arranged for the Internationale Beleuchtungskommission (International Illumination Commission) to meet at the Reichsanstalt in 1913.[109] In addition to hosting such international metrological standards conferences, Warburg's Reichsanstalt, thanks to Geiger, helped set national and international standards for radioactive preparations.

## Testing, Testing, Testing

The Technical Section, still directed by Ernst Hagen, continued to consist of four laboratories: Precision-Mechanical, Electrical, Heat and Pressure, and Optics. (All of these laboratories were still served by the

Chemical Laboratory and the workshop.) By 1913 the section had a staff of thirty-three major technical workers and a small army of technical assistants and administrative personnel. Although the staff did some research, most of its workers devoted most of their time to doing routine testing work for industry and, to a far lesser extent, for government and academic scientific institutes.

Like the Scientific Section, the Technical Section during Warburg's tenure required some reforms. Above all, the rise of the Electrical Laboratory to become the largest of all Reichsanstalt laboratories, which was the outstanding feature of the section during Warburg's presidency, led to a need for new, larger facilities to meet the electrical industry's ever greater and changing demands on the Reichsanstalt. As we shall see presently, Warburg managed to have new facilities built for the section's electrical and magnetic work.

That Warburg did so must in part be attributed to the serious interest that he, in contrast to Helmholtz and especially Kohlrausch, developed in the field of applied or technical physics.[110] Following his appointment to the presidency, he quickly came to understand technology's needs and, thanks especially to his numerous former students who had gone into industry, made countless connections in the industrial world.[111] He developed much closer relations between the Reichsanstalt and industry than his predecessors had, often visiting individual manufacturers' plants.[112] He became an active member of a number of technological associations, later receiving honors from several of them: In 1907 and 1908 he chaired Berlin's Elektrotechnischer Verein and later received its Siemens-Stephan-Gedankplatte.[113] He developed much closer relations with the Verein Deutscher Ingenieure – a group with which his predecessors were sometimes at odds – employing a Verein engineer at the Reichsanstalt and arranging for two of the latter's members to sit on the Verein's scientific advisory committee. He cofounded the Beleuchtungstechnische Gesellschaft (Society for Illumination Technology) in 1912 and remained its chairman until 1919. He received an honorary doctorate of engineering degree in 1917 from the Technische Hochschule Charlottenburg, simultaneously receiving praise from every major German technological association for his "personal and official services" aimed at maintaining "the intimate ties of physics and technology."[114] And he cofounded the Deutsche Gesellschaft für Technische Physik (German Society for Technical Physics) in 1919, becoming its first honorary chairman.[115] "There is absolutely no doubt," Lewald told his superiors in the ministry, "that President Warburg has excellently managed both the physical and the technical side [of the Reichsanstalt]."[116] Warburg grew with the office.

Arnold Leman headed the Precision-Mechanical Laboratory, still the smallest of the section's four laboratories, just as he had since the Reichsanstalt's founding. In 1913, Leman and his chief associates, Arnold Blaschke and A. Werner, tested and certified more than 200 items: for example, precision gauge blocks, plug gauges, screw spindles, tuning forks, tachometers, centrifuges, and micrometer screws. Owing to the increased demands of industry, more than a third of the laboratory's work was devoted to testing and certifying the precision gauge blocks. Consequently, its projected scientific studies had "made only scant progress."[117]

The Electrical Laboratory's growth during Warburg's reign to become the largest of all Reichsanstalt laboratories reflected, as already mentioned, the continued growth of the German electrical industry. In 1912, for example, the laboratory conducted about one-half of all of the Technical Section's work.[118] The Electrical Laboratory still consisted of three sublaboratories: High-Voltage, Low-Voltage, and Magnetics; in addition, it supervised the electrical testing offices around Germany. Its nineteen major scientific or technical workers outnumbered those in the entire Scientific Section and constituted more than one-half of the Technical Section's major workers and more than one-third of the entire Reichsanstalt's major workers.

Within the Electrical Laboratory, the High-Voltage Laboratory conducted the lion's share of the work; indeed, with its fourteen principal scientific and technical workers this sublaboratory alone was larger than any single Reichsanstalt laboratory, except of course the Electrical Laboratory itself. From 1904 to 1913, it was under the command of Ernst Orlich and from 1913 to 1918, under Karl Willy Wagner, one of Germany's outstanding telegraph engineers.

During 1913 the High-Voltage Laboratory tested more than 900 different items or pieces of material, including more than 600 pieces of direct- and alternating-current measuring apparatus (transformers, resistors, condensers, and so on); 16 generators, motors, and ventilators; more than 150 pieces of insulation materials and apparatus for conveying and distributing electrical energy; and more than 130 items for the laboratory's own use.[119]

The laboratory's close cooperation with the electrical industry was also manifested in its work for or with the Verband Deutscher Elektrotechniker. Harald Schering, for example, helped develop testing procedures for insulating material up to 500 volts and worked on the influence of arc lamps on such material. Moreover, he and Rudolf Schmidt helped improve standards for electrical meters. In addition, Schering and Schmidt determined, "in accordance with technology's wish," the limits of error for transformers, and Schering and Egon Alberti continued to develop, at the request of the Verband, methods for testing transformers.[120]

Apart from its routine testing duties and its work for the Verband, the High-Voltage Laboratory also conducted a number of more research-oriented investigations. Alberti and Volkmar Vieweg used two different methods for determining the magnetizing current in transformers, and Walter Rogowski used a magnetic voltmeter to measure the magnetic parameter of a number of electromagnetic devices, including those used in electrical streetcar systems. His colleagues Vieweg and August Wetthauer used a dynamometer to determine the torsion of gyrating waves, while Robert Lindemann and W. Hüter completed experiments on the resistance of vibrating cylindrical and disk coils. Finally, Adolf Güntherschulze investigated the behavior of electrolytic valves and the losses in a mercury rectifier as a function of its load, frequency, and curvature.[121]

All of the High-Voltage Laboratory's testing and research work occurred in increasingly cramped, inadequate quarters. As early as 1907, that is, less than a decade after the Technical Section's new facilities had been completed, the inadequacy of the section's electrical facilities began seriously to affect its ability to conduct electrical studies and testing. In particular, the alternating-current work done in the High-Voltage Laboratory lacked the necessary space to meet an increased workload and "to prevent dangers to the life and health of the P.T.R.'s officials."[122] Moreover, it lacked appropriate, up-to-date equipment. Thanks to the enormous growth and rapid technological changes in the electrical industry (particularly the rise of high-voltage, alternating-current systems during the 1890s and 1900s), industry's demands on the Reichsanstalt's electrical resources increasingly outstripped its ability to respond adequately. Lack of sufficient space and proper equipment also placed serious limitations on the Reichsanstalt's research capacities. The board declared:

The present space is not even adequate for the regular setup of all necessary measuring arrangements used in testing work. Scientific work must be executed in the same rooms as the testing work, usually in the unofficial evening hours and often after removal or modification of the setups used for testing purposes.[123]

The hindrance to or preclusion of research by heavy testing demands was a constant refrain at and concern to the Reichsanstalt. That concern was now exacerbated by the construction of new, well-equipped electrical laboratories at several Technische Hochschulen, which threatened to reduce the Reichsanstalt to a second-rate research center in matters of advanced electrical technology.[124] The High-Voltage Laboratory needed more than a remodeling of its extant facilities; it needed an entirely new building.[125]

It took nearly four years (from 1907 to 1911) before Warburg and Lewald could convince the government and the legislature of the necessity of a new electrical building and before construction could begin. Official approval was delayed in part because of Warburg's concern that the new high-voltage work (especially that for wireless telegraphy) to be conducted by the Technical Section would harm the Scientific Section's low-voltage or precision-electrical studies, rendering the latter's facilities inferior to those of the Technical Section or of certain foreign laboratories.[126] As a consequence, Warburg asked the government to construct two new electrical laboratories, one for each section: that for the Technical to be devoted to high-voltage work, that for the Scientific to low-voltage work or precision-electrical measurements (especially studies on the fundamental electrical units, induction, and capacitance).[127] The secretary of the treasury, however, opposed funding a second new laboratory and would approve funding the Technical Section's new High-Voltage Laboratory only after Warburg agreed to give up his plans to construct a precision-electrical laboratory for the Scientific Section. Warburg pointed out to the secretary that the Reichsanstalt had razed the Scientific Section's machine house in order to save the costs of purchasing new land for the Technical Section's new building. Thus the Scientific Section had sustained an absolute loss of resources. The secretary, however, was unimpressed with this argument; Warburg was forced to abandon his plans for a new laboratory for the Scientific Section and to accept the secretary's ultimatum.[128] As partial compensation, however, the government did fund a new wing for the Scientific Section's *Observatorium.*[129]

In 1914, after three years of construction, the new High-Voltage Laboratory, which cost 767,000 marks to build and equip, opened its doors.[130] With its new equipment the laboratory could undertake current measurements up to 15,000 amperes and, "in conformity with the needs of technology," voltage measurements up to 400,000 volts.[131] With the Reich's financial support, Warburg had managed to build a first-class, high-voltage laboratory capable of meeting industry's testing demands and research needs.

As in the High-Voltage Laboratory, routine testing work was the main preoccupation of the Low-Voltage Laboratory. With the death of Lindeck in 1911, Eduard Grüneisen, a former student of Warburg's and a physicist in the Scientific Section's Electrical Laboratory from 1899 to 1911, became the new head of the Low-Voltage Laboratory.[132] Aided by Erich Giebe, in 1913 Grüneisen tested more than 300 electrical items or pieces of material: for example, conduction and resistance materials, resistance assemblies, Clark cells, Weston standard cells, and accumulators. Moreover, Grüneisen developed resistances of 1,000 and 10,000 ohms. Finally, the two men sought to determine ever more precisely the absolute value of the ohm.[133]

Ernst Gumlich directed the Magnetic Laboratory throughout the imperial period and beyond. In 1913, with his sole aide August Steinhaus, Gumlich tested more than 200 pieces of magnetic material, more than half of which were pieces of dynamo sheet iron. In addition, the two men conducted five major research investigations: First, they completed their studies on the influence of vibrations on the magnetic properties of dynamo sheet iron, a topic of great importance to both the electrical industry and the German navy. (That work began after the Verband Deutscher Elektrotechniker gave the Reichsanstalt 5,000 marks to undertake studies on the magnetic properties of iron, especially dynamo sheet iron.)[134] Second, they produced hysteresis-free magnetization curves. Third, they completed laborious experiments on determining initial and reversible permeability by means of a magnetometer. Fourth, they studied the influence of chemical composition and heat treatment on iron alloys – the study having been requested by the Verband and the alloys supplied by Krupp (although, in previous years, other firms had supplied alloys, too.)[135] Fifth, they investigated supposedly nonmagnetic nickel steel, determining that the disturbances in compasses on warships were caused by magnetic steel contained within armored parts near the compasses. (Karl Rottok, the German navy's representative on the board, continued to advocate the navy's interest in studies of magnetic materials and their properties. The navy supplied the necessary materials.)[136]

The increase in the Magnetic Laboratory's testing and research work during Warburg's reign led Warburg to have a new, disturbance-free Magnetic Laboratory built. Begun in 1911 and completed in 1913, the laboratory, which was located on the Telegraphenberg near Potsdam, cost 111,500 marks to build. Most of that money, 100,000 marks, had been provided in 1901 as part of the Reichsanstalt's settlement with the Berlin-Charlottenburger Strassenbahngesellschaft.[137] Without that unusual source of funding it is most unlikely that the laboratory would have been built. For Warburg's plan to construct a constant-temperature building, wherein work requiring measurements at constant temperatures and studies on fundamental electrical units could be undertaken, and his plan to expand a Reichsanstalt building in order to conduct low-temperature measurements were both rejected by the Reich. His proposals came too close on the heels of the recent appropriations for the High-Voltage and Magnetic laboratories.[138]

Karl Feussner still supervised the work of the electrical testing offices, just as he had done since their founding in 1901. By 1913 there were seven offices (Bremen, Chemnitz, Frankfurt am Main, Hamburg, Ilmenau, Munich, and Nürnberg). Above all, the offices tested, improved, standardized, and certified electrical meters. Moreover, they tested various electrical toll apparatus, timers, galvanometers, voltmeters, wattmeters, and so on. Despite the seemingly large number of tests – 47,548 to be exact – the offices still remained limited in scope, for their testing work was only a small fraction of that undertaken by the (private) electrical firms themselves. A breakdown of the offices' activities shows, moreover, that the Munich office, the only one attached to a local municipal electrical works, conducted more than one-half of all the offices' tests, and the Frankfurt am Main office did more than a fourth. Thus only two offices really had much work to do, the other five languishing.[139] They languished, as they had in the past, because of opposition by industry. The Vereinigung der Elektrizitätswerke and the Verband Deutscher Elektrotechniker initiated, Hagen told the board, "a lively agitation" against the right of the offices to set standards for and to test electrical meters.[140] Johannes Görges, professor of electrotechnology at the Technische Hochschule Dresden, told his fellow board members "that within the industry an active sense of alarm exists over the mandatory standardization of electrical meters that was presumably being planned." Wilhelm Kohlrausch agreed, claiming that the industry already had stronger specifications than the law required. The board thus opposed forcing standards for electrical meters upon the industry.[141]

The Heat and Pressure Laboratory was the Technical Section's second largest laboratory during Warburg's presidency. The death in 1912 of the laboratory's original leader, Wiebe, a man of unsurpassed technical knowledge in thermometry and of extensive personal contacts throughout the thermometer industry, led Warburg to transfer Scheel from the Scientific Section's Heat Laboratory to that of the Technical Section.[142] Scheel was aided in his new duties by a handful of senior associates – Friedrich Gruetzmacher, Guido Moeller, Johann Disch, Alfred Schulze, and Paul Hebe – along with numerous subordinate personnel.

In 1913 the laboratory tested nearly 11,000 thermometers, electrical and optical-temperature gauges, petroleum apparatus, and similar equipment. About two-thirds of its work consisted of thermometer testing. With its testing of more than 2,000 medicinal thermometers in Charlottenburg and its supervision of the testing of another 86,000 medicinal thermometers in Ilmenau and still another 28,000 in Gehlberg, the laboratory contributed immeasurably to the German health profession and German health care. Yet all of this testing work left little time for research.[143]

As to the laboratory's nonthermometric work, Schulze and an aide tested nearly 1,700 electrical and optical-temperature gauges; the bulk of their work involved testing thermocouples. Scheel, Disch, and Hebe together prepared official testing regulations for barometers, manometers, and indicators. The setting of limits of error for barometers – and probably for the manometers and indicators as well – was determined "in agreement with the interested parties." Finally, after various requests the laboratory began testing hygrometers, and in response to requests from the German branch of the International Petroleum Committee it began investigations on petroleum apparatus.[144]

The Optical Laboratory of 1913, the third largest of the section's four laboratories, conducted exactly the same types of optical tests and studies that it had always performed. As in years past, Brodhun, Lie-

Figure 5.3. Cartoon from *Simplicissimus* (1908): "Taxing Electricity." Reprinted from *Simplicissimus 13* (1908): 389.

benthal, and Schönrock executed this work, taking on only one new senior official, August Wetthauer. During 1913 Brodhun, Liebenthal, and Wetthauer performed more than 700 routine photometric tests on Hefner lamps, carbon-filament lamps, tungsten lamps, and others.[145] As a result of the new illuminant tax law of 1909, the laboratory was burdened with new bureaucratic duties.[146] In 1913 Brodhun and Liebenthal had to perform more than 300 optical tests for government tax officials (Figure 5.3). They also helped develop, in cooperation with the Lichtnormalien-Kommission (Light Standards Commission) of the Verband Deutscher Elektrotechniker, regulations for measuring the luminous intensity of tubular light sources and participated in meetings of the Internationale Lichtmesskommission (International Photometric Commission), held that year at the Reichsanstalt.[147]

Most of the remainder of the laboratory's work fell to Schönrock, who continued to devote himself to saccharimeter testing and investigations. During 1913 he tested fifteen saccharimeter quartz plates and made calculations on a Zeiss sugar refractometer widely used in the sugar trade.[148] Both Zeiss and the International Committee for Unitary Methods in Sugar Investigation, in which Schönrock was a leading scientific figure, were interested in developing trustworthy tables showing the dependency of the refraction of sugar on its content.[149]

As the above account of the Technical Section's work in 1913 indicates, during Warburg's tenure the section dedicated most of its resources to executing tests for German industry; the pressures to fulfill industry's constant testing demands simply left the section with little time for research. The small amount of research that it did conduct was aimed largely at establishing procedures for or improving the quality of its tests.

The Reichsanstalt's leaders were naturally aware of this state of affairs and its dangers. In 1907, for example, Warburg reported that, despite the establishment of new testing bureaus in England and America, the Technical Section had experienced an increase in its testing activities since 1904.[150] Most of the section's workers and resources were "absorbed by testing activity."[151] Hagen, for his part, had serious concerns about the implications of this heavy testing load for his section. He told the board that the section needed to do more than mere testing work. If it did not, its "[scientific] officials would become mechanics. It is precisely the additional scientific work," Hagen asserted, "that he [i.e., the scientific official] needs in order to recover from the monotonous work of completing testing orders. Without the additional scientific work it would be impossible for the section to retain qualified officials."[152] Hagen's remarks were a revelation for several board members and an indication of another needed reform at the Reichsanstalt. Deadening testing work dominated the section's activities and lowered its morale.

Warburg's and Hagen's concerns about the enormous testing demands on the Technical (and, to a lesser extent, the Scientific) Section are supported by some quantitative data. As Table 5.2 shows, between 1887 and 1918 German industry paid the Reichsanstalt 1.6 million marks in testing fees, an enormous sum that reflected industry's enormous demands on the Reichsanstalt's resources. In 1913 alone, the Reichsanstalt earned 115,338 marks in testing fees, more than twice as much as Kohlrausch's Reichsanstalt of 1903 had earned and nearly eight times as much as Helmholtz's of 1893 had earned.[153] As Figure

Table 5.2. *Reichsanstalt's income from testing fees, 1887–1918*

| Year | Amount (marks) | Year | Amount (marks) |
|------|---------------|------|---------------|
| 1887/88 | 4,359 | 1903 | 55,942 |
| 1888/89 | 8,292 | 1904 | 59,760 |
| 1889/90 | 10,478 | 1905 | 63,230 |
| 1890/91 | 11,257 | 1906 | 67,683 |
| 1891/92 | 12,267 | 1907 | 79,414 |
| 1892/93 | 14,867 | 1908 | 83,950 |
| 1893/94 | 15,279 | 1909 | 90,453 |
| 1894/95 | 18,570 | 1910 | 89,745 |
| 1895/96 | 20,454 | 1911 | 90,000 |
| 1896/97 | 28,791 | 1912 | 91,664 |
| 1897/98 | 29,688 | 1913 | 115,338 |
| 1898 | 33,439 | 1914 | 105,655 |
| 1899 | 40,334 | 1915 | 41,489 |
| 1900 | 42,890 | 1916 | 46,691 |
| 1901 | 51,724 | 1917 | 53,979 |
| 1902 | 60,022 | 1918 | 57,090 |
| | | Total  1887–1918 | 1,594,794 |

*Sources:* K. Strecker, "Die Personalverhältnisse der wissenschaftlichen Beamten der Physikalisch-Technischen Reichsanstalt" (29 August 1911), Anlage 9, BA 265, PTB-IB; "Uebersicht über die Einnahmen und Ausgaben der Physikalisch-Technischen Reichsanstalt seit ihrer Errichtung," n.d. (probably 1915), BA 240, ibid.; and [Emil Warburg], "Denkschrift über die Tätigkeit der Physikalisch-Technischen Reichsanstalt von Anfang 1912 bis Ende 1920," *SBVR 365* (1920/21): 880–911, on 881.

5.4 shows, there was a nearly continuous increase in income from testing between 1887 and 1914. That fact in turn issued from two others: First, resources permitting, the Reichsanstalt had a legal obligation to fulfill all authorized testing requests; and second, it maintained an egalitarian testing policy – it did not matter whether a testing request came from a one-man mechanical workshop or the firm of Friedrich Krupp, Aktiengesellschaft. The Reichsanstalt's testing policy aimed to serve all of German industry and to have tests conducted on a first-come-first-served basis, with dispatch (normally within two to four weeks), and for a low fee.

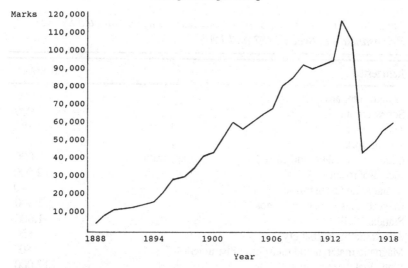

Figure 5.4. The Reichsanstalt's income from testing fees, 1887-1918. For sources see Table 5.2.

Several specialized measures of the testing demands on the Reichsanstalt help illuminate the meaning of these facts for the Reichsanstalt's evolution. Table 5.3 presents a breakdown of the total number of tests (principal items only) conducted between 1887 and 1905. All told, the Reichsanstalt conducted more than 325,000 tests through 1905. On average, the Reichsanstalt as a whole conducted about 20,000 tests annually.[154] The Heat and Pressure Laboratory, to cite figures for one laboratory, though admittedly one that had by far the greatest testing demands on it, tested more than 300,000 thermometers between 1887 and 1915.[155] (If the testing work done on so-called fever thermometers at the thermometric control stations in Ilmenau and Gehlberg between 1887 and 1913 is included, then that figure explodes to more than 1.5 million.)[156] Finally, Figures 5.5, 5.6, and 5.7 provide three further selected measures of the Institute's testing load: Between 1887 and 1918 the Reichsanstalt often conducted 1,000 or more resistance measurements per year, annually tested hundreds of direct- and alternating-current apparatus, and annually tested thousands of nonmedicinal thermometers.

German industry avidly sought Reichsanstalt testing – and the research behind that testing – for two reasons. First, testing served as an external quality control for a manufacturer's products and as a means

Table 5.3. *Number of tests (principal items only) conducted by the Reichsanstalt between 1887 and 1905*

| Item tested | Number |
|---|---|
| Length standard | 1,000 |
| Screw thread | 900 |
| Tachometer | 180 |
| Tuning fork | 2,700 |
| Current-, voltage-, and capacity-measuring apparatus | 1,800 |
| Electrical meter | 2,000 |
| Dynamo and transformer | 40 |
| Electrical resistance and rheostat | 2,500 |
| Standard cell | 1,600 |
| Accumulator and battery | 650 |
| Magnetic material and measuring instrument | 800 |
| Medicinal thermometer | 217,000 |
| Other thermometer (mostly for very precise measurements) | 30,000 |
| Pyrometric thermocouple | 5,000 |
| Barometer | 450 |
| Technical manometer and indicator spring | 250 |
| Safety device for boiler | 50,000 |
| Apparatus for analyzing petroleum | 3,200 |
| Standard lamp for luminous intensity (Hefner lamp) | 1,300 |
| Quartz plate for determining sugar content | 250 |
| Electrical lamp | 3,500 |
| Other lamp (gas, oil, acetylene, alcohol, etc.) | 1,800 |
| Total | 326,920 |

*Source:* Ernst Hagen and Karl Scheel, "Die Physikalisch-Technische Reichs-anstalt," in *Ingenieurwerke in und bei Berlin: Festschrift zum 50 Jährigen Bestehen des VDI,* ed. Verein Deutscher Ingenieure (Berlin, 1906), 60–67, on 67. Cf. (for 1887–1903) [Friedrich Kohlrausch], *Die bisherige Tätigkeit der Physikalisch-Technischen Reichsanstalt: Aus dem Reichstage am 19. Februar 1904 überreichten Denkschriften . . .* (Braunschweig, 1904), 5–6, reprinted in *GAFK, 1,* 1032–47.

by which he might compare his products with those of other manufac-turers. Second, an item tested by the Reichsanstalt and provided with one of its certification seals helped German industry win consumer confidence for its products, thereby raising the value of German goods. In particular, manufactured products certified with a quality-control

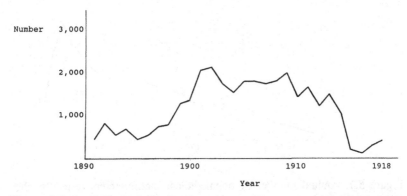

Figure 5.5. Annual number of resistance measurements conducted by the Reichsanstalt between 1891 and 1918. Adapted from "Abteilung II für Elektrizität und Magnetismus," *Forschung und Prüfung: 50 Jahre Physikalisch-Technische Reichsanstalt*, ed. Johannes Stark (Leipzig, 1937), 190.

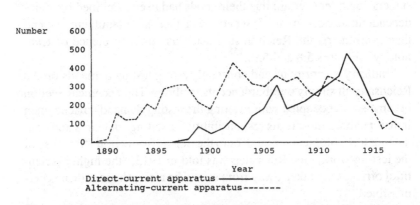

Figure 5.6. Annual number of direct- and alternating-current apparatus tested by the Reichsanstalt between 1887 and 1918. Adapted from "Abteilung II für Elektrizität und Magnetismus," *Forschung und Prüfung: 50 Jahre Physikalisch-Technische Reichsanstalt*, ed. Johannes Stark (Leipzig, 1937), 203.

seal from the Reichsanstalt helped German industry gain the confidence of foreign consumers, many of whom were often unfamiliar with German trade names.[157] German industry thereby gained a larger share of the hotly contested world market of the post-1870 era. Having a product emblazoned with a Reichsanstalt seal, foreign and domestic con-

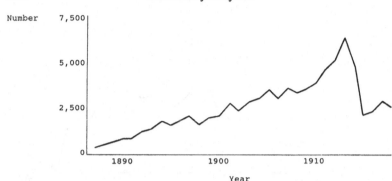

Figure 5.7. Annual number of nonmedicinal thermometers tested by the Reichsanstalt between 1887 and 1918. Adapted from "Abteilung III für Wärme und Druck," *Forschung und Prüfung: 50 Jahre Physikalisch-Technische Reichsanstalt,* ed. Johannes Stark (Leipzig, 1937), 263.

sumers could feel certain that their goods had been examined by an independent, noncommercial agency. For German manufacturers and their consumers, the Reichsanstalt acted as "the conscience of technology" (Figures 5.8 and 5.9).[158]

Industry's enormous and constantly rising testing demands on the Reichsanstalt simply overburdened its facilities. The Technical Section became so preoccupied with meeting the testing demands placed upon it that it shifted most of its responsibility for setting standards and conducting technological research onto the Scientific Section. "Because of the testing work," the Reichstag was told in 1913, "the higher [scientific] officials are taken away from scientific research more than is permissible."[159]

Although the testing demands that it placed on the Reichsanstalt were minute compared with those of industry, academic science also profited greatly from the Reichsanstalt's work. Before the establishment of the Reichsanstalt, routine testing and not so routine standards work had necessarily been the province of the academic physical

Figure 5.8 *(facing page)*. Official Reichsanstalt certificate of testing results on a so-called laboratory thermometer. From National Bureau of Standards, Entry 64: Records of the Heat Division, Temperature Physics Section, 1888-1911, Record Group 167, National Archives, Washington, D.C.

# Beglaubigungsschein

für

## das Thermometer

PTR *55530.*

Das in *zehntel* Grade C geteilte Thermometer ist als *Laboratorium-* Thermometer geprüft und beglaubigt worden; seine Angaben sind zur Zeit bei:

*0 Grad um 0,04*
*10 „ „ 0,05*
*20 „ „ 0,06* } *Grad zu hoch,*
*30 „ „ 0,06*
*40 „ ohne wesentlichen Fehler,*
*50 „ um 0,04 Grad zu hoch,*
*60 „ „ 0,08 „ „ „*

Die vorstehenden auf *fünfzigstel* Grade abgerundeten Prüfungsergebnisse sind auf das Gasthermometer bezogen.

Als Kennzeichen der vollzogenen Beglaubigung ist auf das Thermometer die amtliche Nummer, PTR, der Reichsadler und die Jahreszahl aufgeätzt worden. Strichmarken befinden sich bei – *10* und + *60* Grad.

Charlottenburg, den *7. März 1913.*

## Physikalisch-Technische Reichsanstalt
### Abteilung II.

*Hagen*

Fabr.-Nr.———

# Deutsche Mechaniker-Zeitung.

**Vereinsblatt der Deutschen Gesellschaft für Mechanik und Optik.**

Herausgegeben vom Vorstande der Gesellschaft.

Erscheint seit 1891.

| Beiblatt zur Zeitschrift | Organ für die gesamte |
| für Instrumentenkunde. | Glasinstrumenten-Industrie. |

Redaktion: A. Blaschke, Charlottenburg 4, Fritsche-Str. 39.

Verlag von Julius Springer in Berlin N.

| Heft 8. | 15. April. | 1910. |

**Nachdruck nur mit Genehmigung der Redaktion gestattet.**

Prüfungsbestimmungen der Physikalisch-Technischen Reichsanstalt.

### Allgemeine Bestimmungen[1]).

§ 1.

*Arbeitsgebiete.*

Die Physikalisch-Technische Reichsanstalt übernimmt die Ausführung von Prüfungen und Beglaubigungen nach den folgenden allgemeinen und den für die einzelnen Arbeitsgebiete erlassenen besonderen Bestimmungen.

Die Arbeitsgebiete umfassen:

*Präzisionsmechanik* (Längenmaße, soweit sie nicht zur Maß- und Gewichtsordnung gehören, Lehren, Kreisteilungen, Schrauben, Libellen, Tachometer, Arbeitszähler, Ausdehnungs- und Dichte-Bestimmungen u. a.);

*Druck* (Barometer, Manometer für hohe und niedrige Drucke, Luftpumpen, Indikatoren u. a.);

*Akustik* (Stimmgabeln, Resonatoren u. a.);

*Wärme* (Flüssigkeits-Thermometer von − 200 ° bis 575 ° C, Widerstands-Thermometer, Thermoelemente nebst Meß- und Registrier-Vorrichtungen für Messungen bis 1600 °, optische Pyrometer, Kalorimeter; Bestimmung von Schmelz- und Siedepunkten, spezifischen Wärmen und Verbrennungswärmen; Abel-Prober, Siedeapparate für Mineralöle, Zähigkeitsmesser u. a.);

*Elektrizität* (Widerstände, Normalelemente, Strom-, Spannungs-, Leistungsmesser, Elektrizitätszähler, Strom- und Spannungswandler, Frequenz- und Phasenmesser, Kapazitäten, Induktivitäten, Wellenmesser; Leitungs-, Widerstands-, Isolationsmaterialien, Dielektrika, Installations-Gegenstände; Primärelemente, Akkumulatoren, Generatoren, Motoren, Transformatoren u. a.);

*Magnetismus* (magnetische Apparate und Materialien);

*Optik* (Hefnerlampen und Glühlampen für photometrische Zwecke, Beleuchtungslampen und Zubehör, photometrische Apparate; Bestimmung der optischen Konstanten von festen und flüssigen Körpern sowie von Linsen, Prismen und dioptrischen Apparaten; Untersuchung von Planflächen und Planparallel-Platten, Quarz-Platten und Saccharimetern u. a.).

Außerdem werden auch Untersuchungen anderer Art ausgeführt, wie die Prüfung von Glassorten auf Verwitterbarkeit und das Verhalten gegenüber chemischen Agentien, die Untersuchung von Metallen hinsichtlich ihrer Verwendbarkeit bei der Anfertigung von Apparaten u. dergl.

§ 2.

*Prüfungsanträge.*

Die Prüfungsanträge sind schriftlich an die Physikalisch-Technische Reichsanstalt — Abteilung II — einzureichen[2]). Ebenso sind alle sonstigen Schreiben, welche

[1]) Vgl. *Centralbl. f. d. D. Reich vom 8. 4. 1910. 38. S. 101. 1910.*
[2]) Adresse: Charlottenburg 2, Werner-Siemens-Str. 8—12; Station für Frachtsendungen: Charlottenburg, Güterbahnhof. Telephon: Amt Charlottenburg Nr. 93.

Figure 5.9. First page of the Reichsanstalt's newly issued testing regulations (1910). Reprinted from *DMZ*, Heft 8 (15 April 1910): 73, in *ZI 30* (1910).

scientist, who lacked, however, sufficient resources for conducting extensive work in measuring physics. After the establishment of the Reichsanstalt, German physical scientists no longer needed to spend precious research time and material resources on standardizing instruments or establishing measuring standards, for example those for heat, electricity, and optics. The Reichsanstalt now conducted such work for them; it institutionalized measuring physics, thereby relieving academic scientists. As Ostwald rightly stated in 1906:

> Many older physicists can still remember how they had to
> devote weeks and sometimes months to calibrating and making
> other studies of their thermometers in order to obtain the possi-
> bility of a more exact (within limits) temperature determina-
> tion. Today they need only let the Reichsanstalt examine their
> instrument; they can use the saved time for real scientific
> work.[160]

But where the academic physicist "saved time" for his research, the Reichsanstalt physicist lost time from his own. And he became dissatisfied, even as he knew that testing and standardization work was part of his official duties.

## Wartime: A Crippled Institute, A Dying Empire

The Reichsanstalt's founders had intended that standards and testing work comprise only one-half of the Institute's work; the other half they reserved for pure, largely experimental physics. At least by the start of Kohlrausch's presidency, however, that bifurcation into equal halves had failed to materialize. The Reichsanstalt's institutional setting within the German physics community and the heavy testing demands placed upon it by German industry effectively forced it to make metrology and testing the Institute's primary function and, as a consequence, to let research in pure physics suffer.

To be sure, throughout the imperial period the Institute did an unprecedented amount of research in measuring physics. Creating and improving standards, instruments, and measuring methods – and making measurements with these – was itself research in pure (as well as applied) physics. Research in measuring physics required much time, extensive material resources, and a large staff of skilled physicists and technicians. It was laborious, often unappreciated work that frequently appeared as "raw" data, mere observational results in reference works or scientific articles. Occasionally, as in the case of measurements on blackbody radiation, it appeared as experimental data marshaled to support scientific arguments. To give a quantitative dimension to the point, between 1887 and 1914 Reichsanstalt physicists published the results of their work in measuring physics in some 170 articles in the *Verhandlungen der Physikalischen Gesellschaft zu Berlin* and the *Verhandlungen der Deutschen Physikalischen Gesellschaft,* in some 185 articles in the *Zeitschrift für Instrumentenkunde,* and in some 200 in the *Annalen der Physik.*[161]

Apart from its general work in measuring physics and its specific scientific discoveries, the Reichsanstalt contributed to the well-being of the German physics community in at least a third major way: It employed physicists and it provided guest physicists with excellent research facilities. To be more precise, between 1887 and 1918 it employed – excluding lower-level technicians – thirty-odd senior professional physicists on a full-time basis and invited some ten others to work there as guests. (In 1910, Kohlrausch thought that the Reichsanstalt probably had as many physicists on its staff as there had been in all of Germany during the 1860s.)[162] As noted earlier, to a certain extent the Reichsanstalt functioned as a kind of postdoctoral institution, allowing young physicists an opportunity to conduct research, to publish, and eventually to transfer into an academic career. The employment and provision of research facilities for German physicists; the participation of Reichsanstalt associates on the boards of editors or as editors of leading physics journals like the *Annalen der Physik,* the *Verhandlungen der Deutschen Physikalischen Gesellschaft,* and the *Zeitschrift für Instrumentenkunde;* the participation of Reichsanstalt physicists in the physics colloquia of the University of Berlin and in the meetings of the Deutsche Physikalische Gesellschaft – these and other activities placed the Reichsanstalt at the center of the Berlin physics community and made it an integral part of the larger German and world physics communities.

Moreover, the Reichsanstalt did a considerable amount of research (as well as testing) for a host of German industries: The gas, electrical, telegraphic, radio, optical, mechanical, steel, iron, porcelain, and sugar industries and, to a lesser extent, the chemical, petroleum, and motor industries all regularly turned to the Reichsanstalt for help in establishing fundamental standards. The establishment of electrical, thermometric, mechanical, radiation, and other standards; the increase in knowledge of the properties of materials; and the testing of nearly all physical apparatus and instruments except for those in the domain of weights and measures – these were prodigious contributions to German industry.

Still, especially after 1905 the Reichsanstalt's leaders became convinced that research – above all in pure physics – was suffering at the Institute and that the immediate cause of that suffering was precisely the burden of testing imposed by industry. Some thought that reform of the Reichsanstalt's overall organization was now needed. As early as 1908 several board members voiced concern that the Reichsanstalt's division into a Scientific and a Technical Section was no longer achieving its goal of fruitful cooperation between science and technology at the Reichsanstalt.[163] Inside each section there had evolved independent, "parallel laboratories" for heat, electricity, and optics, those of the Scientific Section being devoted principally to research and those of the Technical Section predominantly to testing work.[164] After several years of consultation with board members and, presumably, Reichsanstalt workers, in 1914 Warburg dissolved the two original sections and replaced them with three new ones: Optics, Electricity, and Heat. Each of these was in turn divided into two subsections: "a pure scientific (Subsection a)" completely free of testing work and a "technical-scientific (Subsection b)" responsible for the testing work. Each section had its own director, who sought, among other things, to integrate science and technology. The president maintained overall authority for the three sections and also supervised the Precision-Mechanical Laboratory, the Chemical Laboratory, the workshop, and the disturbance-free Magnetic Laboratory near Potsdam.[165] The object was to achieve a better integration of science and technology and, no less, to help retain more of the skilled scientists who lacked close contact with research and suffered from monotonous testing work.

Unfortunately, the new organizational structure took effect on 1 October 1914, less than two months after the outbreak of World War I. The war obliterated any potential benefits of the new organization. By 1915 fully one-half of all Reichsanstalt personnel, including twenty-two senior scientists, had either joined the German armed services – where they were sent to the front or attached to a military-technical unit – or were dispatched to conduct war-related work in factories.[166] Those who did stay in Charlottenburg spent most of their time conducting military-related work: for example, testing lights and lamps, electrical equipment, meteorological and medicinal apparatus, nonmagnetic steel alloys (for the navy) and doing research on artillery and ballistics problems, on wireless telegraphy, and so on.[167] Although Warburg offered to place the Reichsanstalt at the disposal of the German Ministry of War and the navy, its war work, in contrast to that conducted by the Americans at the Bureau of Standards and the English at the National Physical Laboratory, was of a minor, testing nature. Warburg later reported that "the Reichsanstalt's plant was for the most part brought to a standstill." "The officials who remained behind had been greatly burdened by testing work and by conducting largely unimportant work for the army."[168] Worse still, the German war effort took the lives of Lindemann and Schultze, as well as of at least six minor Reichsanstalt officials.[169] Apart from Warburg's own work in quantitative photochemistry, one of the few signs of scientific life at the Reichsanstalt during the war was Einstein's and W. J. de Haas's – they were guest scientists there in 1914-1915 – experimental testing of André-Marie Ampère's hypothesis that magnetism was due to circular molecular currents within magnets. Magnetization, they showed in what subsequently became known as the Einstein-de Haas effect, induced rotation in an iron rod.[170] The restructuring of the Reichsanstalt was a stillborn reform.

The coming of the war killed other plans too. Warburg had expected to host an international conference on the temperature scale at the Reichsanstalt at 1914; the war prevented this.[171] More important for Warburg's program to increase research at the Reichsanstalt, the war largely prevented the use of monies from the new Helmholtz Fund and the Rathenau Foundation. Grants from the Helmholtz Fund were first awarded for 1914-1915: Schultze and Giebe received 750 marks each for trips to Kammerlingh Onnes's low-temperature laboratory at Leiden and the National Physical Laboratory at Teddington, respectively; Geiger received 3,000 marks to purchase apparatus to conduct radioactivity research, as opposed to testing work financed by the Reich; Holborn received 2,000 marks for an air pump; and Schönrock received 1,500 marks as a first contribution for constructing a spectrometer and 2,000 marks to purchase fluorite to be used in light-dispersion experiments. All told, in 1914-1915 Warburg awarded 10,000 marks for scientific research and travel.[172] With the coming of the war, the awards became largely useless; only about 2,500 – all for radioactivity – of the 10,000 marks awarded was actually disbursed. In August 1915 Warburg reported that the Helmholtz Fund had 15,000 marks available; he purchased war bonds with it.[173] In 1916, however, he did award 6,000 marks for scientific research: 3,000 marks to Schönrock for a spectrometer; 2,000 marks for constructing a blackbody radiator that would, it was still hoped, help establish the basis for a unit of luminous intensity; and 1,200 to 1,400 marks for an assistant for Einstein.[174] As for funds from the Rathenau Foundation, it awarded only one grant during the entire war: an annual payment of 1,500 marks to Jaeger.[175]

Political forces far beyond Warburg's control – in particular, German nationalism and imperialism – that had once helped the Reichsanstalt come into being, and that had maintained its well-being, now prevented him from fully reforming the Reichsanstalt. Indeed, they crippled his institute: Four years of war and neglect left the Reichsanstalt, including Warburg's program to increase the Reichsanstalt's scientific strength, in shambles. As the empire died around him in 1918 and as the new Weimar Republic struggled for survival, the seventy-two-year-old Warburg could only hope to rebuild the Reichsanstalt to its prewar strength and to leave his more ambitious plans for increased scientific research to his successors. For four years he became a caretaker president, until at last in 1922 a successor was found.

There was more than a little irony in the fate of the Reichsanstalt during these war years; for the Institute's work had done much to further some essential political goals of the new German Reich: Its research and testing for industry had helped increase Germany's domestic and foreign trade, which in turn strengthened the Reich economically and politically. Its development of physical units and standards for all of Germany helped unify the Reich politically and so helped diminish the centuries-old German problem of political particularism. German scientists, industrialists, and government officials, whether in Munich, Hamburg, Leipzig, or elsewhere in the Reich, turned to the Reichsanstalt in Berlin for authoritative answers to their metrological problems. Moreover, by functioning as the official German institution responsible on the international plane for physical metrology, the Institute increased the young Reich's status as a national power. Finally, the Institute's numerous individual scientific discoveries helped make Berlin a center of German and international science, and so increased the cultural standing of the Reich's capital. In short, imperial Germany's Reichsanstalt had abundantly demonstrated the roles and functions that physics and technology could perform in building a modern society and nation-state. But where the Franco-Prussian War had entailed a brilliant new institution, World War I proved too great an ordeal for imperial German science to withstand intact.

# Notes

## Introduction

1. Karl-Heinz Manegold, *Universität, Technische Hochschule und Industrie: Ein Beitrag zur Emanzipation der Technik im 19. Jahrhundert unter besonderer Berücksichtigung der Bestrebung Felix Kleins* (Berlin, 1970); and idem, "Felix Klein als Wissenschaftsorganisator: Ein Beitrag zum Verhältnis von Naturwissenschaft und Technik im 19. Jahrhundert," *Technikgeschichte* 35 (1968): 177-204.

2. "Frankfurter Bezirksverein," *Zeitschrift für angewandte Chemie 20* (1907): 603-8; Emil Fischer and Ernst Beckmann, *Das Kaiser-Wilhelm-Institut für Chemie Berlin-Dahlem* (Braunschweig, 1913), 8; Lothar Burchardt, *Wissenschaftspolitik im Wilhelminischen Deutschland: Vorgeschichte, Gründung und Aufbau der Kaiser-Wilhelm-Gesellschaft zur Förderung der Wissenschaften* (Göttingen, 1975), 18, 24-28, passim; Günter Wendel, *Die Kaiser-Wilhelm-Gesellschaft 1911-1914: Zur Anatomie einer imperialistischen Forschungsgesellschaft* (Berlin, 1975), 37-38, 57-71, 266-67, 324-27, passim; and Walter Ruske, *Hundert Jahre Materialprüfung in Berlin: Ein Beitrag zur Technikgeschichte* (Berlin, 1971), 277- 326.

3. Oliver Lodge, "Section A. Mathematics and Physics. Opening Address by Prof. Oliver J. Lodge, . . . , President of the Section," *Nature 44* (1891): 383-87; Douglas Galton, "President's Address," *Report of the Sixty-fifth Meeting of the British Association for the Advancement of Science Held at Ipswich in September 1985 65* (1895): 3-35, quote on 34; and idem, "Physics at the British Association," *Nature 52* (1895): 532-36, on 532.

4. "A National Physical Laboratory," *Nature 55* (1897): 368-69. For further evidence on the Reichsanstalt as a model for the National Physical Laboratory see "The National Physical Laboratory," *Nature 55* (1897): 385-86; "Report on a National Physical Laboratory," *Nature 58* (1898): 548-49; and R. T. Glazebrook, "The Aims of the National Physical Laboratory of Great Britain," *Annual Report of the Board of Regents of the Smithsonian Institution . . . for the Year Ending June 30, 1901* (Washington, D.C., 1902), 341-57.

See also the following letters in ZStAP, RI, Sig. 13144/4: Frank C. Lascelles to His Excellency Herr von Bülow (copy), 19 March 1898, Bl. 36-37; Lascelles to Bülow (copy), 2 February 1900, Bl. 71-72; Friedrich Kohlrausch to Staatssekretär des Innern, 20 February 1900, Bl. 73-74; and *National Physical Laboratory: Report of the Committee Appointed by the Treasury to Inquire Generally into the Work Now Performed at the National Physical Laboratory* . . . (London, 1908), in Sig. 13185/1, Bl. 103 ff.

5. "The National Physical Laboratory," *Nature 58* (1898): 565-66; and "The National Physical Laboratory," *Engineering 71* (1901): 707-8. On the origins of the laboratory see Russell Moseley, "Science, Government, and Industrial Research: The Origins and Development of the National Physical Laboratory, 1900-1975" (Dissertation, University of Sussex, 1976), Chap. 1; and idem, "The Origins and Early Years of the National Physical Laboratory: A Chapter in the Pre-History of British Science Policy," *Minerva 16* (1978): 222-50.

6. See, e.g., Francis Crocker, "The Precision of Electrical Engineering," *TAIEE 14* (1897): 237-49, esp. 240; Edward W. Morley, "Visits to Scientific Institutions in Europe," *American Architect and Building News 59* (1898): 12-13; Henry S. Carhart, "The Imperial Physico-Technical Institution in Charlottenburg," *TAIEE 17* (1900): 555-83, including remarks (582) on the Reichsanstalt by Arthur Edwin Kennelly; and Senate Committee on Commerce, *National Standar[d]izing Bureau,* to accompany S. Report 4680, 56th Cong., 2d sess., 1901.

7. Henry S. Pritchett, "The Story of the Establishment of the National Bureau of Standards," *Science 15* (1902): 281-84, on 282.

8. Hale to Stratton (draft copy), 7 February 1900, National Academy of Sciences, Washington, D.C., George Ellery Hale Papers, Roll 34, pp. 51-54.

9. Ames to Stratton, 13 January 1900, Massachusetts Institute of Technology, Libraries, Institute Archives and Special Collections, Samuel Wesley Stratton Papers, MC 8, Box 1.

10. "Scientific Notes and News," *Science 16* (1902): 437. For further statements on the effectiveness of the Reichsanstalt and calls for an American version of it, see House of Representatives, Committee on Coinage, Weights, and Measurements, *National Standardizing Bureau,* 56th Cong., 1st sess., 1900, H. Document No. 625, Serial 3997; Senate, Subcommittee of the Committee on Commerce, *Hearing Before the Subcommittee of the Committee on Commerce, United States Senate, upon the Bill (S. 4680) to Establish a National Standardizing Bureau,* 8, 18, 34, 40, 46, 48, 56th Cong., 2d sess., 1900-1, S. Document No. 70, Vol. 5; Chap. 6, "A Typical Scientific Institution: The Physical and Technical Institute at Charlottenburg," in Ray Stannard Baker's *Seen in Germany* (London, 1902), 161-93; and Rexmond Cochrane, *Measures for Progress: A History of the National Bureau of Standards* (Washington, D.C., 1966), 38-39, 44, 60-61.

11. W. Kaufmann, book review of H. Pellat, *Die Physikalisch Technischen Staatslaboratorien* (six-page pamphlet), in *PZ 1* (1899-1900): 486; and Itakura Kiyonobu and Eri Yagi, "The Japanese Research System and the Establishment of the Institute of Physical and Chemical Research," in *Science and Society in Modern Japan: Selected Historical Sources,* ed. Nakayama Shigeru, David L. Swain, and Eri Yagi (Cambridge, Mass., 1974), 158-201, esp. 161-63.

12. K. Strecker, "Die Personalverhältnisse der wissenschaftlichen Beamten der Physikalisch-Technischen Reichsanstalt" (29 August 1911), BA 265, PTB-IB.

13. These fall into three categories: (1) Older, panegyrical accounts: Karl Scheel, "Werner von Siemens und die Physikalisch-Technische Reichsanstalt," *Dinglers Polytechnisches Journal 331* (1916): 405-8; Emil Warburg, "Werner Siemens und die Physikalisch-Technische Reichsanstalt," *NWN 4* (1916): 793-97; and Johannes Zenneck, "Werner v. Siemens und die Gründung der Physikalisch- Technischen Reichsanstalt," *Deutsches Museum, Abhandlungen und Berichte 3* (1931): 1-26. (2) Science-policy accounts: Frank Pfetsch, "Scientific Organisation and Science Policy in Imperial Germany, 1871-1914: The Foundation of the Imperial Institute of Physics and Technology," *Minerva 3* (1970): 557- 80, which is virtually identical to Chapter 3 of Pfetsch's *Zur Entwicklung der Wissenschaftspolitik in Deutschland 1750-1914* (Berlin, 1974), 109-28; and Lothar Burchardt, "Der Weg zur PTR: Die Projektierung und Gründung hochschulferner physikalischer Forschungsstätten im Kaiserreich," *PB 32* (1976): 289-97. (3) Marxist accounts: Gisela Buchheim, "Initiativen zur Gründung der Physikalisch-Technischen Reichsanstalt (1887)," *NTM 11* (1974): 33-43; idem, "Reichstagsdebatten über die Gründung der Physikalisch-Technischen Reichsanstalt," *NTM 12* (1975): 1-13; idem, "Die Wechselbeziehungen zwischen Industrie, Staat und Wissenschaft gezeigt am Beispiel der Gründung der Physikalisch-Technischen Reichsanstalt (1887)," *Acta Historiae Rerum Naturalium Necnon Technicarum,* special issue, 8 (1976): 189-97; idem, "Die Entwicklung des elektrischen Messwesens und die Gründung der Physikalisch-Technischen Reichsanstalt," *NTM 14* (1977): 16-32; idem, "Die Gründungsgeschichte der Physikalisch-Technischen Reichsanstalt von 1872 bis 1887," *Dresdener Beiträge zur Geschichte der Technikwissenschaften,* Heft 3 (1981) and Heft 4 (1982), which incorporates Buchheim's previous studies; and Günter Wendel, "On the History of the Founding in 1887 in Germany of the Imperial Institute of Physics and Technology" (in Russian), *Voprosy istorii estestvoznaniia i tekhniki* No. 3 (52) (1976), 66-70. I have presented my own views on the Reichsanstalt's origins in "Werner Siemens and the Origin of the Physikalisch-Technische Reichsanstalt, 1872-1887," *HSPS 12* (1982): 253-83; and in Chapters 1 and 2, this volume.

14. Johannes Stark, ed., *Forschung und Prüfung: 50 Jahre Physikalisch-*

*Technische Reichsanstalt* (Leipzig, 1937); H. Moser, ed., *Forschung und Prüfung: 75 Jahre Physikalisch-Technische Reichsanstalt* (Braunschweig, 1962); and J. Bortfeldt, W. Hauser, and H. Rechenberg, ed., *Forschen-Messen-Prüfen: 100 Jahre Physikalisch-Technische Reichsanstalt/Bundesanstalt 1887-1987* (Weinheim, 1987).

15. See, however, notes 18 and 19 below.

16. "Berichte von Universitäten, Hochschulen und wissenschaftlichen Einrichtungen," *PB 2* (1946): 85-91, on 88; Ernst Brüche, "Das Schicksal der PTR," *PB 6* (1950): 422-26; and idem, "Physikalisch-Technische Reichsanstalt, Berlin-Charlottenburg," *PB 8* (1952): 429.

17. G. Schmid, "Die Verluste des ehemaligen Reichsarchivs im zweiten Weltkrieg," in *Archivar und Historiker: Studien zur Archiv- und Geschichtswissenschaft: Zum 65. Geburtstag von Heinrich Otto Meissner*, ed. Staatliche Archivverwaltung im Staatssekretariat für Innere Angelegenheiten (Berlin, 1956), 176-207, on 188.

18. David Cahan, "The Physikalisch-Technische Reichsanstalt in Imperial Germany: A Study in the Relations of Science, Technology and Industry in Imperial Germany" (Ph.D. dissertation, The Johns Hopkins University, 1980).

19. David Cahan, *Meister der Messung: Die Physikalisch-Technische Reichsanstalt im Deutschen Kaiserreich* (Weinheim, 1988).

20. Hans Kangro, *Vorgeschichte des Planckschen Strahlungsgesetzes: Messungen und Theorien der spektralen Energieverteilung bis zur Begründung der Quantenhypothese* (Wiesbaden, 1970).

## Chapter 1. Physics and empire

1. [Debray, Jamin, Daubrée, and Laboulaye], "Discours prononcés par des membres de l'Académie aux funérailles de M. Regnault," *Comptes rendus hebdomadaires des séances de l'Académie des sciences 86* (1878): 131-43, esp. 134, 142; J. H. Gladstone, ["Henri Victor Regnault,"] *Journal of the Chemical Society 33* (1878): 235-39; J.-B. Dumas, "Eloge historique de Henri-Victor Regnault, membre de l'Académie des sciences de l'institut de France," *Mémoires de l'Académie des sciences de l'Institut de France 42* (1883): 37-72, esp. 54, 60-65, 70-72, quote on 72; F. Henning, "Henri Victor Regnault," *PZ 11* (1910): 770-74, on 773-74; and Robert Fox, *The Caloric Theory of Gases from Lavoisier to Regnault* (Oxford, 1971), 295-300.

2. Fox, *The Caloric Theory of Gases*, 281-318, esp. 314-18.

3. Emil DuBois-Reymond, "Der deutsche Krieg," in his *Reden*, 2nd enlarged ed., ed. Estelle DuBois-Reymond, 2 vols. (Leipzig, 1912), *1*, 393-420.

4. Emil DuBois-Reymond, "Das Kaiserreich und der Friede," in ibid., *1*, 421-30, quote on 421.

5. Ibid., 429.

6. J. L. Heilbron, *The Dilemmas of an Upright Man: Max Planck as Spokesman for German Science* (Berkeley, Calif., 1986).

7. Rudolf Virchow, ["Ueber die Aufgaben der Naturwissenschaften in dem neuen nationalen Leben Deutschlands,"] *Tageblatt der 44. Versammlung Deutscher Naturforscher und Aerzte in Rostock 1871* No. 5 (1871): 73-81, quotes on 75, 77.

8. K. Schwarzschild, "Präzisionstechnik und wissenschaftliche Forschung," *DMZ*, Heft 14 (15 July 1914): 149-53; Heft 15 (1 August 1914): 162-65, on 151, in *ZI 34* (1914).

9. Kenneth O. May, "Gauss, Karl Friedrich," *DSB 5* (1972): 298-315, on 305-6; A. E. Woodruff, "Weber, Wilhelm Eduard," *DSB 14* (1976): 203-9, on 204; and Christa Jungnickel and Russell McCormmach, *Intellectual Mastery of Nature: Theoretical Physics from Ohm to Einstein*, 2 vols., Vol. 1: *The Torch of Mathematics* (Chicago, 1986), 22-23, 63-77, 89-93, 101-5, 130-48.

10. Wilhelm Jaeger, *Die Entstehung der internationalen Masse der Elektrotechnik (= Geschichtliche Einzeldarstellungen aus der Elektrotechnik,* ed. Elektrotechnischer Verein, Band 4) (Berlin, 1932).

11. One measure of the growing importance of precision physics was the establishment in 1889 of a separate "Abteilung für Instrumentenkunde" within the Versammlung deutscher Naturforscher und Aerzte ("62. Versammlung deutscher Naturforscher und Aerzte zu Heidelberg in den Tagen von 17. bis 23. September 1889," *ZI 9* [1889]: 224). On the tasks of precision physics see Emil Warburg, "Verhältnis der Präzisionsmessungen zu den allgemeinen Zielen der Physik," in *Physik (= Die Kultur der Gegenwart: Ihre Entwicklung und ihre Ziele*, Teil 3, *Mathematik, Naturwissenschaften, Medizin*, Abteilung 3, *Anorganische Naturwissenschaften*, ed. P. Hinneberg), ed. Emil Warburg (Leipzig, 1915), 653-60.

12. David S. Landes, *The Unbound Prometheus: Technological Change and Industrial Development in Western Europe from 1750 to the Present* (Cambridge, 1972), 4, 235-36, 249-326.

13. S. Bradbury, *The Evolution of the Microscope* (Oxford, 1967), 240-47; N. Günther, "Abbe, Ernst," *DSB 1* (1970): 6-9; and Kei-ichi Tsuneishi, "On the Abbe Theory (1873)," *Japanese Studies in the History of Science 12* (1973): 79-91.

14. Bradbury, *Evolution*, 259-67; and Felix Auerbach, *The Zeiss Works and the Carl-Zeiss Stiftung in Jena: Their Scientific, Technical, and Sociological Development and Importance Popularly Described*, trans. Siegfried Paul and Frederic J. Cheshire (London, 1904), 19-21.

15. C. Faulhaber, "Die Optische Industrie," in *Handbuch der Wirtschaftskunde Deutschlands*, 4 vols. (Leipzig, 1904), *3*, 455-72, on 469-70.

16. Paul Wildner, "Die Glasindustrie," in ibid., 263-92, on 279-80, 283-84.

17. Of the extensive literature on the technical innovations in and history of the electrical industry during the late nineteenth and early twentieth centuries,

see especially Alfred Cruse, "Die elektrische Industrie," in ibid., 949-1008; Georg Dettmar, *Die Entwicklung der Starkstromtechnik in Deutschland* (Berlin, 1940); Arthur A. Bright, Jr., *The Electric Lamp Industry: Technological Change and Economic Development from 1800 to 1947* (New York, 1949); Harold C. Passer, *The Electrical Manufacturers, 1875-1900* (Cambridge, Mass., 1953); Peter Czada, *Die Berliner Elektroindustrie in der Weimarer Zeit: Eine regionalstatistisch-wirtschaftshistorische Untersuchung* (Berlin, 1969); and Thomas P. Hughes, *Networks of Power: Electrification in Western Society, 1880-1930* (Baltimore, Md., 1983).

18. Czada, *Die Berliner Elektroindustrie*, 52-54.

19. Georg Dettmar, "Die Elektrizitäts-Industrie," in *Deutschland unter Kaiser Wilhelm II*, ed. P. Zorn and H. v. Berger, 3 vols. (Berlin, 1914), 2, 559-78, on 575.

20. Joseph Loewe, "Die elektrotechnische Industrie," in *Die Störungen im deutschen Wirtschaftsleben während der Jahre 1900 ff.*, ed. Verein für Sozialpolitik, 8 vols. in 4 (Leipzig, 1903), 3, 75-155, on 93-94.

21. Ibid., 81.

22. Bright, *The Electric Lamp Industry*, 159; I. C. R. Byatt, "Electrical Products," in *The Development of British Industry and Foreign Competition, 1875-1914*, ed. Derek H. Aldcroft (London, 1968), 238-73, on 240, 267; and Landes, *The Unbound Prometheus*, 290.

23. Hajo Holborn, "The Prusso-German School: Moltke and the Rise of the General Staff," in *Makers of Modern Strategy from Machiavelli to the Nuclear Age*, ed. Peter Paret (with Gordon A. Craig and Felix Gilbert) (Princeton, N.J., 1986), 281-95, on 284, 287.

24. Wilhelm Foerster, *Lebenserinnerungen und Lebenshoffnungen (1832 bis 1910)* (Berlin, 1911), 115, 161-62.

25. Gunter E. Rothenberg, "Moltke, Schlieffen and the Doctrine of Strategic Envelopment," in *Makers of Modern Strategy*, ed. Peter Paret, 296-325, on 303-4; Paul Eversheim, "Die Physik im Kriege," in *Deutsche Naturwissenschaft, Technik und Erfindung in Weltkriege*, ed. B. Schmid (Munich, 1919), 57-79; and Willem D. Hackmann, "Sonar Research and Naval Warfare 1914-1954: A Case Study of a Twentieth-Century Establishment Science," *HSPS 16* (1986): 83-110, on 90-99.

26. In general, see G. Neumayer, "Die Kriegsflotten und die wissenschaftliche Forschung," in *Beiträge zur Beleuchtung der Flottenfrage*, 4th series (Munich, 1900), 27-34.

27. Konrad H. Jarausch, *Students, Society and Politics in Imperial Germany: The Rise of Academic Illiberalism* (Princeton, N.J., 1982), 136.

28. For further details see David Cahan, "The Institutional Revolution in German Physics, 1865-1914," *HSPS 15:2* (1985): 1-65, on 44-48, 50-56.

29. Ibid., 48-50.

30. Ibid., 3-12.

31. Ibid., 15-21.

32. Ibid., 21-25, 42.

33. E. Warburg, "Ueber den Aufschwung der modernen Naturwissenschaften," *Berichte der Naturforschenden Gesellschaft zu Freiburg i. B. 6* (1892): 18-29, on 23.

34. Friedrich Kohlrausch, "Antrittsrede," *SB* (1896): (II) 736-43, on 737, reprinted in *GAFK, 1*, 1024-31, on 1024-25.

35. *Karl Schellbach: Rückblick auf sein wissenschaftliches Leben, nebst zwei Schriften aus seinem Nachlass und Briefen von Jacobi, Joachimstahl und Weierstrass* (= *Abhandlungen zur Geschichte der mathematischen Wissenschaften mit Einschluss ihrer Anwendungen, 20:1*), ed. Felix Müller (Leipzig, 1905), 14, 19, 26, 28-29, 32-34. Schellbach inspired a love for science and technology in the antimilitarist, antichauvinist crown prince, whose confidant he became (Schellbach, *Erinnerungen an den Kronprinzen Friedrich Wilhelm von Preussen* [Breslau, 1890], 22- 23, passim; Foerster, *Lebenserinnerungen* [Berlin, 1911], 128-29, 133, 160-61).

36. The memorandum is printed in Gisela Buchheim, "Die Denkschrift vom 30.7.1872 von K.-H. Schellbach: Ein Nachtrag zur Vorgeschichte der Physikalisch-Technischen Reichsanstalt," *NTM 23* (1986): 99-101; draft letter of the Akademie's response to the Kultusminister (Falk), 27 March 1873, in *Verhandlungen [= Protokolle] Physikalisch-mathematische Klasse*, 1872-73, Zentrales Archiv der Akademie der Wissenschaften der DDR, Historische Abteilung, Abschnitt II, Sign. II-VI, 60, Bl. 55-61, quote on 58; see also II-V, 50, 118.

37. Foerster, *Lebenserinnerungen*, 74-76, 160-61.

38. See notes 23-25; P. Guthnick, "Wilhelm Foerster," *Vierteljahrsschrift der Astronomischen Gesellschaft 50* (1924): 5- 13, on 7.

39. Foerster to Morozowicz, 27 October 1873, SAA 61/Lc 973; Manfred Messerschmidt, "Die politische Geschichte der preussisch-deutschen Armee," in *Handbuch zur deutschen Militärgeschichte 1648-1939*, Band 2, Abschnitt 4, *Militärgeschichte im 19. Jahrhundert 1814-1890*, ed. F. Forstmeier and Hans Meier-Welcker (Munich, 1975), 317.

40. "Denkschrift betreffend die Begründung eines mechanischen Instituts," "Vorschläge zur Hebung der wissenschaftlichen Mechanik und Instrumentenkunde," "Grundzüge der Organisation des mechanischen Instituts," in (Prussian) Landtag, *Anlagen zu den Stenographischen Berichten über die Verhandlungen des Hauses der Abgeordneten*, 1876, Band 1, Aktenstück 53, pp. 531-36; Morozowicz to Moltke, 3 November 1873, SAA 61/Lc 973.

41. It consisted of Morozowicz (chairman), DuBois-Reymond, Foerster, Helmholtz, Georg Neumayer, Franz Reuleaux, Schellbach, Siemens, and other governmental, military, and academic officials.

42. Minutes of meeting, 8 December 1873, SAA 61/Lc 973.

43. Minutes of meeting, 20 December 1873, ibid.

44. Published in (Prussian) Landtag *Anlagen*, quote on 531.

45. Ibid., 535-36.

46. Falk (Kultusministerium) to Siemens, 15 July 1875, SAA 61/Lc 973.

47. (Prussian) Landtag, Haus der Abgeordneten, *Stenographische Berichte über die Verhandlungen*, 1876, Band 1, 17 March 1876, pp. 732-46. Cf. Gert Schubring, "Mathematics and Teacher Training: Plans for a Polytechnic in Berlin," *HSPS 12* (1981): 161-94.

48. "Conferenz zur Berathung von Vorschlägen in Bezug auf Förderung der wissenschaftlichen Mechanik im Sitzungszimmer des Kultusministeriums," First Session, 30 November 1882, copy in SAA 61/Lc 973.

49. Ibid.

50. Cf. Gisela Buchheim, "Die Entwicklung des elektrischen Messwesens und die Gründung der Physikalisch-Technischen Reichsanstalt," *NTM 14* (1977): 16-32.

## Chapter 2. Rift in the foundations

1. Gossler to Heinrich von Boetticher (Reich minister of the interior), 23 April 1884, ZStAP, RI, Nr. 13144/1, Bl. 21-28Rs.

2. The committee included Foerster, Helmholtz, Siemens, Paalzow, and Reuleaux from the older committees and several new recruits: Carl Bamberg and Rudolf Fuess, instrumentmakers; Hans Landolt, professor of chemistry at the Landwirtschaftliche Hochschule in Berlin; H. C. Vogel, director of the Astrophysikalisches Observatorium in Potsdam; and other governmental and military representatives.

3. "Conferenz zur Berathung von Vorschlägen in Bezug auf Förderung der wissenschaftlichen Mechanik im Sitzungszimmer des Kultusministeriums," First Session, 30 November 1882, copy in SAA 61/Lc 973; cf. minutes of the Second Session, "Conferenz zur Berathung . . . ," 7 December 1882, copy in ibid. This and subsequent quotations from individuals at committee or board meetings are given as reported by committee or board secretaries.

4. Second Session, "Conferenz zur Berathung . . . ," ibid.

5. Ibid.

6. Ibid.

7. The committee's report ("Denkschrift betreffend die Begründung eines Instituts . . . Vom 16. Juni 1883" and ff.) and all other public documents concerning the Reichsanstalt's founding are published together as part of the following ordered collection: "Begründung der Vorschläge zur Errichtung einer 'physikalisch-technischen Reichsanstalt' für die experimentelle Förderung der exakten Naturforschung und der Präzisionstechnik. 'Vorbemerkungen' [1-6], 'Aufgaben der ersten (wissenschaftlichen) Abtheilung der physikalisch-technischen Reichsanstalt (Ausgearbeitet von Dr. von Helmholtz)' [6-9], 'Aufgaben der zweiten (technischen) Abtheilung der physikalisch-tech-

nischen Reichsanstalt (Ausgearbeitet von Dr. Foerster)' [10-13], 'Organisationsplan' [14-49, including five Anlagen]"; Beilage: "Denkschrift betreffend
die Begründung eines Instituts für die experimentelle Förderung der exakten
Naturforschung und der Präzisionstechnik (Physikalisch- mechanisches Institut). Vom 16. Juni 1883" [1-22], Beilage I. "Votum des Herrn Geheimen Regierungs-Raths Dr. Werner Siemens (April 1883)" [23-27], Beilage II.
"Votum des Chefs der trigonometrischen Abtheilung der königlichen Landesaufnahme, Herrn Oberstlieutenant Schreiber (Mai 1883)" [29-32], Beilage
III. "Votum des Herrn Geheimen Regierungs-Rathes, Prof. Dr. von Helmholtz (Juni 1883). Ueber die Aufgaben der wissenschaftlichen Abtheilung des
in Aussicht genommenen Physikalisch-Mechanischen Instituts" [33-36], in
*Drucksachen zu den Verhandlungen des Bundesraths des Deutschen Reichs,*
1886, Band 1, Aktenstück No. 50 (Berlin, 1886), 1-49, 1-36, respectively. The
quotations cited are from the "Denkschrift betreffend die Begründung eines
Instituts . . . Vom 16. Juni 1883," 2-3.

8. "Denkschrift betreffend die Begründung eines Instituts . . . Vom 16. Juni
1883," 4-9.

9. Ibid., 12.

10. Beilage III, "Votum des Herrn Geheimen Regierungs-Rathes, Prof. Dr.
von Helmholtz," (note 7), 33-36.

11. Beilage I, "Votum des Herrn Geheimen Regierungs-Raths Dr. Werner
Siemens (April 1883)" (note 7), 23-27; reprinted as "Votum betreffend die
Gründung eines Instituts für die experimentelle Förderung der exakten Natur-
forschung und der Präzisionstechnik," in Siemens's *Wissenschaftliche und
Technische Arbeiten,* 2nd ed., 2 vols. (Berlin, 1889 and 1891), *2,* 568-75.

12. Werner Siemens, *Lebens-Erinnerungen* (Leipzig, 1943), 38-41.

13. Ibid., 79-98.

14. August Kundt, "Gedächtnisrede auf Werner von Siemens," *APAWB*
(1893): 1-21, on 3, reprinted in *PPII,* 111- 23.

15. Jürgen Kocka, *Unternehmensverwaltung und Angestelltenschaft am
Beispiel Siemens 1847-1914: Zum Verhältnis von Kapitalismus und Bürokratie in der deutschen Industrialisierung* (Stuttgart, 1969), 204-6.

16. Peter Czada, *Die Berliner Elektroindustrie in der Weimarer Zeit: Eine
regionalstatistisch-wirtschafthistorische Untersuchung* (Berlin, 1969), 38-43.

17. Siemens, *Lebens-Erinnerungen,* 58.

18. Ibid., 372-73. Cf. Siemens's remarks in "Die Elektrizität im Dienste des
Lebens," *EZ 1* (1880): 16-22, on 16, reprinted in his *Arbeiten 2,* 374-87; and
his "Antrittsrede," *Monatsberichte der Preussischen Akademie der Wissenschaften zu Berlin* (1874): 464-67, reprinted in *PPII,* 106-8.

19. Siemens, *Lebens-Erinnerungen,* 57-58, 262, 368-69, 406; Wilhelm von
Siemens, "Werner Siemens und sein Wirkungsfeld," *NWN 4* (1916): 759-71,
on 760.

20. Siemens, *Arbeiten*, Vol. 1: *Wissenschaftliche Abhandlungen und Vorträge;* Hermann von Helmholtz et al., "Nr. 8. Wahlvorschlag für Werner von Siemens (1816-1892) zum OM. Berlin, 20 Oktober 1873," in *PPI*, 84-86; and Kundt, "Gedächtnisrede auf Siemens."

21. Gustav Mie, "Werner Siemens als Physiker," *NWN 4* (1916): 771-76, on 772.

22. Kundt, "Gedächtnisrede auf Siemens," 21. Cf. E. Warburg, "Ueber den Aufschwung der modernen Naturwissenschaft," *Berichte der Naturforschenden Gesellschaft zu Freiburg i. B. 6* (1892): 18-29, on 22.

23. Heinrich Hertz noted that at the Naturforscherversammlung at Heidelberg in September 1889 "there were at least 100 members there, the best physicists, Helmholtz, Kundt, Kohlrausch, Siemens, etc., almost everybody who has a name" (Hertz, *Erinnerungen, Briefe, Tagebücher,* ed. Johanna Hertz [Leipzig, 1927], 222).

24. Siemens, *Lebens-Erinnerungen*, 262-63, 378-80, 392 (quotes on 262, 378). Cf. Siemens to Henry Villard, 27 February 1887, in *Aus einem reichen Leben: Werner von Siemens in Briefen an seine Familie und an Freunde,* 2nd ed., ed. Friedrich Heintzenberg (Stuttgart, 1953), 330.

25. Siemens, *Lebens-Erinnerungen*, 263-72; Richard Ehrenberg, "Werner Siemens in seiner Bedeutung für die deutschen Volkswirtschaft," *NWN 4* (1916): 823-27.

26. Siemens to Gossler, 7 July 1883, ZStA, Merseburg, Rep. 76 Vb Sekt. 1 Tit. X Nr. 4 Band 1, Bl. 402-4.

27. Ibid.

28. Gossler to Siemens, 13 July 1883, SAA 61/Lc 973.

29. Siemens to Gossler, 12 January 1884, ibid., printed in *Werner Siemens: Ein kurzgefasstes Lebensbild nebst einer Auswahl seiner Briefe,* ed. Conrad Matschoss, 2 vols. (Berlin, 1916), 2, 805-6. Siemens later increased the amount of land offered to the state to 19,800 square meters valued at 566,157 marks (see Chapter 3, "New Quarters, Slow Science"). Siemens had originally intended to bequeath a sizable sum of money to the state for the establishment of a natural scientific research institute; his inheritance now allowed him to accelerate his plan (Siemens, *Lebens-Erinnerungen*, 394-95).

30. Gossler to Siemens, 24 January 1884, SAA 61/Lc 973.

31. Siemens to Spieker, 19 February 1884, ibid.; partially printed in *Werner Siemens,* ed. Matschoss, 2, 810-11.

32. Siemens to Boetticher, 20 March 1884, copy in SAA 61/Lc 973; partially printed in "Begründung der Vorschläge zur Errichtung einer 'physikalisch-technischen Reichsanstalt' für die experimentelle Förderung der exakten Naturforschung und der Präzisionstechnik. 'Vorbemerkungen,'" 2-4 (note 7), reprinted in Siemens, *Arbeiten 2,* 576-80.

33. Gossler to Boetticher, 23 April 1884, ZStAP, RI, Nr. 13144/1, Bl. 28; Boetticher to Burchard (minister of the treasury), 8 May 1884, copy in BAK,

Nr. R2 12375.

34. Boetticher and Burchard to Bismarck, 21 July 1884, ZStAP, Reichs-kanzlei, Nr. 975, Bl. 63-74; Bismarck to Boetticher, 30 July 1884, copy in BAK, Nr. R2 12375; and Boetticher to Siemens, 13 August 1884, SAA 61/Lc 973.

35. "Kommission zur Berathung der Organisation und des Kostenan-schlags für das mechanisch-physikalische Institut, 15 Oktober 1884," SAA 61/Lc 973. In addition to Bamberg, Foerster, Fuess, Helmholtz, and Siemens, the committee consisted of Interior officials Lieber, Spieker, Wehrenpfen-ning, and Weymann.

36. Minutes of committee and subcommittee meetings, "Kommission zur Berathung," October 1884 to December 1885, ibid.; and ZStAP, RI, Nr. 13144/8, 13144/9.

37. Minutes, subcommittee meeting, 3 November 1884, ZStAP, RI, Nr. 13144/9, Bl. 66-74.

38. Emil DuBois-Reymond, "Gedächtnisrede auf Hermann von Helmholtz (1821-1894)," *APAWB* (1896): Gedächtnisrede 2, 1-50, reprinted in *PPII*, 68-99, on 96.

39. Siemens to Mayor Rosenthal, 17 September 1884, in *Werner Siemens,* ed. Matschoss, 2, 821.

40. Siemens to Helmholtz, 7 September 1885, SAA 61/Lc 973, partially printed in *Aus einem reichen Leben,* ed. Heintzenberg, 322-23, and in *Werner Siemens,* ed. Matschoss, 2, 855-56.

41. Siemens to Alexander Siemens, 12 March 1886, SAA 2/Li 536. Bis-marck had no interest in science; see A. J. P. Taylor, *Bismarck: The Man and the Statesman* (New York, 1967), 136.

42. Siemens to Rühlmann, 20 February 1886, SAA 61/Lc 973, printed in *Werner Siemens,* ed. Matschoss, 2, 878-80, and cited in Gisela Buchheim, "Initiativen zur Gründung der Physikalisch-Technischen Reichsanstalt (1887)," *NTM 11* (1974): 33-43, on 37. Cf. Siemens, *Lebens-Erinnerungen,* 386.

43. Siemens to Hammacher, 22 February 1886, SAA 2/Li 548.

44. Siemens to [Hermann?] Wedding, chairman of the Verein's Technical Committee, 1 July 1886, ibid.

45. "Bitte des Vorstandes des Vereines Deutscher Ingenieure, betreffend die Errichtung einer 'physikalisch-technischen Reichsanstalt,'" to the Bun-desrat, 12 October 1886, ZStAP, RI, Nr. 13144/2, Bl. 12-16Rs, printed in *Zeit-schrift des Vereins deutscher Ingenieure 30* (1886): 890-91. Cf. Buchheim, "Initiativen," 41; and, on the Verein, Theodore Peters, *Geschichte des Vereines deutscher Ingenieure in zeitlicher Auseinanderfolge* (Berlin, 1912).

46. Deutsche Gesellschaft für Mechanik und Optik to Bismarck, Novem-ber 1886, ZStAP, RI, Nr. 13144/2, Bl. 43-45, signed by Bamberg, Fuess, H. Haensch, Max Hildebrand, L. A. Steinheil, Karl Zeiss, et al.

47. Ministerium für Handel und Gewerbe; Ministerium der öffentlichen Arbeiten; Ministerium der geistlichen, Unterrichts- und Medicinal-Angelegenheiten to Bismarck, 26 December 1886, ibid., Bl. 70-71; report of the ministerial committee, 9 December 1886, copy, ibid., Bl. 72-73. Ultimately two additional (making a total of four) board seats were allotted to engineers; the board also agreed to meet the engineers' demands as far as personnel and money allowed (see the section on "Compromises," this chapter, and "The Formation of a Scientific Bureaucracy," Chapter 3).

48. Foerster, *Lebenserinnerungen*, 192 (quote). Only the Bavarian government denied support to the Reichsanstalt, on the grounds that the support of science was the exclusive, legal responsibility of the individual states, not of the Reich (*Protokolle über die Verhandlungen des Bundesraths des Deutschen Reichs* [Berlin, 1886], 25th Sitzung, 4 June 1886, §348, p. 264; Karl Griewank, "Aus den Anfängen gesamtdeutscher Wissenschaftspflege," in *Volkstum und Kulturpolitik: Eine Sammlung von Aufsätzen Gewidmet Georg Schreiber zum fünfzigsten Geburtstage*, ed. H. Konen and J. P. Steffes [Cologne, 1932], 208-36, on 234; and Gisela Buchheim, "Reichstagsdebatten über die Gründung der Physikalisch-Technischen Reichsanstalt," *NTM 12* [1975]: 1-13, on 2, 4).

49. *SBVR 93* (1886-87), 297-311, on 297-98; Buchheim, "Reichstagsdebatten," 5.

50. Georg Siemens, *Geschichte des Hauses Siemens*, 3 vols. (Munich, 1947), *1*, 167.

51. "Vereins-Angelegenheiten," *EZ 8* (1887): 2-3, signed by Foerster, Rühlmann, Siemens et al.; Foerster, *Die Physikalisch-Technische Reichsanstalt: Ein Beitrag zur Verständigung* (Berlin, 1887). Foerster, *Lebenserinnerungen*, 193, and Buchheim, "Reichstagsdebatten," 6-8.

52. Foerster, *Lebenserinnerungen*, 193; cf. ibid., 129, and Georg Siemens, *Geschichte des Hauses Siemens*, *1*, 167.

53. *SBVR 93*, 298-310, quotes on 298, 308.

54. Ibid., 301-2.

55. Ibid., 311; Buchheim, "Reichstagsdebatten," 5, 12.

56. *SBVR 95* (1887), 299.

57. Ibid.; and ibid., Aktenstück Nr. 31, Anlageband 1, *97* (1887), 294.

58. Kohlrausch to Siemens, 10 April 1887, SAA 61/Lc 973.

59. *Geschäftsordnung für die Physikalisch-Technische Reichsanstalt* (Berlin, 1888), 1. Cf. "Anlage B zum Protokoll vom 9. August 1887," Board Session 3, 9 August 1887, SAA 61/Lc 973.

60. "Kommission zur Vorberathung," "I. Sitzung der Subkommission," 24 October 1884, in ZStAP, RI, Nr. 13144/9, Bl. 38-42.

61. Friedrich Kohlrausch, "Denkschrift über die Tätigkeit der Physikalisch-Technischen Reichsanstalt von Anfang 1900 bis Ende 1903," *SBVR 206* (1903-4): 1139.

62. Cf. Board Session 3, 9 August 1887, SAA 61/Lc 973, where Siemens noted "that in yesterday's discussion the idea may have taken hold that the Reichsanstalt might develop into an institute competitive with private institutes or into a privileged institute of discovery [*Erfindungsinstitut*]." Both possibilities," he admonished, "are unconditionally excluded [from the purposes of the Reichsanstalt]. Moreover, there must not even be the *appearance* that such goals are being pursued." Siemens made a similar declaration in "Anlage B zum Protokoll vom 9. August 1887," ibid. See also Board Session 4, 10 August 1887, ibid.

63. *Geschäftsordnung*, 1.

64. Board Session 1, 6 August 1887, SAA 61/Lc 973.

65. Ibid.

66. Ibid.

67. Ibid. The committee consisted of Abbe, Bamberg, Foerster, Grashof, and A. Mensing and was aided by the counsel of Loewenherz. One of its first acts was to reject Helmholtz's appointee to head the Technical Section's Mechanics Laboratory.

68. Board Session 3, 9 August 1887, ibid.

69. Ibid.

70. "Anlage A zum Protokoll vom 10. August 1887. Arbeitsplan für die Physikalisch-technische Reichsanstalt für die Zeit vom 1. Oktober 1887 bis 1. April 1888," ibid.

71. Board Session 4, 10 August 1887, ibid.

72. Ibid.

73. Board Session 5, 26 March 1888, ibid. The committee consisted of Abbe, Clausius, Wilhelm Dietrich, Foerster, Helmholtz, Repsold, and Hugo von Seeliger.

74. *Geschäftsordnung*, 1-2.

## Chapter 3. Between charisma and bureaucracy

1. Leo Koenigsberger, *Hermann von Helmholtz*, 3 vols. (Braunschweig, 1902-3), *3*, 96-97.

2. Siemens to Helmholtz, 7 September 1885, partially printed in *Werner Siemens: Ein kurzgefasstes Lebensbild nebst einer Auswahl seiner Briefe*, ed. Conrad Matschoss, 2 vols. (Berlin, 1916), *2*, 855-56.

3. Unless otherwise noted, the following account of Helmholtz's life and work is based on Emil DuBois-Reymond, "Gedächtnisrede auf Hermann von Helmholtz (1821- 1894)," *APAWB* (1896): Gedächtnisrede 2, 1-50, reprinted in *PPII*, 68-99; Koenigsberger, *Hermann von Helmholtz;* E. Warburg, "Helmholtz als Physiker," in E. Warburg, M. Rubner, and M. Schlick, *Helmholtz als Physiker, Physiologe und Philosoph: Drei Vorträge* (Karlsruhe, 1922), 3-14; and R. Steven Turner, "Helmholtz, Hermann von," *DSB 6* (1972): 241-53.

4. Thomas S. Kuhn, "Energy Conservation as an Example of Simultaneous Discovery," in *Critical Problems in the History of Science*, ed. Marshall Clagett (Madison, Wis., 1959), 321-56, reprinted in Kuhn's *The Essential Tension: Selected Studies in Scientific Tradition and Change* (Chicago, 1977), 66-104.

5. Rosalie Braun-Artaria, *Anna von Helmholtz: Ein Erinnerungsblatt* (n.p., 14 December 1899), p. 5 of unpaginated pamphlet.

6. Rowland to his mother, January 1876, Henry A. Rowland Papers, MS. 6, Special Collections, Milton S. Eisenhower Library, The Johns Hopkins University; cf. Rowland to his sister Arura, 21 February 1876, ibid.

7. Michael Pupin, *From Immigrant to Inventor* (New York, 1960), 230.

8. Max Planck, "Die Physikalische Gesellschaft um 1890," *PB 6* (1950): 433-34, on 433.

9. Eugen Goldstein, "Aus vergangenen Tagen der Berliner Physikalischen Gesellschaft," *NWN 13* (1925): 39-45, on 40.

10. Wilhelm Ostwald, *Lebenslinien: Eine Selbstbiographie*, 3 vols. (Berlin, 1926-27), *1*, 187; *2*, 370.

11. Hermann von Helmholtz, "Ueber das Verhältniss der Naturwissenschaften zur Gesammtheit der Wissenschaft: Akademische Festrede gehalten zu Heidelberg beim Antritt des Prorectorats 1862," in Helmholtz's *Vorträge und Reden*, 4th ed., 2 vols. (Braunschweig, 1896), *1*, 159-85, quotes on 180-82.

12. Anna von Helmholtz to her sister Ida, 3 June 1888, in *Anna von Helmholtz: Ein Lebensbild in Briefen*, ed. Ellen von Siemens- Helmholtz, 2 vols. (Berlin, 1929), *2*, 9-10.

13. DuBois-Reymond, "Gedächtnisrede auf Helmholtz," 96.

14. Helmholtz to Hertz, 15 December 1888, in Heinrich Hertz, *Erinnerungen, Briefe, Tagebücher*, ed. Johanna Hertz (Leipzig, 1927), 261; and Warburg, "Helmholtz als Physiker," 12.

15. "II. Sitzung der Subkommission," 25 October 1884, ZStAP, RI, Nr. 13144/9, Bl. 51-58.

16. "Vierte Sitzung der Gesamt-Kommission," 3 November 1885, ibid., Bl. 137-47.

17. "Fünfte Sitzung der Gesamt-Kommission," 13 November 1885, ibid., Bl. 148-52.

18. Burchard to Boetticher, 8 December 1885, BAK, R2 12375.

19. Gossler to Bismarck, 15 January 1886, ibid.

20. See, e.g., *HEDR* (1893-94), 4-5, 32-33, 40-41, on 4.

21. Burchard to Boetticher, 23 January 1886, BAK, R2 12375.

22. Boetticher to Burchard, 29 January 1886 and enclosure to same, ibid.

23. Hermann von Helmholtz, "Spezifizierung meiner dienstlichen Einnahmen" (copy) 4 April 1887; and Boetticher to Jacobi (Burchard's successor as minister of the treasury), 18 April 1887, ibid.

24. Boetticher to Jacobi, ibid.

25. Jacobi to Boetticher, 24 April 1887, ibid.

26. Reichskanzlei to Boetticher, 9 June 1887, ibid., R 43 F/2365, 3/8; and Boetticher's report to "Seiner Durchlaucht," 6 June 1887, ibid.

27. Hermann to Anna, 7 August 1887, in *Anna von Helmholtz: Ein Lebens-bild*, ed. Siemens-Helmholtz, *1*, 307-8.

28. See Helmholtz's correspondence with various Interior officials in ZStAP, RI, Nr. 13144/3.

29. On the former see Helmholtz to Max Harrwitz, 11 February 1892, SPK, Slg. Darmstaedter, F i a 1847: Helmholtz; on the latter, Helmholtz to the Gesellschaft für Erdkunde, 20 November 1893, ibid., F i a 1847 (2): Helmholtz.

30. [Hermann von Helmholtz], "Denkschrift über die bisherige Thätigkeit der Physikalisch-Technischen Reichsanstalt," *SBVR 122* (1890-91): 1368-80, on 1370.

31. DuBois-Reymond, "Gedächtnisrede auf Helmholtz," 97. Cf. Johannes Pernet, "Hermann von Helmholtz. 31. August 1821 bis 8. September 1894: Ein Nachruf," *Neujahrsblatt der Naturforschenden Gesellschaft in Zürich 97* (1895): 1-36, on 31-32; Gustav Wiedemann, "Hermann von Helmholtz, Wissenschaftliche Abhandlungen," *AP 290* (1895): i-xxiv, on xxi; Koenigsberger, *Helmholtz, 3*, 1-2, 31, passim; and Warburg, "Helmholtz als Physiker," 10.

32. Wiedemann, "Helmholtz," xxi.

33. Pernet, "Hermann von Helmholtz," 32; cf. Koenigsberger, *Helmholtz, 3*, 1-2, and Wilhelm Wien, "Ein Rückblick," in *Wilhelm Wien: Aus dem Leben und Wirken eines Physikers*, ed. Karl Wien (Leipzig, 1930), 1-50, on 14.

34. Cf. Hertz, entry for 17 November 1878, in his *Erinnerungen*, ed. Hertz, 75; and Rowland to his mother, January 1876, Rowland Papers.

35. Wien, "Ein Rückblick," 18; and, for *"Forschergeist,"* Hermann Ebert, *Hermann von Helmholtz* (Stuttgart, 1949), 170-71.

36. Emil Fischer, *Aus meinem Leben* (Berlin, 1922), 142 (quote).

37. Braun-Artaria, *Anna von Helmholtz*, p. 7 of unpaginated pamphlet; Marie von Bunsen, *Zur Erinnerung an Frau Anna von Helmholtz* (n.p., 10 December 1899), p. 7 of unpaginated pamphlet; and *Anna von Helmholtz: Ein Lebensbild*, ed. Siemens-Helmholtz, *2*, 10-11. Cf. Pupin, *From Immigrant to Inventor*, 231.

38. *SBVR 284* (1912), 914.

39. Wilhelm Spielmann, *Handbuch der Anstalten und Einrichtungen zur Pflege von Wissenschaft und Kunst in Berlin* (Berlin, 1897), 4-5; and F. A. Medicus, *Das Reichsministerium des Innern* (Berlin, 1940), 12-16.

40. *Geschäftsordnung für die Physikalisch-Technische Reichsanstalt* (Berlin, 1888), 2-3 (hereafter cited as *Geschäftsordnung*).

41. Board Session 1, 6 August 1887, SAA 61/Lc 973; and Board Session 1, 7 March 1903, SAA 4/Lk 28.

42. Board Session 1, 6 August 1887, SAA 61/Lc 973.

43. *Geschäftsordnung*, 3. Unless otherwise noted, the term "president" hereafter refers to the president of the Reichsanstalt.

44. Board Session 2, 16 March 1891, SAA 61/Lc 973.

45. Posadowsky-Wehner to (Kaiser) Wilhelm/Reichskanzler, 18 November 1897; and Wilhelm to Hohenlohe-Schillingsfürst (?), 29 November 1897, ZStAP, RI, Nr. 13147/2, Bl. 98-105.

46. Helmholtz to Weymann, 25 May 1887, ibid., Nr. 13146/1, Bl. 15-24 (quote on 19).

47. Ibid.; and "A. Kundt," *ZI 14* (1894): 258.

48. Helmholtz to Weymann, 25 May 1887, Bl. 21.

49. Max Planck, "Persönliche Erinnerungen aus alten Zeiten," in Planck's *Vorträge und Erinnerungen*, 5th ed. (Stuttgart, 1949), 1-14, on 9.

50. H. Rubens, "A. Paalzow, Gedächtnisrede," *VDPG 10* (1908): 451-62, on 461.

51. F. Fraunberger, "Quincke, Georg Hermann," *DSB 11* (1975): 241-42, on 242; and Weymann to Boetticher, 12 November 1892, ZStAP, RI, Nr. 13146/1, Bl. 209-10.

52. F. Kohlrausch, "Gustav Wiedemann: Nachruf," *VDPG 1* (1899): 155-67, reprinted in *GAFK, 1,* 1064-76, on 1067. On Wiedemann's work for the German Electrical Standards Committee see Gisela Buchheim, "Die Entwicklung des elektrischen Messwesens und die Gründung der Physikalisch-Technischen Reichsanstalt," *NTM 14* (1977): 16-32.

53. Weymann to Boetticher, 21 February 1889, ZStAP, RI, Nr. 13146/1, Bl. 98-99.

54. "Kommission zur Vorberathung des Planes für eine physikalisch-technische Reichsanstalt," "Dritte Sitzung der Gesamt-Kommission," 2 November 1885, ZStAP, RI, Nr. 13144/9, Bl. 128-36.

55. One point favoring Quincke, for example, was that he taught in Baden (Weymann to Boetticher, 12 November 1892, ZStAP, RI, Nr. 13146/1, Bl. 209-10).

56. For their services, members received only reimbursement for their travel expenses (*Geschäftsordnung*, 3).

57. Weymann to Boetticher, 15 July 1892, ZStAP, RI, Nr. 13146/1, Bl. 197-203.

58. The president did, however, require post facto approval for any changes exceeding one-quarter of the total annual funds at his disposal (*Geschäftsordnung*, 3-4).

59. There were only two instances – one in 1906 and one in 1914 – in which Reichsanstalt employees attempted to patent inventions on the basis of work that allegedly originated from official Reichsanstalt work. The second of these attempts led the board, which was ever vigilant about conflicts of interest between the Reichsanstalt and private industry, to prohibit completely the appli-

cation for or selling of patents by Reichsanstalt workers. (See W. C. Heraeus to Emil Warburg [copy], 17 November 1906; Warburg to Heraeus, 14 December 1906; "Besprechung der §.37 der Geschäftsordnung der Physikalisch-Technischen Reichsanstalt"; Board Session 3, 15 March 1907; Board Session 2, 10 March 1914, all in BA 240, PTB-IB; and cf. *Geschäftsordnung für die Physikalisch-Technische Reichsanstalt* [Berlin, 1917], 7.)

60. *Geschäftsordnung,* 3-9 (quotes on 6, 9).

61. Board Session 2, 16 March 1891, SAA 61/Lc 973.

62. Hermann von Helmholtz, "Vorrede," *WAPTR 1* (1894): n.p.

63. Board Session 3, 17 March 1891, SAA 61/Lc 973.

64. See the title pages of the *Annalen 237* (1877); *290* (1895); *306* (1900); *326* (1906); *337* (1910); *367* (1920).

65. Planck to Wien, 23 June 1907, SPK, Nachlass W. Wien: Max Planck.

66. Planck to Wien, 1 July 1907, ibid. On Planck's editing of the *Annalen,* see Lewis Pyenson, *The Young Einstein: The Advent of Relativity* (Bristol, 1985), 194-214; and Christa Jungnickel and Russell McCormmach, *Intellectual Mastery of Nature: Theoretical Physics from Ohm to Einstein,* 2 vols., Vol. 2: *The Now Mighty Theoretical Physics 1870-1925* (Chicago, 1986), 309-12.

67. Board Session 2, 16 March 1891, SAA 61/Lc 973.

68. "An unsere Leser," *ZI 1* (1881): n.p.

69. Leopold Loewenherz, "Betrifft die *Zeitschrift für Instrumentenkunde,*" 22 December 1888, in ZStAP, RI, Nr. 13150/1, Bl. 22-27.

70. Deutsche Gesellschaft für Mechanik und Optik to Bismarck, November 1886, ZStAP, RI, Nr. 13144/2, Bl. 43-45. Fuess and Bamberg later explicitly stated that they meant the *Zeitschrift für Instrumentenkunde;* see their letter to Weymann, 6 April 1887, ibid., Bl. 126-30.

71. R. Fuess, E. Abbe, and W. Foerster, "Anlage B zum Protokoll vom 23. März 1888. Antrag, betreffend Beihülfe an die Zeitschrift für Instrumentenkunde," SAA 61/Lc 973.

72. Board Session 4, 10 August 1887, ibid.

73. Fuess, Abbe, and Foerster, "Anlage B"; and Board Session 3, 23 March 1888, ibid.

74. Title page, *ZI 8* (1888); and Die Herausgeber, ibid., 1.

75. Das Redaktionskuratorium, "An unsere Leser!" *ZI 15* (1895): 1.

76. Board Session 2, 16 March 1891, SAA 61/Lc 973; and Helmholtz, "Vorrede."

77. Helmholtz to Boetticher, 9 October 1888, ZStAP, RI, Nr. 13144/3, Bl. 30-41, on 31-32.

78. Rudolf Denke, "Das Siemens-Grundstück: Ein Beitrag zu seiner Geschichte," *Der Bär von Berlin. Jahrbuch des Vereines für die Geschichte Berlins,* 6th series (1956): 108- 34, on 108; K. Haemmerling, *Charlottenburg: Das Lebensbild einer Stadt, 1905-1955* (Berlin, 1955), 113; and F. Leyden,

"Charlottenburg um 1880," in *Berlin,* ed. J. J. Haesslin (Munich, 1971), 322-24.

79. Hertz to parents, 22 June 1880, in Hertz's *Erinnerungen,* ed. Hertz, 97.

80. F. Leyden, *Gross-Berlin: Geographie der Weltstadt* (Breslau, 1933), 207.

81. Haemmerling, *Charlottenburg,* 18; Leyden, "Charlottenburg um 1880," 324.

82. "Taxé des von dem Kaiserlichen Geheimen Regierungsrathes Herrn Dr. Siemens dem deutschen Reiches übereigneten Grundstückes," 16 November 1885, signed Bohl, Königlicher Baurath, in "Kommission zur Vorberathung des Planes für eine physikalisch-technische Reichsanstalt," June 1883 to November 1885, ZStAP, RI, Nr. 13144/8, Bl. 259-62.

83. "III. Sitzung der Subkommission," Berlin, 29 October 1884, SAA 61/Lc 973. See also Kohlrausch's report of 9 August 1887 and Board Session 4, 25 March 1889, ibid.

84. *HEDR* (1892-93), 4-5, 32-35, 40-41, on 40-41.

85. Siemens to Spieker, 19 February 1884, SAA 61/Lc 973, partially reprinted in *Werner Siemens,* ed. Matschoss, *2,* 810-11.

86. Siemens to Weymann, 14 May 1887, SAA 61/Lc 973.

87. See Spieker to Siemens, 17 February 1884; Spieker to Siemens, 20 February 1884; Gossler to Siemens, 29 December 1884; and Helmholtz to Spieker, 28 March 1885, all in ibid.; Helmholtz to ?, 12 January 1885, and (two letters of) Helmholtz to Spieker, 28 March 1885, copies in "Kommission zur Vorberathung," ZStAP, RI, Nr. 13144/8, Bl. 150-51, 161-62, 178, respectively.

88. "VIII. Sitzung der Subkommission, Berlin, 1. Mai 1885"; and "IX. Sitzung der Subkommission, Berlin, 13. Juli 1885," SAA 61/Lc 973.

89. Siemens to Boetticher, 21 July 1886, ZStAP, RI, Nr. 13144/2, Bl. 4-6; and Weymann to Maybeth, 3 August 1886, ibid., Bl. 8-10.

90. [Helmholtz], "Denkschrift über die bisherige Thätigkeit der Reichsanstalt," 1369; and Johannes Pernet, "Ueber die physikalisch-technische Reichsanstalt zu Charlottenburg und die daselbst ausgeführten elektrischen Arbeiten," *Schweizerische Bauzeitung 18* (1891): 1-6, on 3.

91. On the (extended) length of time needed to win approval for and to construct academic physics institutes, see David Cahan, "The Institutional Revolution in German Physics, 1865- 1914," *HSPS 15:2* (1985): 1-65, on 17-19.

92. See Siemens's remarks in "IV. Sitzung der Subkommission, Berlin, den 3.11.1884," ZStAP, RI, Nr. 13144/9, Bl. 69-74; and in "VIII. Sitzung der Subkommission, Berlin, 1. Mai 1885," SAA 61/Lc 973.

93. "Erste Sitzung der Kommission zur Berathung der Organisation und des Kostenanschlags für das mechanisch-physikalische Institut, abgehalten im Reichsamt des Innern, Mittwoch den 15. Oktober 1884 Morgens 10 Uhr," SAA 61/Lc 973.

94. Helmholtz to Boetticher, 7 November 1887, ZStAP, RI, Nr. 13144/2, Bl. 255; "Bericht über die Einrichtung und die bisherige Thätigkeit der physikalisch-technischen Reichsanstalt," 17 January 1888, SAA 61/Lc 973; [Helmholtz], "Denkschrift über die bisherige Thätigkeit der Reichsanstalt," 1369; and "Die Thätigkeit der Physikalisch-Technischen Reichsanstalt bis Ende 1890," *ZI 11* (1891): 149-70, on 150.

95. Hermann von Helmholtz, Reichsanstalt agenda for 1889/90: "Anlage 1 zum Protokoll vom 22. März 1889," SAA 61/Lc 973; Otto Lummer, "Ueber die Ziele und die Thätigkeit der Physikalisch-Technischen Reichsanstalt," *Verhandlungen des Vereines zur Beförderung des Gewerbfleisses 73* (1894): 151- 84, on 159.

96. "Begründung der Vorschläge zur Errichtung einer 'physikalisch-technischen Reichsanstalt' für die experimentelle Förderung der exakten Naturforschung und der Präzisionstechnik. 'Vorbemerkungen' [1-6], 'Aufgaben der ersten (wissenschaftlichen) Abtheilung der physikalisch-technischen Reichsanstalt (Ausgearbeitet von Dr. von Helmholtz)' [6-9], 'Aufgaben der zweiten (technischen) Abtheilung der physikalisch-technischen Reichsanstalt (Ausgearbeitet von Dr. Foerster)' [10-13], 'Organisationsplan' [14-49, including 5 Anlagen] . . . ," in *Drucksachen zu den Verhandlungen des Bundesraths des Deutschen Reichs,* Band 1, 1886, Aktenstück No. 50, 1-49, on 16-18; and "IV. Gebäude für die Verwaltungsbehörden des Deutschen Reiches. 11. Die Physikalisch-Technische Reichsanstalt in Charlottenburg," in *Berlin und seine Bauten,* ed. Architekten-Verein zu Berlin und Vereinigung Berliner Architekten, 3 vols. in 2 (Berlin, 1896), *2,* 80-84, on 82-83.

97. "IV. Gebäude für die Verwaltungsbehörden," *2, 81.*

98. "Begründung der Vorschläge," 18.

99. Paul Spieker, "E. Sternwarten und andere Observatorien," in *Handbuch der Architektur,* ed. Joseph Durm et al. (Darmstadt, 1888), Teil 4, Halbband 6, Heft 2, 474-567, on 564-66; Pernet, "Ueber die physikalisch-technische Reichsanstalt," 2; and "IV. Gebäude für die Verwaltungsbehörden," *2,* 83.

100. "Begründung der Vorschläge," 17.

101. This description of the *Observatorium* is based on Pernet, "Ueber die physikalisch-technische Reichsanstalt," 2-3; Ernst Hagen and Karl Scheel, "Die Physikalisch-Technische Reichsanstalt," *Ingenieurwerke in und bei Berlin: Festschrift zum 50 Jährigen Bestehen des VDI,* ed. Verein Deutscher Ingenieure (Berlin, 1906), 60-67, on 60-65; and "IV. Gebäude für die Verwaltungsbehörden," *2,* 82-83.

102. Sir Douglas Galton, "On the Reichsanstalt, Charlottenburg, Berlin," *Report of the Sixty-fifth Meeting of the British Association for the Advancement of Science Held at Ipswich in September 1895 65* (1895): 606-8, quote on 606; Helmholtz to Boetticher, 1 March 1888 and Staatssekretär des Innern to Helmholtz, 19 March 1888, ZStAP, RI, Nr. 13144/2, Bl. 260-62, 265, respectively; Board Session 1, 20 March 1888, SAA 61/Lc 973; and letters of 4

October 1894, 23 and 29 November 1894, 24 December 1894, 22 January 1895, and 23 February 1895 between Anna von Helmholtz and various officials in the Ministry of the Interior, in ZStAP, RI, Nr. 13144/3, Bl. 117-20, 123-26, 128, 147, 161-62, respectively.

103. "IV. Gebäude für die Verwaltungsbehörden," 2, 83-84.

104. *HEDR* (1887-88), 2-3, 32-35, on 33.

105. "IV. Gebäude für die Verwaltungsbehörden," 2, 81.

106. Ibid.

107. Extract from Otto Warburg's taped autobiography, quoted in Hans Krebs (with Roswitha Schmid), *Otto Warburg: Zellphysiologe, Biochemiker, Mediziner, 1883-1970* (Stuttgart, 1979), 94. Cf. A. G. Webster, "A National Physical Laboratory," *Pedagogical Seminary 2* (1892): 90-101, on 98.

108. *HEDR* (1887-88), 33.

109. Ibid.; and "IV. Gebäude für die Verwaltungsbehörden," 2, 81-82.

110. *HEDR* (1887-88), 33; *HEDR* (1891-92), 4-5, 30-37, on 36; Pernet, "Ueber die physikalisch-technische Reichsanstalt," 3; Hagen and Scheel, "Die Physikalisch-Technische Reichsanstalt," 61-62; and "IV. Gebäude für die Verwaltungsbehörden," 2, 82.

111. *HEDR* (1887-88), 33; Henry S. Carhart, "The Imperial Physico-Technical Institution in Charlottenburg," *TAIEE 17* (1900): 555-83, on 562.

112. During the early years the Reichsanstalt did not always or consistently use the term "laboratory" when referring to a group of its scientists who together concentrated on a given field (e.g., electricity); instead, it sometimes used the term "work group."

113. See Heinrich Hertz to ?, 26 August 1891, DM 3217.

114. *HEDR* (1892-93), 32; and *HEDR* (1893-94), 32.

115. *J. C. Poggendorff's Biographisch-Literarisches Handwörterbuch zur Geschichte der exacten Wissenschaften. . . 3* (1898): 1336 (hereafter cited as *Poggendorff* [title varies]); *Poggendorff 6* (1923-31): 2645; and Weymann to Siemens, 22 July 1890, SAA 61/Lc 973.

116. [Ernst Hagen], "5ter Bericht über die Thätigkeit der Physikalisch-Technischen Reichsanstalt. (Dezember 1892 bis Februar 1894)," *ZI 14* (1894): 261-79, 301-16, on 261-63 (hereafter cited as "Thätigkeit").

117. Board Session 2, 18 March 1892, SAA 61/Lc 973.

118. Lummer, "Ueber die Ziele," 163.

119. *Poggendorff 6* (1923-31): 1216-17.

120. Board Session 4, 25 March 1889, SAA 61/Lc 973.

121. See, e.g., Der Vorsitzender des Elektrotechnischen Vereins to Durchlauchtigster Fürst, Gnädigster Fürst und Herrn (RI), 27 February 1886, ZStAP, RI, Nr. 13158/2, Bl. 154-59.

122. H. v. Helmholtz, "Ueber die Berathungen des Pariser Kongresses, betreffend die elektrischen Masseinheiten," *EZ 2* (1881): 482-89, on 482; Board Session 3, 17 March 1891, SAA 61/Lc 973; "Die Thätigkeit der Physi-

kalisch-Technischen Reichsanstalt in den Jahren 1891 und 1892," *ZI 13* (1893): 113-40, on 117; and Helmholtz to Boetticher, 21 February 1893, ZStAP, RI, Nr. 13159/3, Bl. 135-41.

123. "Thätigkeit," 263-66; E. Warburg, "Die Physikalisch-Technische Reichsanstalt in Charlottenburg," *Zeitschrift des Oesterreichischen Ingenieur- und Architekten-Vereines 60* (1908): 513-17, 529-33, on 529-31; and K. Scheel, "Die Physikalisch-Technische Reichsanstalt in Charlottenburg," *Akademische Rundschau 1* (1913): 221-27, on 224.

124. Pernet, "Ueber die physikalisch-technische Reichsanstalt," 3; and Lummer, "Ueber die Ziele," 170-71.

125. Hydrographisches Amt des Reichs-Marine-Amts to Helmholtz, 13 November 1891, BA 658, PTB-IB.

126. Fritz Reiche, "Otto Lummer," *PZ 27* (1926): 459-67; [Erich Waetzmann and Fritz Reiche], *Akademische Reden: Zum Gedächtnis an Otto Lummer gehalten in Breslau 1925* (Braunschweig, 1928); E. Buchwald, "Erinnerung an Otto Lummer," *PB 6* (1950): 313-16; Clemens Schaefer, "Otto Lummer zum 100. Geburtstag," *PB 16* (1960): 373-81; and Armin Hermann, "Lummer, Otto Richard," *DSB 8* (1973): 551-52. On the hiring of Robert von Helmholtz at the Reichsanstalt, see Board Session 3, 23 March 1889, SAA 61/Lc 973. For the quote see Lummer to [Leo Koenigsberger ?], 11 January 1902, SPK, Slg. Darmstaedter F 2 c 1890.

127. "Biographische Notizen," *WAPTR 1* (1894): n.p.; and W. Steinhaus, "Ernst Gumlich zum Gedächtnis," *ZTP 11* (1930): 129-31.

128. Hans Kangro, "Wien, Wilhelm Carl Werner Otto Fritz Franz," *DSB 14* (1976): 337-42.

129 F. Henning, "Ferdinand Kurlbaum † [Obituary]," *ZTP 8* (1927): 525-27, on 525.

130. "Thätigkeit," 266-71.

131. Board Session 1, 20 March 1889, SAA 61/Lc 973; and Weymann to Boetticher, 28 May 1889, ZStAP, RI, Nr. 13159/3, Bl. 30-31. See also O. Lummer and E. Brodhun, "Photometrische Untersuchungen. I. Ueber ein neues Photometer," *ZI 9* (1889): 41-50, on 42.

132. Vorstand des Deutschen Vereins von Gas- und Wasserfachmänner to RI, 3 August 1888, ZStAP, RI, Nr. 13158/2, Bl. 185-90.

133. Helmholtz to Boetticher, 12 November 1888, ibid., Bl. 193- 96.

134. Board Session 2, 21 March 1888; and Board Session 2, 22 March 1889, SAA 61/Lc 973.

135. O. Lummer and E. Brodhun, "Photometrische Untersuchungen. IV. Die photometrische Apparate der Reichsanstalt für den technischen Gebrauch," *ZI 12* (1892): 41-50, on 41; and Lummer, "Ueber die Ziele," 174-82 (quote on 174).

136. H. Helmholtz, "Die Beglaubigung der Hefnerlampe. (Mittheilung aus der Physikalisch-Technischen Reichsanstalt)," *ZI 13* (1893): 257-67; and note

189 below.

137. "Thätigkeit in den Jahren 1891 und 1892," 120.

138. Lummer, "Ueber die Ziele," 176-80.

139. O. Lummer and F. Kurlbaum, "Bolometrische Untersuchungen," *AP* 282 (1892): 204-24, on 204-5. In 1892 Lummer and Kurlbaum constructed a greatly improved bolometer capable of yielding highly precise comparisons of two heat or light sources – e.g., two electric lamps – with one another. Their goal was "to compare the light radiation of a flame directly with the radiation of a constant heat source and to base the unit of luminous intensity on an absolute unit of heat-radiation" (O. Lummer and F. Kurlbaum, "Ueber die Herstellung eines Flächenbolometers," *ZI 12* [1892]: 81-89, on 81). For further remarks on the industrial uses of the Reichsanstalt's work on the unit of luminous intensity see "Thätigkeit in den Jahren 1891 und 1892," 114, 120

140. Lummer and Kurlbaum, "Bolometrische Untersuchungen," 208.

141. "Thätigkeit," 266.

142. O. Lummer and F. Kurlbaum, "Bolometrische Untersuchung für eine Lichteinheit," *SB* (1894): (I) 229-38, on 231, 238.

143. "Thätigkeit," 266.

144. Board Session 2, 16 March 1891, SAA 61/Lc 973; and "Thätigkeit in den Jahren 1891 und 1892," 122. On Landolt's general interest in advancing precision mechanics and optics see O. Schönrock, "Hans Heinrich Landolt," *ZI 30* (1910): 92-96, on 92.

145. Schönrock, "Landolt," 94-96.

146. J. H. van't Hoff, "Gedächtnisrede auf Hans Heinrich Landolt," *APAWB: Physikalisch-mathematische Klasse* (1910): Gedächtnissrede 2, 1-13, on 8, reprinted in *PPII*, 130-37.

147. "Thätigkeit," 270.

148. "Thätigkeit in den Jahren 1891 und 1892," 114. Cf. Wiedemann, "Helmholtz," xxi.

149. [Friedrich Kohlrausch], "Denkschrift über die Thätigkeit der Physikalisch-Technischen Reichsanstalt vom Anfang des Jahres 1893 bis Ostern 1895," *SBVR 151* (1895-97): 373-91, on 373.

150. [Helmholtz], "Denkschrift über die bisherige Thätigkeit der Reichsanstalt," 1369; and Hans Ebert, "Baugeschichte und Wissenschaftsentwicklung: Zur Geschichte der TH/TU Berlin," *Technische Universität Berlin: Baugeschichte-Bauplanung. TUB Dokumentation aktuell 1/1977*, ed. Jürgen Dietrich Besch et al. (Berlin, 1977), 29-98, on 46.

151. Board Session 3, 9 August 1887, and Board Session 4, 25 March 1889, SAA 61/Lc 973.

152. Board Session 4, 25 March 1889, ibid.

153. [Helmholtz], "Denkschrift über die bisherige Thätigkeit der Reichsanstalt," 1369; cf. Pernet, "Ueber die physikalisch-technische Reichsanstalt," 10-11.

154. Weymann to Boetticher, 22 March 1888, ZStAP, RI, Nr. 13146/1, Bl. 83-85; *HEDR* (1892-93), 41.

155. Loewenherz to Helmholtz, 19 August 1887, Zentrales Archiv der Akademie der Wissenschaften der DDR, Nachlass Helmholtz.

156. R. Fuess, "Gedenkfeier für Dr. Leopold Loewenherz," *ZI 13* (1893): 177-91; and the notice by the editors and publisher of the *Zeitschrift für Instrumentenkunde, ZI 12* (1892): 401.

157. R. Fuess and C. Bamberg to Weymann, 6 April 1887, ZStAP, RI, Nr. 13144/2, Bl. 126-30.

158. Board Session 1, 6 August 1887, SAA 61/Lc 973.

159. Helmholtz to Boetticher, 19 March 1891, ZStAP, RI, Abt. III, Loewenherz L 460. See also Helmholtz to Boetticher, 19 June 1891 (draft); Boetticher to Helmholtz, 25 June 1891; and Helmholtz to Boetticher, 23 January 1893, ibid.

160. *HEDR* (1892-93), 32.

161. [Kohlrausch], "Denkschrift über die Thätigkeit der Reichsanstalt 1893 bis 1895," 373.

162. Ministerium der geistlichen, Unterrichts- und Medicinal-Angelegenheiten to Helmholtz, 19 July 1880, DM 1964-3; E. Gumlich, "Nachruf für Ernst Hagen," *PZ 24* (1923): 145-49; Karl Scheel, "Ernst Hagen," *Deutsches Biographisches Jahrbuch 5* (1930): 146-48; and Rolf Sonnemann et al., *Geschichte der Technischen Hochschule Dresden: 1828-1978* (Berlin, 1978), 82.

163. Hertz to his parents, 14 and 24 October 1880, *Erinnerungen,* ed. Hertz, 102-4, on 103, and 104-5, on 105, respectively; Weymann to Boetticher, 24 March 1894, ZStAP, RI, Nr. 13147/2, Bl. 7-9; Koenigsberger, *Helmholtz, 3,* 134; and Heinrich Kayser, *Erinnerungen aus meinem Leben* (n.p., 1936), MS at the American Philosophical Society, 70;

164. Weymann to Boetticher, 24 March 1894.

165. Loewenherz to Helmholtz, 19 August 1887. See also Loewenherz's "Die Aufgaben der zweiten (technischen) Abtheilung der physikalisch-technischen Reichsanstalt," *ZI 8* (1888): 153-57.

166. Der Direktor der Königlichen Normal Eichungs-Kommission (Nieberding) to Boetticher, 29 March 1887, ZStAP, RI, Nr. 13144/2, Bl. 123-24.

167. Board Session 2, 8 August 1887, SAA 61/Lc 973; and death notice of Leman by the *Zeitschrift*'s Board, publisher, and editors, *ZI 34* (1914): 341.

168. "Thätigkeit," 272-75.

169. Ibid., 273.

170. Board Session 1, 20 March 1889, and Board Session 1, 17 March 1892, SAA 61/Lc 973.

171. Board Session 1, 20 March 1888, and Board Session 1, 20 March 1889, ibid. The Reichsanstalt accelerated its tuning-fork investigations in response to the "urgent wish" of the military and the Prussian Ministry of In-

struction (Board Session 1, 20 March 1889, ibid).

172. Staatssekretär to Nieberding, 1 October 1887, ZStAP, RI, Nr. 13144/2, Bl. 228-30; A. Boettcher, "Nachruf auf Geheimen Regierungsrat Prof. Dr. H. F. Wiebe und Regierungsrat Dr. J. Domke," *DMZ*, Heft 20 (15 October 1913): 209-13, on 209, in *ZI 33* (1913); and W. Meissner, "Hermann F. Wiebe," *Petroleum: Zeitschrift für die gesamten Interessen der Petroleum-Industrie und des Petroleum-Handels 8* (1912): 69-71, on 71.

173. Letter from fifteen Thuringian thermometer manufacturers to RI, 11-15 June 1888, ZStAP, RI, Nr. 13169/1, Bl. 109-109a; Boetticher to Helmholtz, 25 June 1888, ibid., Bl. 110; and Helmholtz to Boetticher, 13 March 1889, ibid., Bl. 136-40.

174. The period covered is actually from 1 Febraury 1893 to 31 January 1894; see "Thätigkeit," 302.

175. Ibid., 301-6.

176. For an explicit statement of the close working relationship between the Reichsanstalt and the Thuringian glass industry see Loewenherz's report on this topic: "Betrifft die Herstellung von Thermometerglas, Anlage 6 zum Protokoll vom 22. März 1889," SAA 61/Lc 973.

177. Board Session 2, 22 March 1889, ibid.

178. "Thätigkeit," 307-9.

179. *Poggendorff 4* (1883-1904): 416.

180. Emil Warburg, "Zum Umlauf in PTR III (Das Original läuft in Abt. I um)," 26 October 1915, BA 816, PTB-IB.

181. "Thätigkeit," 275-79. Cf. Pernet's remarks ("Ueber die physikalisch-technische Reichsanstalt," 4) on the importance of the testing work of the Technical Section's Electrical Laboratory for the electrical industry.

182. "Thätigkeit," 275; W. Jaeger, "Die Physikalisch-Technische Reichsanstalt: Fünfundzwangzig Jahre ihrer Tätigkeit. 3. Elektrizität," *NWN 1* (1913): 273- 79, on 275; and Warburg, "Zum Umlauf in PTR III."

183. "Thätigkeit," 275-78.

184. Berliner Elektricitäts-Werke to RI, 3 August 1892, ZStAP, RI, Nr. 13159/3, Bl. 79-81. Helmholtz, for his part, agreed fully on the need for such certifications (Helmholtz to Boetticher, 12 November 1892, ibid., Bl. 114-17).

185. "Thätigkeit," 279.

186. *Poggendorff 6* (1923-31): 337; and E. Gehrcke, "Eugen Brodhun † [Obituary]," *Das Licht 8* (1938): 260 (for the quote).

187. *Poggendorff 4* (1883-1904): 884; and Emil Liebenthal, in ZStAP, RI, Abt. III, L 458.

188. "Thätigkeit," 309-12.

189. Helmholtz, "Die Beglaubigung der Hefnerlampe," 257, 259; Helmholtz to Boetticher, 30 March 1893, ZStAP, RI, Nr. 13159/3, Bl. 123; and Deutscher Verein von Gas- und Wasserfachmännern zu Dresden, "Bericht der Lichtmess-Commission," *Journal für Gasbeleuchtung und Wasserversor-*

*gung 36* (1893): 449-51, on 451.

190. "Thätigkeit," 309-10.

191. Ibid., 310-12.

192. K. Strecker, "Die Personalverhältnisse der wissenschaftlichen Beamten der Physikalisch-Technischen Reichsanstalt" (29 August 1911), Anlage 9, BA 265, PTB-IB.

193. Board Session 4, 25 March 1889, SAA 61/Lc 973.

194. Board Session 2, 16 March 1891, and Staatssekretär to Siemens, 10 October 1891, ibid.

195. *HEDR* (1892-93), 41; and *HEDR* (1893-94), 40-41. This appropriation was later reduced by 47,500 marks to 1,910,500 marks (*HEDR* [1895-96], 4-5, 34-41, on 41).

196. Hagen and Scheel, "Die Physikalisch-Technische Reichsanstalt," 62

197. "IV. Gebäude für die Verwaltungsbehörden," 2, 84.

198. Ibid.

199. *HEDR* (1893-94), 41.

200. Hagen and Scheel, "Die Physikalisch-Technische Reichsanstalt," 65.

201. Ibid.; and *HEDR* (1893-94), 41.

202. *HEDR* (1893-94), 41; and [Friedrich Kohlrausch], "Denkschrift über die Thätigkeit der Physikalisch-Technischen Reichsanstalt vom Frühjahr 1895 bis zum Sommer 1897," *SBVR 163* (1897-98): 870-87, on 870.

203. Hagen and Scheel, "Die Physikalisch-Technische Reichsanstalt," 62. This description of the Technical Section's main building relies exclusively on ibid., 62-64.

204. Ibid., 65; and *HEDR* (1893-94), 41.

205. Hagen to Kayser, 25 May 1911, SPK, Slg. Darmstaedter, F i c 1902 (6), Bessel-Hagen. Hagen's reference was to his and Rubens's important set of studies confirming the validity of Maxwell's electromagnetic theory of light in the infrared part of the spectrum (see Chapter 4, "Measuring for Marks").

206. As remarked above (note 195), during the fiscal year 1895-96 the legislature reduced the Reichsanstalt's appropriations by 47,500 marks, which accounts for the difference between the two figures (880,190 and 927,690) given above. Exactly which, if any, categories were specifically reduced, is unknown.

207. This grand total is based on the figures given in *HEDR* for the years 1893 to 1898 along with those provided by Carhart, "The Imperial Physico-Technical Institution," 561-62.

208. Strecker, "Die Personalverhältnisse der wissenschaftlichen Beamten," Anlage 1; and "Uebersicht über die Einnahmen und Ausgaben der Physikalisch-Technischen Reichsanstalt seit ihrer Errichtung" (n.d., probably 1915), BA 240, PTB-IB.

209. *Verzeichniss der Veröffentlichungen aus der Physikalisch-Tech-*

*nischen Reichsanstalt 1887 bis 1900* (Berlin, 1901), 17-19.

210. Koenigsberger, *Helmholtz, 3,* 2, 24, 70, 80.

211. Ibid., 24, 63, 65.

212. Ibid., 72; and Pernet, "Hermann von Helmholtz," 34.

213. *Vorschläge zu gesetzlichen Bestimmungen über elektrische Massein-heiten, entworfen durch das Curatorium der Physikalisch-Technischen Reichsanstalt: Nebst kritischem Bericht über den wahrscheinlichen Werth des Ohm nach dem bisherigen Messungen verfasst von Dr. Friedrich Ernst Dorn* (Berlin, 1893).

214. Anna von Helmholtz to Ellen von Siemens-Helmholtz, 14 October 1893, *Anna von Helmholtz: Ein Lebensbild,* ed. Siemens-Helmholtz, 2, 75-77.

215. Koenigsberger, *Helmholtz, 3,* 93-101; and Pernet, "Hermann von Helmholtz," 35.

216. Koenigsberger, *Helmholtz, 3,* 122-23; and Hagen to Kayser, 25 November 1894, SPK, Slg. Darmstaedter, F i c 1902: Bessel-Hagen.

217. Death announcement and burial notice by Helmholtz's family, 8 September 1894, SPK, Slg. Darmstaedter, F i a 1847: Helmholtz; on the flag's being at half-mast, Julius Reiner, *Hermann von Helmholtz* (Leipzig, 1905), 53.

218. Max Planck, "Persönliche Erinnerungen," in Planck's *Physikalische Abhandlungen und Vorträge,* 3 vols. (Braunschweig, 1958), *3,* 358-63, on 362.

219. Hagen to Kayser, 25 November 1894.

### Chapter 4. Masters of measurement

1. See the Introduction.

2. Wilhelm von Bezold et al., "Nr. 13. Wahlvorschlag für Friedrich Kohlrausch (1840-1910) zum OM. Berlin, 30 Mai 1895," in *PPI,* 98.

3. Unless otherwise noted, the following account of Kohlrausch's life and work is based on Adolf Heydweiller, "Friedrich Kohlrausch," in *GAFK, 1,* xxxv-lxviii; L. Holborn, "Friedrich Kohlrausch † [Obituary]," *EZ 31* (1910): 211-12; Eduard Riecke, "Friedrich Kohlrausch † [Obituary]," *PZ 11* (1910): 73-76; idem, "Friedrich Kohlrausch," *Nachrichten von der Königlichen Gesellschaft der Wissenschaften zu Göttingen: Geschäftliche Mitteilungen aus dem Jahre (1910),* Heft 1, 71-85; Heinrich Rubens, "Gedächtnisrede auf Friedrich Kohlrausch," *APAWB: Physikalisch-Mathematische Klasse* (1910): Gedächtnisrede 1, 3-11, reprinted in *PPII,* 177-82; K. Scheel, "Friedrich Kohlrausch † [Obituary]: Nachruf," *Naturwissenschaftliche Rundschau 25* (1910): 153-54; Emil Warburg, "Friedrich Kohlrausch, Gedächtnisrede, . . . ," *VDPG 12* (1910): 911-38; and W. Wien, "Friedrich Kohlrausch † [Obituary]," *AP 31* (1910): 449-54. On Kohlrausch's early career making and his style and methods of research see David Cahan, "Instruments, Institutes, and

Scientific Innovation: Friedrich Kohlrausch's Route to a Law of Electrolytic Conductivity," *Osiris 5* (forthcoming 1989).

4. Kohlrausch quoted in Warburg, "Kohlrausch," 920.

5. Rubens, "Gedächtnisrede auf Kohlrausch," 6-7; and on Kohlrausch's lack of interest in theoretical matters, Warburg, "Kohlrausch," 929.

6. For an analysis of Kohlrausch's route to his law see Cahan, "Instruments, Institutes, and Scientific Innovation."

7. Wien, "Kohlrausch," 452 (quote); Warburg, "Kohlrausch," 920.

8. *GAFK, 1*, xvii-xxxiv.

9. Heydweiller, "Kohlrausch," lvi.

10. Kohlrausch to Emil Fischer, 13 July and 15 October 1894; and 28 July 1906, EFP; Kohlrausch to Wilhelm Wien, 15 September 1906, DM 2452; Heydweiller, "Kohlrausch," lvi; and Wien, "Kohlrausch," 450.

11. Kohlrausch to Fischer, 18 October 1894, EFP; and Kohlrausch to Fischer, 15 November 1894, ibid.

12. Kohlrausch to Eurer Hochwohlgeboren (probably the curator of the University of Strasbourg), 7 March 1895, Dossier personnel de Friedrich Kohlrausch, AL 103/102, Archives du Bas-Rhin, Strasbourg, France.

13. Ibid. (quote); Wien, "Kohlrausch," 451; Heydweiller, "Kohlrausch," lvi; and Rubens, "Gedächtnisrede auf Kohlrausch," 5.

14. Heydweiller, "Kohlrausch," lvi; Rubens, "Gedächtnisrede auf Kohlrausch," 5.

15. Riecke, "Kohlrausch," 81.

16. Heydweiller, "Kohlrausch," lvi.

17. Kohlrausch to Fischer, 13 July 1894.

18. Wien, "Kohlrausch," 450-51.

19. Boetticher to Posadowsky-Wehner, 17 December 1894, BAK, R2 12376.

20. Boetticher to Posadowsky-Wehner, 24 November 1894, ibid.

21. Boetticher to Posadowsky-Wehner, 17 December 1894, ibid.

22. Posadowsky-Wehner to Boetticher, 22 December 1894, ibid.

23. Staatssekretär des Innern to Staatssekretär des Reichsschatzamts, 30 November 1904, ZStAP, RI, Personalakte Warburg, Nr. 645, Bl. 3-5.

24. Kohlrausch to Verehrte Freundin! 28 March 1897, DM, 1932- 17/32.

25. See Table 3.2.

26. Paul Forman, John Heilbron, and Spencer Weart, "Physics *circa* 1900: Personnel, Funding, and Productivity of the Academic Establishments," *HSPS 5* (1975): 1-185, on 92, 95.

27. [Friedrich Kohlrausch], "Denkschrift über die Thätigkeit der Physikalisch-Technischen Reichsanstalt vom Frühjahr 1895 bis zum Sommer 1897," *SBVR 163* (1897-98): 870-87, on 871.

28. "The National Physical Laboratory," *Nature 58* (1898): 565-66, on 565.

29. *HEDR* (1903), 2-3, 26-29, on 26-28; "Uebersicht über die Einnahmen

und Ausgaben der Physikalisch-Technischen Reichsanstalt seit ihrer Errichtung" (n.d., probably 1915), BA 240, PTB-IB; and K. Strecker, "Die Personalverhältnisse der wissenschaftlichen Beamten der Physikalisch-Technischen Reichsanstalt" (29 August 1911), Anlage 10, BA 265, PTB-IB.

30. Strecker, "Die Personalverhältnisse," Anlage 2, BA 265, PTB-IB.

31. Heydweiller, "Kohlrausch," xxxviii.

32. Arrhenius to Ostwald, 19 December 1886, in *Aus dem wissenschaftlichen Briefwechsel Wilhelm Ostwalds, Teil 2, Briefwechsel mit Svante Arrhenius und Jacobus Hendricus Van't Hoff*, ed. Hans-Günther Körber, 2 vols. (Berlin, 1969), *2*, 26-27, on 27.

33. Arrhenius to Ostwald, 20 January 1887, ibid., 28-30, on 30.

34. W. Wien, "Ein Rückblick," in *Wilhelm Wien: Aus dem Leben und Wirken eines Physikers*, ed. Karl Wien (Leipzig, 1930), 1-50, on 18; cf. Wien, "Kohlrausch," 451.

35. Scheel, "Kohlrausch," 153.

36. Kohlrausch to Arrhenius, 31 May 1893, SAC; cf. Kohlrausch to Carl Barus, 7 July 1892, CBP, and Kohlrausch to Fischer, 17 June 1894, EFP.

37. Heydweiller, "Kohlrausch," lvi.

38. Kohlrausch to Barus, 30 May 1896, CBP.

39. Kohlrausch to Arrhenius, 1 February 1900, SAC; cf. Kohlrausch to Barus, 23 May 1901, CBP.

40. Kohlrausch to Arrhenius, 21 February 1903, SAC.

41. Friedrich Kohlrausch, "Antrittsrede," *SB* (1896): (II) 736-43, reprinted in *GAFK, 1*, 1024-31, and in *PPII*, 172-76.

42. [Friedrich Kohlrausch], "Die Tätigkeit der Physikalisch-Technischen Reichsanstalt im Jahre 1903," *ZI 24* (1904): 133-47, 167-80, on 133-34 (hereafter cited as "Tätigkeit, 1903").

43. [Ernst Hagen], "Die Thätigkeit der Physikalisch-Technischen Reichsanstalt in der Zeit vom 1. März 1894 bis 1. April 1895," *ZI 15* (1895): 283-300, 324-43, on 297-98; and [Friedrich Kohlrausch], "Die Thätigkeit der Physikalisch-Technischen Reichsanstalt in der Zeit vom 1. Februar 1898 bis 31. Januar 1899," *ZI 19* (1899): 206-16, 240-56, on 210-11.

44. Board Session 1, 8 March 1902, BA 239, PTB-IB.

45. Board Session 4, 12 March 1902, ibid.

46. "Protokoll der Kommissionssitzung des Kuratoriums der Physikalisch-Technischen Reichsanstalt für die Beratung über die spezifische Wärme der Gase," Session 1, 10 March 1902, ibid.

47. Board Session 1, 7 March 1903, ibid.

48. "Tätigkeit, 1903," 133-35.

49. Carl Linde, *Aus meinem Leben und von meiner Arbeit: Aufzeichnungen für meine Kinder und meine Mitarbeiter* (Munich, 1916); Max Jakob, "C. Lindes Lebenswerk," *NWN 5* (1917): 417-23; O. Knoblauch, "Carl von Linde zum 11. Juni 1932," *ZTP 13* (1932): 25-52; J. Zenneck, "Carl von Lindes

Wesen und Wirken," *Technische Ueberwachung 3* (1942): 93-97; and Friedrich Klemm, "Linde, Carl von," *DSB 8* (1973): 365-66.

50. Linde to Kohlrausch, 3 November 1895; 24 March, and 13, 15, and 17 April 1896; Holborn to Kohlrausch, 8, 14, and 30 April and 6 May 1896, all in the archive of the Physikalisch-Technische Bundesanstalt, Braunschweig; Linde, *Aus meinem Leben,* 89; and L. Holborn and W. Wien, "Ueber die Messung tiefer Temperaturen," *AP 295* (1896): 213-28.

51. "Tätigkeit, 1903," 135.

52. Kohlrausch to Wilhelm von Siemens, 15 March 1904, SAA 4/Lk 28. On Scheel, see also [Emil Warburg], "Denkschrift über die Tätigkeit der Physikalisch-Technischen Reichsanstalt von Anfang 1904 bis Ende 1906," *SBVR 239* (1907): 446-63, on 461; H. Ebert, "Nachruf auf Karl Scheel," *Zeitschrift für Physik 104* (1936): i-iii; idem, "Karl Scheel zum siebzigsten Geburtstag," *ZTP 17* (1936): 65-66; Ernst Brüche, "Aus der Vergangenheit der Physikalischen Gesellschaft. 5. Die alten Gesellschaftszeitschriften," *PB 17* (1961): 225-32, on 232; and idem, "Unser Geheimrat Scheel," *PB 22* (1966): 121-28.

53. *Deutsche Senioren der Physik,* ed. Bild- und Filmsammlung Deutscher Physiker (Leipzig, 1936), ix-xi, 1, on x.

54. "Tätigkeit, 1903," 137.

55. On Diesselhorst's expertise in these fields see, H. Diesselhorst, "Wärmeleitung," in *Encyklopädie der mathematischen Wissenschaften mit Einschluss ihrer Anwendungen,* ed. A. Sommerfeld (Leipzig, 1903-21), *Physik,* Teil 1, Vol. 5, 208-31; Diesselhorst to Sommerfeld, 30 April 1900, AIP; and Kohlrausch to Wien, 13 March 1899, DM 2436, 26 November 1900, DM 2440, and 17 December 1900, DM 2442.

56. "Tätigkeit, 1903," 137.

57. Ibid., 135-38.

58. Ibid., 134.

59. [Friedrich Kohlrausch], "Die Tätigkeit der Physikalisch-Technischen Reichsanstalt im Jahre 1902," *ZI 23* (1903): 113-25, 150-57, 171-84, on 117 (hereafter cited as "Tätigkeit, 1902").

60. Fischer to Jaeger (quote), and Fischer to Warburg, 2 June 1906, EFP.

61. Emil Fischer, "Die Begründung einer chemischen Reichsanstalt," *Verhandlungen des Vereins zur Beförderung des Gewerbefleisses* (1906): 113-23, on 113-14.

62. Of the many studies of the origins of quantum physics, see above all Martin Klein, "Max Planck and the Beginnings of the Quantum Theory," *Archive for History of Exact Sciences 1* (1962): 459-79; Hans Kangro, *Vorgeschichte des Planckschen Strahlungsgesetzes: Messungen und Theorien der spektralen Energieverteilung bis zur Begründung der Quantenhypothese* (Wiesbaden, 1970); and Thomas Kuhn, *Black-Body Theory and the Quantum Discontinuity, 1894-1912* (New York, 1978).

63. Kangro, *Vorgeschichte*, 41, n. 31.

64. Hans Kangro, "Rubens, Heinrich (Henri Leopold)," *DSB 11* (1975): 581-85.

65. On the essential lack of research and research facilities at the Technische Hochschule Charlottenburg in the 1890s see *Berlin und seine Bauten*, ed. Architekten-Verein zu Berlin und Vereinigung Berliner Architekten, 3 vols. in 2 (Berlin, 1896), *1*, 15 and *2*, 287-95, on 295; Wilhelm Westphal, "Heinrich Rubens," *NWN 10* (1922): 1017-20, on 1017; idem, "Dem Andenken an Heinrich Rubens," *PB 8* (1952): 323-25, on 323; idem, "Das Physikalische Institut der TU Berlin," *PB 11* (1955): 554-58, on 555; Ernst Brüche, "Aus der Vergangenheit der Physikalischen Gesellschaft. VI. Die Jahre zwischen Quanten- und Relativitätstheorie," *PB 17* (1961): 400-10, on 405; and Jost Lemmerich, *Zur Geschichte der Physik an der Technischen Hochschule Berlin-Charlottenburg*, Vol. 3 of *Berliner Beiträge zur Geschichte der Naturwissenschaften und der Technik*, ed. Friedrich G. Rheingans and Edgar Swinne (Berlin, 1986), 13, 17, 21.

66. On the importance to industry of establishing scientifically grounded thermal and optical measures see, e.g., F. v. Hefner- Alteneck, "Vorschlag zur Beschaffung einer constanten Lichteinheit," *Journal für Gasbeleuchtung und Wasserversorgung 27* (1884): 73-79; idem, "Zur Frage der Lichteinheit," ibid., *27* (1886): 3-8; "Literatur: Beleuchtungswesen," ibid., *36* (1893): 703; "Herstellung gewisser Einheiten für Lichtmessungen," March 1889 to 30 April 1894, ZStAP, RI, Sig. 13159/3, esp. the "Bericht über die Verhandlungen der Delegirtenkammer des Elektrikercongresses in Chicago, 21-26 August 1893," by one Budde to the Minister der geistlichen, Unterrichts- und Medizinal-Angelegenheiten, Bl. 190-98; H. Remané, "Die geschichtliche Entwickelung, die Herstellung, die physikalischen Eigenschaften und die Anwendungen der elektrischen Glühlampen," *DMZ*, Heft 23 (1 December 1899): 209-13, 221- 27, in *ZI 19* (1899). The Verein zur Beförderung des Gewerbefleisses was so interested in the topic of the heat and light radiation of burned gases that it established an award in the late 1880s for the best work on the subject. The prizewinner, the young Dutch physicist W. H. Julius, characterized this virtually unexplored topic as one that "simultaneously has great practical importance and enters deeply into the theory of matter" (W. H. Julius, *Die Licht- und Wärmestrahlung verbrannter Gase* [Berlin, 1890], 1 [quote], 29). American industry, as well as that of other nations, was equally anxious to find a scientifically acceptable unit and temperature scale; see, e.g., A. Macfarlane, "On the Units of Light and Radiation," *TAIEE 12* (1895): 85-98; W. M. Stine, "Macfarlane on Units of Light and Radiation," ibid. *12* (1895): 152-53; and Ed. L. Nichols, Clayton Sharp, and Chas. Matthews, "Standards of Light: Preliminary Report of the Sub-Committee of the Institute," ibid. *13* (1896): 135- 207, esp. the statement of J. W. Howell on 201.

67. W. Wien, "Eine neue Beziehung der Strahlung schwarzer Körper zum

zweiten Hauptsatz der Wärmetheorie," *SB* (1893): (I) 55-62, on 62.

68. Willy Wien, "Temperatur und Entropie der Strahlung," *AP 288* (1894): 132-65, quote on 133.

69. Wien, "Ein Rückblick," 18.

70. O. Lummer and F. Kurlbaum, "Bolometrische Untersuchungen für eine Lichteinheit," *SB* (1894): (I) 229-38; O. Lummer and F. Kurlbaum, "Ueber die neue Platinlichteinheit der Physikalisch-technischen Reichsanstalt," *Verhandlungen der Physikalischen Gesellschaft zu Berlin 14* (1895): 56-70.

71. [Friedrich Kohlrausch], "Die Thätigkeit der Physikalisch-Technischen Reichsanstalt in der Zeit vom 1. April 1895 bis 1. Februar 1896," *ZI 16* (1896): 203-18, 233-49, on 208.

72. W. Wien and O. Lummer, "Methode zur Prüfung des Strahlungsgesetzes absolute schwarzer Körper," *AP 292* (1895): 451-56, on 455.

73. Ibid., 451-52 (emphasis added).

74. [Kohlrausch], "Thätigkeit vom 1. April 1895 bis 1. Februar 1896," 209.

75. W. Wien, "Ueber die Energievertheilung im Emissionsspectrum eines schwarzen Körpers," *AP 294* (1896): 662-69; and F. Paschen, "Ueber Gesetzmässigkeiten in den Spectren fester Körper," *AP 294* (1896): 455-92.

76. Max Planck, "Ueber irreversible Strahlungsvorgänge," *AP 306* (1900): 69-122, which summarizes Planck's results on radiation theory reported between 1897 and 1899.

77. Otto Lummer, "Light and Its Artificial Production," in *Annual Report of the Board of Regents of the Smithsonian Institution . . . to July, 1897* (Washington, D.C., 1898), 273-99, esp. 275-77, 278 (quote), 287-92, 297 (quote), 299 (quote).

78. F. Paschen, "Ueber Gesetzmässigkeiten in den Spectren fester Körper," *AP 296* (1897): 662-723; and O. Lummer and E. Pringsheim, "Die Strahlung eines 'schwarzen' Körpers zwischen 100 und 1300°C," *AP 299* (1897): 395-410.

79. [Kohlrausch], "Thätigkeit Februar 1898 bis Januar 1899," 256; and O. Lummer and F. Kurlbaum, "Der electrisch geglühte 'absolute schwarze' Körper und seine Temperaturmessung," *Verhandlungen der Physikalischen Gesellschaft zu Berlin 17* (1898): 106-11.

80. [Kohlrausch], "Thätigkeit Februar 1898 bis Januar 1899," 213-14.

81. [Friedrich Kohlrausch], "Die Thätigkeit der Physikalisch-Technischen Reichsanstalt in der Zeit vom Februar 1899 bis Februar 1900," *ZI 20* (1900): 140-50, 172-86, on 147- 48 (quote on 147).

82. F. Paschen, "Ueber die Vertheilung der Energie im Spectrum des schwarzen Körpers bei niederen Temperaturen," *SB* (1899): (I) 405-20 (II) 959-76.

83. O. Lummer and E. Pringsheim, "1. Die Vertheilung der Energie im Spectrum des schwarzen Körpers und des blanken Platins; 2. Temperaturbestimmung fester glühender Körper," *VDPG 1* (1899): 215-35, quotes on 230-

33. (This article is a continuation of work described in their "Die Vertheilung der Energie im Spektrum des schwarzen Körpers," ibid., 23-41.)

84. [Kohlrausch], "Thätigkeit Februar 1899 bis Februar 1900," 150.

85. O. Lummer and E. Pringsheim, "Ueber die Strahlung des schwarzen Körpers für lange Wellen," *VDPG 2* (1900): 163-80.

86. Lord Rayleigh [John William Strutt], "Remarks upon the Law of Complete Radiation," *Philosophical Magazine 49* (1900): 539-40.

87. O. Lummer and E. Jahnke, "Ueber die Spectralgleichung des schwarzen Körpers und des blanken Platins," *AP 308* (1900): 283-97; and Max Thiesen, "Ueber das Gesetz der schwarzen Strahlung," *VDPG 2* (1900): 65-70.

88. E. F. Nichols and Heinrich Rubens, "Ueber Wärmestrahlen von grosser Wellenlänge," *Naturwissenschaftliche Rundschau 11* (1896): 545-49; and idem, "Beobachtungen elektrischer Resonanz an Wärmestrahlen von grosser Wellenlänge," *SB* (1896): (II) 1393-1400. On Rubens's work see H. Kangro, "Ultrarotstrahlung bis zur Grenze elektrisch erzeugter Wellen: Das Lebenswerk von Heinrich Rubens," *Annals of Science 26* (1970): 235-59, *27* (1971): 165- 200; and his *Vorgeschichte*, 49-60, 125-29, 160-64, 200-6.

89. H. Rubens and F. Kurlbaum, "Ueber die Emission langwelliger Wärmestrahlen durch den schwarzen Körper bei verschiedenen Temperaturen," *SB* (1900): (II) 929-41; see also their "Anwendung der Methode der Reststrahlen zur Prüfung des Strahlungsgesetzes," *AP 309* (1901): 649-66.

90. Max Planck, "Ueber eine Verbesserung der Wien'schen Spectralgleichung," *VDPG 2* (1900): 202-4; see also idem, *Wissenschaftliche Selbstbiographie,* 2nd ed. (Leipzig, 1948), 27.

91. Planck, *Wissenschaftliche Selbstbiographie, 27.*

92. Max Planck, "Zur Theorie des Gesetzes der Energieverteilung im Normalspectrum," *VDPG 2* (1900): 237-45; idem, "Ueber das Gesetz der Energieverteilung im Normalspectrum," *AP 309* (1901): 553-63. For an analysis of Planck's route to the idea of a quantum of action see the references in note 62 above.

93. O. Lummer and E. Pringsheim, "Temperaturbestimmung mit Hilfe der Strahlungsgesetze," *PZ 3* (1901-2): 97-100, quote on 98.

94. See Svante Arrhenius's report of 1908 to the Nobel Committee for Physics: *Kommittéutlåtande, Nobelkommittén för Fysik/Kemi* (1908), The Nobel Archives, Royal Academy of Sciences, Stockholm, cited in Elisabeth Crawford, *The Beginnings of the Nobel Institution: The Science Prizes, 1901-1915* (Cambridge, 1984), 165.

95. See, e.g., E. Pringsheim, "Die Strahlungsgesetze und ihre Anwendungen," *Naturwissenschaftliche Rundschau 15* (1900): 1- 2, 17-19; L. Holborn and F. Kurlbaum, "Ueber ein optisches Pyrometer," *SB* (1901): (II) 712-19; Otto Lummer, *Die Ziele der Leuchttechnik* (Munich, 1903); and W. Nernst, "Ueber die Helligkeit glühender schwarzer Körper und über ein einfaches Py-

rometer," *PZ 7* (1906): 380-83. For a summary of the radiation laws and their application to the determination of temperature and evaluation of the efficiency of gas- and electric-illumination substances and devices see Georg Gehlhoff, "Die Strahlungsgesetzte und ihre Anwendungen," *Helios: Fach-Zeitschrift für Elektrotechnik 16* (1910): 137-41, 149-55, 175-76, 181-86, 381-86, 393-96, 405-10, 417-20.

96. [Friedrich Kohlrausch], "Die Thätigkeit der Physikalisch-Technischen Reichsanstalt im Jahre 1901," *ZI 22* (1902): 110-24, 143-60, on 120-21; see also O. Lummer and E. Gehrcke, "Ueber den Bau der Quecksilberlinien; ein Beitrag zur Auflösung feinster Spektrallinien," *SB* (1902): (I) 11-17; "Tätigkeit, 1903," 139-41.

97. "Tätigkeit, 1903," 139.

98. Paul Forman, "Paschen, Louis Carl Heinrich Friedrich," *DSB 10* (1974): 345-50, on 346.

99. See note 65 above.

100. Wilhelm Wien, "Physik und Technik," in Wien's *Aus der Welt der Wissenschaft: Vorträge und Aufsätze* (Leipzig, 1921), 235-63, on 253-54. Lummer said that he pursued the radiation laws in order to achieve scientific understanding, but he gladly welcomed their practical utility, too. (See his "Light and Its Artificial Production"; *Die Ziele der Leuchttechnik,* 1-2; and "Wissenschaftliche Grundlagen zur ökonomischen Lichterzeugung," *Zeitschrift für Beleuchtungswesen 10* [1904]: 1-3.) On Lummer's dual interest in science and technology, see Fritz Reiche, "Otto Lummer," *PZ 27* (1926): 459-67, esp. 463-64; and [Erich Waetzmann and Fritz Reiche], *Akademische Reden: Zum Gedächtnis an Otto Lummer gehalten in Breslau 1925* (Braunschweig, 1928), 15-17, 19, 23.

101. F. Henning, "Ferdinand Kurlbaum," *PZ 29* (1928): 97- 104, on 102.

102. In 1902 Lummer habilitated at the University of Berlin; he was presumably interested in entering an academic career. He and Kohlrausch occasionally came into conflict, according to Clemens Schaefer, "Otto Lummer zum 100. Geburtstag," *PB 16* (1960): 373-81, on 379-80.

103. Hans Kangro, "Pringsheim, Ernst," *DSB 11* (1975): 149-50, on 149.

104. Kangro, "Rubens, Heinrich," 581.

105. Cf. Weymann to Staatssekretär des Innern, 13 February 1899, ZStAP, RI, Sig. 13147/2, Bl. 147-49.

106. [Friedrich Kohlrausch], "Denkschrift über die Tätigkeit der Physikalisch-Technischen Reichsanstalt von Anfang 1900 bis Ende 1903," *SBVR 206* (1903-4): 1139-59, on 1159; *DMZ,* Heft 1 (1 January 1898): 1, in *ZI 18* (1898); and *DMZ,* Heft 2 (15 January 1904): 17, in *ZI 24* (1904).

107. "Tätigkeit, 1903," 141.

108. [Friedrich Kohlrausch], "Denkschrift über die Tätigkeit der Physikalisch-Technischen Reichsanstalt vom Anfang des Jahres 1893 bis Ostern 1895," *SBVR 151* (1895-97): 373-91, on 382 (quote); Kohlrausch to Barus, 19

September 1895, CBP.

109. Henning, "Kurlbaum," 102-3.

110. "Tätigkeit, 1903," 141-42.

111. Ibid., 142-43.

112. Ibid., 144-45.

113. Ernst Orlich, "Stephan August Lindeck," *ZI 31* (1911): 329-31.

114. *DMZ*, Heft 5 (1 March 1901): 45, in *ZI 21* (1901); and *DMZ*, Heft 4 (15 February 1904): 38, in *ZI 24* (1904).

115. Board Session 2, 9 March 1903, SAA 4/Lk 28.

116. "Tätigkeit, 1903," 146-47.

117. "Entwurf eines Gesetzes, betreffend die elektrischen Masseinheiten," *SBVR 164* (1897-98): 1735-41, on 1736 (quote), 1739.

118. Ibid., 1735-36.

119. Ibid., 1739-40.

120. [Friedrich Kohlrausch], "Denkschrift über die Thätigkeit der Physikalisch-Technischen Reichsanstalt vom Sommer 1897 bis Ende 1899," *SBVR 176* (1898-1900): 3449-63, on 3449; and [F. Kohlrausch], *Die bisherige Tätigkeit der Physikalisch-Technischen Reichsanstalt: Aus dem Reichstage am 19. Februar 1904 überreichten Denkschrift. Mit einem Verzeichnis der Veröffentlichungen aus den Jahren 1901-1903* (Braunschweig, 1904), 14 (quote), reprinted in *DMZ*, Heft 13 (1 July 1904): 121-25, Heft 14 (15 July 1904): 134-37, in *ZI 24* (1904), and in *GAFK, 1*, 1032-47. See also "Herstellung gewisser Einheiten für Lichtmessungen: Entwurf eines Gesetzes betreffend die elektrischen Masseinheiten," May 1896 to November 1898, ZStAP, RI, Nr. 13160; and *Der Verband Deutscher Elektrotechniker, 1893-1918* (Berlin, [1918]), 34, 60.

121. Board Session 2, 10 March 1902, BA 239, PTB-IB; and cf. Board Session 1, 7 March 1903, ibid.

122. "Tätigkeit, 1903," 145-46.

123. Ibid., 167-71.

124. Ibid., 171-73.

125. Ibid., 173-75.

126. Ibid., 175; E. Hagen and H. Rubens, "Ueber Beziehungen zwischen dem Reflexionsvermögen der Metalle und ihrem elektrischen Leitvermögen," *SB* (1903): (I) 269-77; Kangro, "Ultrarotstrahlung bis zur grenze elektrisch erzeugter Wellen," 242-48; and for its role in the discovery of superconductivity, Per Dahl, "Kamerlingh Onnes and the Discovery of Superconductivity: The Leyden Years, 1911-1914," *HSPS 15:* (1984): 1-37, on 17-18.

127. "Uebersicht über die Einnahmen und Ausgaben der Physikalisch-Technischen Reichsanstalt" (n.d., probably 1915), BA 240, PTB-IB; and Strecker, "Die Personalverhältnisse," Anlage 9, BA 265, ibid.

128. Kohlrausch to Arrhenius, 17 November 1898, SAC.

129. "Vermischtes," *Centralblatt der Bauverwaltung 15* (1895): 276. Carl

Junk, "Physikalische Institute," in *Handbuch der Architektur*, 2nd ed., ed. Eduard Schmitt (Stuttgart, 1905), Teil 4, Halbband 6, Heft 2a,1, pp. 164-236, reported (205- 6) that the side effects of electrical streetcars could produce measuring errors of 100 to 140 times beyond normal.

130. See, e.g., Helmholtz to Boetticher (copy), 27 January 1893, ZStAP, RI, Nr. 13144/3, Bl. 138-40, along with numerous other letters in ibid.

131. Kohlrausch to Weymann (copy), 17 January 1895, ibid., 136-37.

132. Kohlrausch to H. Wagner, 8 May 1895, Niedersächsische Staats- und Universitätsbibliothek Göttingen, Handschriftenabteilung, Sig. H. Wagner 29.

133. Friedrich Kohlrausch, "Diskussion über die Frage der Störungen wissenschaftlicher Institute durch electrische Bahnen," *EZ 16* (1895): 427-29, 444-45, reprinted in *GAFK, 1*, 886-901, quotes on 899-901. For the electrotechnologists' own concerns about and investigations of this problem see *Die ersten 25 Jahre des Elektrotechnischen Vereins*, ed. Emil Naglo (Berlin, 1904), 58-59, 76-79; *Der Verband Deutscher Elektrotechniker*, 41- 42; and, esp. *50 Jahre Elektrotechnischer Verein: Festschrift* . . . ed. Hans Görges (Berlin, 1929), 262-70 (and the references therein).

134. Kohlrausch to Barus, 19 September 1895, CBP.

135. John P. McKay, *Tramways and Trolleys: The Rise of Urban Mass Transport in Europe* (Princeton, N.J., 1976), 37-39, 70-81, 99-100, 121-24, passim.

136. Eduard Buchmann, *Die Entwicklung der Grossen Berliner Strassenbahn und ihre Bedeutung für die Verkehrsentwickelung Berlins* (Berlin, 1910).

137. L. Holborn, "Elektrische Strassenbahnen und physikalische Institute," *Preussische Jahrbücher 81* (1895): 177-84, quote on 180. See also the relevant letters in ZStAP, RI, Nr. 13144/3.

138. "Vermischtes: Einfluss elektrischer Strassenbahn auf Galvanometer," *Centralblatt der Bauverwaltung 13* (1893): 551; "Vermischtes," ibid. *15* (1895): 276; Junk, "Physikalische Institute," 205-6; Felix Auerbach, *Ernst Abbe: Sein Leben, sein Wirken, seine Persönlichkeit*, 2nd ed. (Leipzig, 1922), 113-14; Wilhelm Wien, "Das Physikalische Institut und das Physikalische Seminar," in *Die wissenschaftlichen Anstalten der Ludwig-Maximilians- Universität zu München. Chronik zur Jahrhundertfeier*, ed. Karl Alexander von Müller (Munich, 1926), 207-11, on 209.

139. Kohlrausch quoted in Heydweiller, "Kohlrausch," lvii.

140. "Besprechung im Reichsamt des Innern am 2. November 1901 betreffend die Anlage eines von magnetischen Störungen freien Hülfslaboratoriums der Physikalisch-Technischen Reichsanstalt," ZStAP, RI, Nr. 13144/4, Bl. 148-51. For details of the agreement between the Reichsanstalt and the Berlin-Charlottenburger Strassenbahngesellschaft see Kohlrausch to Staatssekretär, RI, 10 February 1902, BAK, R2 12376. The entire battle can be fol-

lowed in greater detail through the materials in ZStAP, RI, Nr. 13144/3. Cf. the disturbances to precision-measurement work at the Physics Institute of the Technische Hochschule Karlsruhe, where after long negotiations the local streetcar company compensated the Institute with 60,000 marks (Otto Lehmann, *Geschichte des Physikalischen Instituts der Technischen Hochschule Karlsruhe: Festgabe der Fridericiana zur 83. Versammlung Deutscher Naturforscher und Aerzte* [Karlsruhe, 1911], 80-81).

141. "Thätigkeit, 1902," 121; and "Tätigkeit, 1903," 138-39.

142. Wien, "Kohlrausch," 451; and Heydweiller, "Kohlrausch," lx, lxvii.

143. Quoted in Heydweiller, "Kohlrausch," lviii.

144. Kohlrausch to Barus, 19 September 1895, CBP; Kohlrausch to Verehrte Freundin! 28 March 1897, DM 1932-17/32.

145. Kohlrausch to Arrhenius, 1 February 1900, SAC.

146. Kohlrausch to Barus, 23 May 1901, CBP.

147. Kohlrausch to Fischer, 29 July 1904, EFP.

148. Quoted in Heydweiller, "Kohlrausch," lviii; cf. Kohlrausch to Barus, 6 July 1905, CBP, where Kohlrausch speculates that part of the failure of his nervous system may have been due to his having worked with radium preparations for four months with his hands unprotected.

149. Kohlrausch to Wien, 23 November 1904, DM 2449.

150. Warburg, "Kohlrausch," 918.

151. Scheel, "Kohlrausch," 153-54.

152. Kohlrausch to Barus, 6 July 1905, CBP.

153. [Friedrich Kohlrausch], "Denkschrift über die Tätigkeit der Physikalisch-Technischen Reichsanstalt von Anfang 1900 bis 1903," *SBVR 206* (1903-4): 1139-59, on 1154- 59.

154. Cf. *SBVR 165* (1898-1900): 509, where this is noted of the Reichsanstalt vis-à-vis academic institutes.

## *Chapter 5. The search for reform*

1. Kohlrausch to Wien, 18 and 23 November 1904, DM 2448 and 2449, respectively; Reichskanzler (Reichsamt des Innern) to Minister der geistlichen und Unterrichtsangelegenheiten, 20 January 1917, ZStAP, RI, Personalakte Warburg, Nr. 645, Bl. 142; *W. C. Röntgen: Briefe an L. Zehnder,* ed. Ludwig Zehnder (Zurich, 1935), 91-92, 98; Otto Glasser, *Dr. W. C. Röntgen* (Springfield, Ill., 1945), 124; and W. Robert Nitske, *The Life of Wilhelm Conrad Röntgen: Discoverer of the X Ray* (Tucson, Az., 1971), 80, 212.

2. K. bayerisches Staatsministerium des Innern für Kirchen- und Schulangelegenheiten to Wilhelm Conrad Röntgen, 3 August 1904, Deutsches Röntgen-Museum, Remscheid; and Kohlrausch to Hochgeehrter Herr Direktor, 21 November 1904, ZStAP, RI, Personalakte Warburg, Nr. 645, Bl. 9-12.

3. Röntgen to Zehnder, 11 October 1904, in *W. C. Röntgen,* ed. Zehnder,

91-92.

4. Peter Paul Koch, "Wilhelm Conrad Röntgen als Forscher und Mensch," *ZTP 4* (1923): 273-79, on 274-76; and Nitske, *The Life of Wilhelm Conrad Röntgen*, 80, 212.

5. The following discussion relies on Kohlrausch to Herr Direktor, 21 November 1904.

6. Ibid. On Voigt see Christa Jungnickel and Russell McCormmach, *Intellectual Mastery of Nature: Theoretical Physics from Ohm to Einstein*, 2 vols., Vol. 2: *The Now Mighty Theoretical Physics, 1870-1925* (Chicago, 1986), 113-19, 268-74. In 1912, Voigt made a number of insufficiently informed and intemperate criticisms of the Reichsanstalt and its first three presidents. (See his *Physikalische Forschung und Lehre in Deutschland während der letzten hundert Jahre: Festrede im Namen der Georg-August Universität zur Jahresfeier der Universität am 5. Juni 1912* [Göttingen, 1912], 20-21; Warburg's reply "Ueber die Ziele der Physikalisch-Technischen Reichsanstalt; zur Abwehr," *PZ 13* [1912]: 1091-93; and Voigt's rejoinder, "Entgegnung," ibid., 1093-95.)

7. Friedrich Kurylo and Charles Susskind, *Ferdinand Braun: A Life of the Nobel Prizewinner and Inventor of the Cathode-Ray Oscilloscope* (Cambridge, Mass., 1981), Chap. 7, esp. 137-55.

8. Kohlrausch to Herr Direktor, 21 November 1904.

9. Staatssekretär des Innern to Staatssekretär des Reichsschatzamts, 30 November 1904, ZStAP, RI, Personalakte Warburg, Nr. 645, Bl. 3-6. Ernst Hagen harbored hopes of becoming the new president or, failing that, of switching into an academic career (Hagen to Heinrich Kayser, 7 October 1904, SPK, Slg. Darmstaedter, F i c 1902: Bessel-Hagen).

10. Kohlrausch to Herr Direktor, 21 November 1904.

11. Warburg to Dr. Richter, Direktor im Reichsamt des Innern, 23 October 1904, ZStAP, RI, Personalakte Warburg, Nr. 645, Bl. 6 ff.

12. Kohlrausch to Herr Direktor, 21 November 1904; Staatssekretär des Innern to Staatssekretär des Reichsschatzamts, 30 November 1904; Staatssekretär (i. A.) des Reichsschatzamts to Staatssekretär des Innern, 1 December 1904; Reichskanzler to Minister der geistlichen pp. Angelegenheiten (draft?), 28 December 1904; Minister der geistlichen, Unterrichts- und Medizinal-Angelegenheiten to Reichskanzler (Reichsamt des Innern), 6 January 1905; Staatssekretär des Innern to Staatssekretär des Reichsschatzamts, 14 January 1905; and Staatssekretär des Innern to Staatssekretär des Reichsschatzamts, 31 January 1905, all in ZStAP, RI, Personalakte Warburg, Nr. 645, Bl. 3-18; and Lewald to Warburg, 10 March 1909, ibid., Sig. 13144/5, Bl. 85.

13. David Farrer, *The Warburgs: The Story of a Family* (New York, 1975), 148-49, 223.

14. Unless otherwise noted, the following account of Warburg's life and work is based on Max Planck et al., "Nr. 24. Wahlvorschlag für Emil Warburg

(1846-1931) zum OM. Berlin, 30. Mai 1895," in *PPI*, 131-32; Emil Warburg, "Antrittsrede," *SB* (1896): (II) 743-45, reprinted in *PPII*, 183-84; Heinrich Rubens, "Adresse an Hrn. Emil Warburg zum 50. Doktorjubiläum," *SB* (1917): (I) 269-71; Ernst Gehrcke, "Warburg als Physiker," *ZTP 3* (1922): 186-92; Albert Einstein, "Emil Warburg als Forscher," *NWN 10* (1922): 823-28; E. Grüneisen, "Emil Warburg zum achtzigsten Geburtstage," *NWN 14* (1926): 203-7; idem, "Emil Warburg," *AP 403* (1931): 521-24; Friedrich Paschen, "Gedächtnisrede des Hrn. Paschen auf Emil Warburg," *SB* (1932): (I) cxv-cxxiii, reprinted in *PPII*, 185-91; and James Franck, "Emil Warburg zum Gedächtnis," *NWN 19* (1931): 993-97.

15. Gehrcke, "Warburg als Physiker," 187-88; and H. Schering, "Emil Warburg und die Technik," *NWN 14* (1926): 208-11, on 208-9.

16. E. Warburg, "Ueber den Aufschwung der modernen Naturwissenschaften," *Berichte der Naturforschenden Gesellschaft zu Freiburg i. B. 6* (1892): 18-29, on 27-28.

17. G. W. Küchler, "Physical Laboratories in Germany," in *Occasional Reports by the Office of the Director-General of Education in India*, No. 4 (Calcutta, 1906), 181-211, on 193. Cf. Georg Gehloff, "Emil Warburg als Lehrer," *ZTP 3* (1922): 193-94, on 193; and R. W. Pohl, "Von den Studien- und Assistentenjahren James Francks: Erinnerungen an das Physikalische Institut der Berliner Universität," *PB 28* (1972): 542-44, on 543.

18. Gehloff, "Emil Warburg," 193 (quote); and Franck, "Emil Warburg zum Gedächtnis," 995-96.

19. A. S. King to H. Kayser, 22 January 1905, SPK, Slg., Darmstaedter F i c 1904.

20. Martin Klein, "Max Planck and the Beginnings of the Quantum Theory," *Archive for History of Exact Sciences 1* (1962): 459-79, on 476.

21. Thomas S. Kuhn, *Black-Body Theory and the Quantum Discontinuity, 1894-1912* (New York, 1978), 216-27, 228-29.

22. An exception was Rubens, who was, it will be recalled, a Reichsanstalt guest between 1895 and 1913 and who in 1906 became professor of experimental physics at the University of Berlin, where he now performed (most of) his experimental work.

23. Hans Krebs, "Two Letters by Wilhelm Conrad Röntgen," *Notes and Records of the Royal Society of London 28* (1973): 83-92; E. Warburg, "Bemerkungen zu der Aufspaltung der Spektrallinien im elektrischen Feld," *VDPG 15* (1913): 1259-66; Warburg to Nobelcomitee für Physik der Königlichen Akademie der Wissenschaften Stockholm, 23 January 1917, Kungl. Vetenskapsakademien, Stockholm (Nobel Archives); Warburg to Herrn Cervai, 17 March 1926, Papers of Emil Warburg privately held by Dr. Peter Meyer-Viol, Lanaken, Belgium; and Richard Wichard Pohl interview with T. S. Kuhn and F. Hund, 25 June 1963, Archive for History of Quantum Physics, AIP.

24. Elisabeth Crawford, *The Beginnings of the Nobel Institution: The Science Prizes, 1901-1915* (Cambridge, 1984), 98-100. Arrhenius was the most successful nominator (ibid.).

25. Board Session 3, 16 June 1905; and Board Session 3, 15 March 1906; both in SAA 4/Lk 29.

26. Board Session 3, 16 June 1905; and Board Session 3, 15 March 1906; both in ibid.

27. Emil Warburg, "Die Physikalisch-Technische Reichsanstalt in Charlottenburg," *Internationale Wochenschrift für Wissenschaft, Kunst und Technik 1* (1907): 537-48, on 538-39, 541 (quote), 547; cf. idem, "Die physikalisch-technische Reichsanstalt in Charlottenburg," *Zeitschrift des Oesterreichischen Ingenieur- und Architekten-Vereines 60* (1908): 513-17, 529-33, on 531.

28. Lewald to Staatssekretär des Innern, 14 August 1907, ZStAP, RI, Nr. 13148/3, Bl. 56-57. See Lewald's similar remarks in Board Session 3, 13 March 1908, BA 240, PTB-IB.

29. On Einstein's appointment to and participation on the board, see *Albert Einstein in Berlin 1913-1933*, Teil I, *Darstellung und Dokumente*, ed. Christa Kirsten and Hans-Jürgen Treder (Berlin, 1979), 10, 50, 53, 158-60; and ibid., Teil II, *Spezialinventar*, 78-83.

30. Einstein to H. Zangger, spring 1912, as cited in Abraham Pais, *'Subtle is the Lord . . .': The Science and the Life of Albert Einstein* (New York, 1982), 209.

31. Board Session 3, 15 March 1906, SAA 4/Lk 29. See also [Emil Warburg], "Denkschrift über die Tätigkeit der Physikalisch-Technischen Reichsanstalt von Anfang 1904 bis Ende 1906," *SBVR 239* (1907): 446-63, on 446.

32. Board Session 1, 10 March 1909, BA 240, PTB-IB.

33. Board Session 1, 8 March 1911, ibid.

34. Board Session 3, 10 March 1911, ibid.

35. K. Strecker, "Die Personalverhältnisse der wissenschaftlichen Beamten der Physikalisch-Technischen Reichsanstalt" (29 August 1911), in BA 265, PTB-IB.

36. Ibid., 1 (quote), 23-24. For confirmation of Strecker's claims see, e.g., Board Session 4, 24 March 1888 and Board Session 2, 22 March 1889, both in SAA 61/Lc 973; Board Session 2, 15 March 1900, BA 239, PTB-IB; and Board Session 3, 10 March 1911, Sonderprotokoll, BA 240, PTB-IB.

37. Strecker, "Die Personalverhältnisse," 2, 5.

38. Ibid., Anlage 7.

39. Ibid., 9, 12, 16.

40. Wien to Hertz, 20 June 1890, DM 3061; and Wien to C. Runge, 24 May 1894, DM 1948-42, where Wien relates his interest in assuming Heinrich Kayser's recently vacated position at the Technische Hochschule Hannover.

41. W. Wien, "Ein Rückblick," in *Wilhelm Wien: Aus dem Leben und*

*Wirken eines Physikers,* ed. Karl Wien (Leipzig, 1930), 1-50, on 19.

42. For example, Ernst Hagen long but unsuccessfully sought to leave the Reichsanstalt for an academic post. (See Kohlrausch to Wien, 31 January 1900, DM 2437 and 2 February 1900, DM 2438; Hagen to Kayser, 7 October 1904; Hagen to Wien, 4 September 1906, DM, Wien Papers; Hagen to Friedrich Althoff, 23 October 1906, ZStA Merseburg, Rep. 92 Althoff B Nr. 60, Bl. 123-24; and Hagen to Althoff, 24 February 1907, ibid., Althoff C Nr. 20, Bl. 13- 15.)

43. Strecker, "Die Personalverhältnisse," Anlage 7.

44. Board Session 4, 24 March 1888, SAA 61/Lc 973.

45. Strecker, "Die Personalverhältnisse," 9.

46. Ibid., Anlage 1.

47. Paul Forman, John L. Heilbron, and Spencer Weart, "Physics *circa* 1900: Personnel, Funding, and Productivity of the Academic Establishments," *HSPS 5* (1975): 1-185, on 42.

48. Strecker, "Die Personalverhältnisse," 8 (quote), 19, Anlagen 3 and 7. The Reichsanstalt's hiring and promotion policy for tenured posts excluded the selection of scientists from outside the Reichsanstalt (ibid., 9).

49. Forman et al., "Physics *circa* 1900," 42; and Strecker, "Die Personalverhältnisse," Anlage 1.

50. Strecker, "Die Personalverhältnisse," 1-2.

51. Ibid., 2-3.

52. Board Session 2, 15 March 1900, BA 239, PTB-IB; cf. Lewald's remarks at Board Session 2, 10 March 1914, BA 240, PTB-IB.

53. Board Session 2, 22 March 1889, SAA 61/Lc 973; Kohlrausch to Wien, 17 December 1900, DM 2442; and A. Güntherschulze, "Die Physikalisch-Technische Reichsanstalt," *Frankfurter Zeitung,* 13 July 1914.

54. Warburg, "Die Physikalisch-Technische Reichsanstalt in Charlottenburg," 538-39, 541 (quote), 547.

55. Güntherschulze, "Die Physikalisch-Technische Reichsanstalt."

56. "Aktenvermerk über das Schreiben des R.A.d.I. vom 3. 8. 1912 Nr. IA 6679 . . . ," BAK, R2 12376.

57. Quoted in Gerhard Becherer, "Die Geschichte der Entwicklung des Physikalischen Instituts der Universität Rostock," *Wissenschaftliche Zeitschrift der Universität Rostock: Mathematisch-Naturwissenschaftliche Reihe 16* (1967): 825-37, on 829, and cited in Forman et al., "Physics *circa* 1900," 103.

58. "Denkschrift betreffend die Vorarbeiten und die Ausarbeitung des Projekts zu einem Neubaue des elektrischen Laboratoriums der Physikalisch-Technischen Reichsanstalt, II. Abteilung," in *HEDR* (1908), Beilage A, Kapitel 3, Titel 8, pp. 60-61; also in Board Session 3, Sonderprotokoll, 15 March 1907, BA 240, PTB-IB. In 1910 Kohlrausch claimed that, given the Reichsanstalt's official duties, the scientific and technical institutes at German

universities and Technische Hochschulen had comparatively greater resources available to them (Board Session 2, 10 March 1910, BA 240, PTB-
IB).

59. Lewald to Staatssekretär des Innern, 16 October 1913, ZStAP, RI, Nr.
13187/1, Bl. 2.

60. *Helmholtz-Fond E.V.,* copy in ZStAP, RI, Nr. 13187/1, Bl. 3-7.

61. "Aufruf," July 1912, in "Helmholtz-Fond," 63/Lh 599, SAA.

62. *Helmholtz-Fond E.V.,* B1. 4.

63. "Geschäftsbericht über die Tätigkeit des Vereins Helmholtz-Fonds
E.V. . . . ," ZStAP, RI, Nr. 13187/1, Bl. 12-13.

64. *Helmholtz-Fond E.V.*; and numerous documents in "Helmholtz-Fond,"
63/Lh 599, SAA. The fund never reached the planned level of 1 million
marks.

65. Lewald to Staatssekretär des Innern, 23 February 1914, including "Emil
Rathenau-Stiftung. Satzung der Emil Rathenau- Stiftung"; *Emil Rathenau-
Stiftung,* ZStAP, RI, Nr. 13188/1, Bl. 12-16, 28-32.

66. *HEDR* (1913), 6-7, 38-43, on 38-42.

67. [Emil Warburg], "Die Tätigkeit der Physikalisch- Technischen Reichsanstalt im Jahre 1913," *ZI 34* (1914): 113-30, 151-64, 184-200, on 197-200
(hereafter cited as "Tätigkeit, 1913").

68. "Feier des 80. Geburtstages des Ehrenmitgliedes der Deutschen Physi-
kalischen Gesellschaft Herrn Geheimrat Professor Dr. Max Planck . . . ,"
*VDPG 19* (1938): 57-76, on 62, reprinted in Planck's *Physikalische Abhand-
lungen und Vorträge,* 3 vols. (Braunschweig, 1958), *3,* 402-16, on 407; and
Walther Meissner interview with T. S. Kuhn, 8 February 1963, Archive for
History of Quantum Physics, AIP.

69. "Tätigkeit, 1913," 123, 125.

70. Ibid., 115-17.

71. The Verein Deutscher Ingenieure had originally requested this work
(see Chapter 4, "High Temperatures and High Science"). Linde, the Verein's
unofficial representative on the board, continued to advocate technology's
need for studies on the behavior of specific heats of gases under high pressures
(Board Session 1, 14 June 1905, SAA 4/Lk 29).

72. "Tätigkeit, 1913," 117-19.

73. Board Session 3, 13 March 1908, BA 240, PTB-IB.

74. "Tätigkeit, 1913," 119-20, 123-25 (quote on 124). On the usefulness to
technology of elastic-hysteresis studies of stress and strain within glass, see
Board Session 1, 13 March 1907, and Board Session 2, 14 March 1912, BA
240, PTB-IB.

75. "Tätigkeit, 1913," 120-23.

76. Ibid., 125. The four firms were the Kautschukwerke Dr. Traun &
Söhne, Schuchardt & Schütte, S. K. F. Kugellagerwerke, and Kugellagerwerke Fichtel & Sachs.

77. Board Session 3, 10 March 1911, BA 240, PTB-IB.

78. *EZ 30* (1909): 138, 167, 360; and "Festschrift des Instituts für Radium-forschung anlässlich seines 40-jährigen Bestandes (1910-1950)," *Sitzungs-berichte der Akademie der Wissenschaften in Wien: Mathematisch- naturwis-senschaftliche Klasse 159:2a* (1950): 1-57.

79. Thaddaeus J. Trenn, "Geiger, Hans (Johannes) Wilhelm," *DSB 5* (1972): 330-33.

80. Rutherford to Bertram Borden Boltwood, 1 and 15 February 1911 and 15 August 1912, in *Rutherford and Boltwood: Letters on Radioactivity,* ed. Lawrence Badash (New Haven, Conn., 1969), 243, 245, 275, respectively.

81. Warburg to Hochgeehrter Herr Geheimrat, 8 January 1912; and Reichs-kanzler (Reichsamt des Innern) to Württembergische Regierung, 17 January 1912, ZStAP, Reichsministerium für Wissenschaft, Erziehung und Volksbil-dung, Personalakte Geiger, Nr. G65, Bl. 6-7, 10-11.

82. Geiger to Rutherford, 13 October 1912, Cambridge University Library, Ernest Rutherford Collection, Add. 7653/G23. (The conversation between Rutherford and Warburg in Brussels probably occurred at or in connection with the first Solvay Conference, held in Brussels between 30 October and 3 November 1911.) Three months later Geiger wrote again to Rutherford, tel-ling him that "Warburg does everything he can to make things easy for me" (Geiger to Rutherford, 20 January 1913, ibid., Add. 7653/G27).

83. Geiger to Rutherford, 13 October 1912.

84. Geiger to Bohr, 12 October 1913, Bohr Scientific Correspondence (2,5), AIP.

85. Warburg to Staatssekretär des Innern, 13 March 1914, Personalakte Geiger, Bl. 15.

86. Sigalia Dostrovsky, "Bothe, Walther Wilhelm (Georg)," *DSB 2* (1970): 337-39; and R. Fleischmann, "Walter Bothe und sein Beitrag zur Atomkern-forschung," *NWN 44* (1957): 457-60.

87. Interview with Sir James Chadwick by Charles Weiner, 15-21 April 1969, AIP (quote); and Harrie Massey and N. Feather, "James Chadwick," *Bi-ographical Memoirs of Fellows of the Royal Society 22* (1976): 11-70, esp. 12-15.

88. "Tätigkeit, 1913," 126-27.

89. [Emil Warburg], "Die Tätigkeit der Physikalisch-Technischen Reichs-anstalt im Jahre 1906," *ZI 27* (1907): 109- 24, 147-60, 184-200, on 124, 155; and idem, "Die Tätigkeit der Physikalisch-Technischen Reichsanstalt im Jahre 1907," *ZI 28* (1908): 101-16, 139-57, 172-89, on 115.

90. "Tätigkeit, 1913," 127-28.

91. Ibid.

92. Ibid., 128.

93. Ibid., 128-29.

94. [Emil Warburg], "Denkschrift über die Tätigkeit der Physikalisch-

Technischen Reichsanstalt von Anfang 1912 bis Ende 1920," *SBVR 365* (1920): 880-911, on 888.

95. Ernest Rutherford, "The British Radium Standard," *Nature 92* (1913): 402-3, on 403.

96. Geiger to Rutherford, 7 January 1914, Cambridge University Library, Ernest Rutherford Collection, Add. 7653/G47.

97. "Tätigkeit, 1913," 129; and Warburg to Regierungsrat Herrn Freiherrn von Stein (in RI), 16 January 1914, ZStAP, RI, Sig. 13144/5, Bl. 158-59.

98. "Tätigkeit, 1913," 129-30; Hans Geiger, "Ueber eine einfache Methode zur Zählung von α- und β- Strahlen," *VDPG 15* (1913): 534-39; Thaddeus J. Trenn, "Die Erfindung des Geiger-Müller-Zählrohres," *Deutsches Museum, Abhandlungen und Berichte 44* (1976): 54-64; and idem, "The Geiger-Müller Counter of 1928," *Annals of Science 43* (1986): 111-35.

99. "Tätigkeit, 1913," 126.

100. Einstein to Warburg, 25 April 1911 or 1912, Warburg File, the Einstein Papers, Department of Manuscripts and Archives, The Hebrew University of Jerusalem, cited in Armin Hermann, *Frühgeschichte der Quantentheorie (1899-1913)* (Mosbach in Baden, 1969), 93.

101. Einstein, "Warburg als Forscher," 823, cited in Paschen, "Gedächtnisrede," cxxi.

102. "Tätigkeit, 1913," 125-26.

103. Grüneisen, "Emil Warburg," 524.

104. Deutscher Verein von Gas- und Wasserfachmänner and Verband Deutscher Elektrotechniker to Reichsamt des Innern, 27 December 1909, ZStAP, RI, Nr. 13166/11, Bl. 61-67. American engineers also still hoped to establish a unit of luminous intensity based upon blackbody radiation. (See C. W. Waidner and G. K. Burgess, "Note on the Primary Standard of Light," *Electrical World 52* [1908]: 625-28, esp. 627.)

105. Warburg to Staatssekretär des Innern, 14 February 1910, ZStAP, RI, Nr. 13166/11, Bl. 69-70. Nernst also strongly advocated conducting radiation research at the Reichsanstalt in order to establish a unit of luminous intensity (see Board Session 2, 15 June 1905, and Board Session 3, 15 March 1906, SAA 4/Lk 29).

106. Board Session 1, 9 March 1910, BA 240, PTB-IB. See also E. Warburg, "Ueber eine rationelle Lichteinheit," *VDPG 19* (1917), 3-10.

107. Friedrich Hoffmann, H. Korte, and H. Willenberg, "Der Werdegang der neuen Lichteinheit," *Das Licht 11* (1941): 107-12.

108. W. Jaeger and St. Lindeck, "Die Internationale Konferenz über elektrische Einheiten und Normale zu London im Oktober 1908," *EZ 30* (1909): 344-48; and Schering, "Warburg und die Technik," 210. On the electrical conferences and the setting of electrical standards see Wilhelm Jaeger, *Die Entstehung der internationalen Masse der Elektrotechnik (= Geschichtliche Einzeldarstellungen aus der Elektrotechnik*, ed. Elektrotechnischer Verein, Band

4) (Berlin, 1932).

109. [Warburg], "Denkschrift 1912 bis 1920," 881.

110. Grüneisen, "Emil Warburg zum achtzigsten Geburtstage," 207; idem, "Emil Warburg," 522; and Franck, "Warburg zum Gedächtnis," 996.

111. Franck, "Warburg zum Gedächtnis," 996.

112. See, e.g., [Warburg], "Die Tätigkeit im Jahre 1906," 110.

113. Copy of Siemens-Stephan-Gedankplatte, Berlin, 1909, in Nachlass Emil Warburg, SPK.

114. "Ein Ehrentag für Emil Warburg," *Stahl und Eisen 37* (1917): 385-86.

115. Schering, "Warburg und die Technik," 209-10; and C. Müller, "Emil Warburg 80 Jahre," *EZ 47* (1926): 317.

116. Lewald to Staatssekretär des Innern, 8 January 1910, ZStAP, RI, Personalakte Warburg, Nr. 645, Bl. 73-74.

117. "Tätigkeit, 1913," 151-52.

118. "Aktenvermerk über die Etatsanmeldungen des R.A.d.I. für 1913 vom Schreiben des R.S.A. vom 22. August 1912, betreffend: Erstdauernde Ausgaben Kap. 13b Titel 2," BAK, R2 12376.

119. "Tätigkeit, 1913," 152-53.

120. Ibid., 153-54.

121. Ibid., 154-55, 157-58.

122. Board Session 3, 15 March 1907, BA 240, PTB-IB.

123. "Denkschrift betreffend die Vorarbeiten und die Ausarbeitung des Projekts zu einem Neubaue des elektrischen Laboratoriums," 60.

124. Ibid.; and "Aktenvermerk über die Etatsanmeldungen des R.A.d.I. für 1913."

125. Board Session 3, 15 March 1907, BA 240, PTB-IB.

126. Board Session 2, 12 March 1908; and Board Session 2, 11 March 1909, ibid.

127. Board Session 2, 11 March 1909, ibid.

128. Board Session 1, 8 March 1911, and Board Session 3, 10 March 1911, ibid.

129. [Emil Warburg], "Die Tätigkeit der Physikalisch-Technischen Reichsanstalt im Jahre 1911," *ZI 32* (1912): 119-35, 155-69, 195-210, on 120.

130. Ibid., 157; and Emil Warburg, "Ueber den Entwicklungsgang der Starkstromtechnik und über deren Beziehungen zur Physikalisch-Technischen Reichsanstalt," *Internationale Monatsschrift für Wissenschaft, Kunst und Technik 8* (1914): 914-29, on 918, 929.

131. [Warburg], "Denkschrift 1912 bis 1920," 880.

132. *DMZ*, Heft 3 (1 February 1912): 32, in *ZI 32* (1912).

133. "Tätigkeit, 1913," 158-61.

134. Board Session 3, 16 June 1905, SAA 4/Lk 29.

135. [Warburg], "Denkschrift Anfang 1904 bis Ende 1906," 447.

136. Board Session 2, 12 March 1908; Board Session 2, 10 March 1910;

and Board Session 1, 8 March 1911, all in BA 240, PTB-IB; and "Tätigkeit, 1913," 162-64.

137. Board Session 2, 12 March 1908; Board Session 1, 8 March 1911, both in BA 240, PTB-IB; and [Emil Warburg], "Die Tätigkeit der Physikalisch-Technischen Reichsanstalt im Jahre 1910," *ZI 31* (1911): 112-25, 148-63, 183-97, on 114.

138. Board Session 1, 8 March 1911, BA 240, PTB-IB; and Schering, "Warburg und die Technik," 211.

139. "Tätigkeit, 1913," 161-62.

140. Board Session 3, 12 March 1909, BA 240, PTB-IB.

141. Board Session 2, 10 March 1910, ibid. Cf. *Der Verband Deutscher Elektrotechniker 1893-1918* (Berlin, [1918]), 46.

142. Editorial, "H. F. Wiebe," *DMZ*, Heft 19 (1 October 1912): 197, in *ZI 32* (1912); and *DMZ*, Heft 22 (15 November 1912): 236, in *ZI 32* (1912).

143. "Tätigkeit, 1913," 184-86.

144. Ibid., 184-88 (quote on 188).

145. Ibid., 189-90.

146. [Emil Warburg], "Die Tätigkeit der Physikalisch-Technischen Reichsanstalt im Jahre 1909," *ZI 30* (1910): 106- 20, 140-60, 174-95, on 184.

147. "Tätigkeit, 1913," 190.

148. Ibid., 190-92.

149. Board Session 3, 13 March 1908, BA 240, PTB-IB.

150. [Warburg], "Denkschrift Anfang 1904 bis Ende 1906," 446-47.

151. Warburg, "Die Physikalisch-Technische Reichsanstalt," 545.

152. Board Session 3, 13 March 1908, BA 240, PTB-IB.

153. *HEDR* (1913), 6.

154. Ernst Hagen and Karl Scheel, "Die Physikalisch-Technische Reichsanstalt," in *Ingenieurwerke in und bei Berlin: Festschrift zum 50 Jährigen Bestehen des VDI*, ed. Verein Deutscher Ingenieure (Berlin, 1906), 60-67, on 67; cf. [Friedrich Kohlrausch], *Die bisherige Tätigkeit der Physikalisch-Technischen Reichsanstalt: Aus dem Reichstage am 19. Februar 1904 überreichten Denkschriften. Mit einem Verzeichnis der Veröffentlichungen aus den Jahren 1901-1903* (Braunschweig, 1904), 6.

155. Karl Scheel, "Die Physikalisch-Technische Reichsanstalt in Charlottenburg," *Akademische Rundschau 1* (1913): 221-27, on 226; and idem, "Die Physikalisch-Technische Reichsanstalt," *Anzeiger für Berg-, Hütten- und Maschinenwesen 38* (1916): 5498-99, 5569-70, on 5570.

156. H. Ebert, "50 Jahre Physikalisch-Technsiche Reichsanstalt: Ein Rückblick," *Glas und Apparat 18* (1937): 203-5, on 205. Cf. Walter Meissner, "Hermann F. Wiebe," *Petroleum: Zeitschrift für die gesamten Interessen der Petroleum-Industrie und des Petroleum-Handels 8* (1912): 69-71, who reports (on 70) that between 1887 and 1913 the Heat and Pressure Laboratory tested more than 250,000 instruments.

157. Scheel, "Die Physikalisch-Technische Reichsanstalt," 225, 5569.

158. Güntherschulze, "Die Physikalisch-Technische Reichsanstalt."

159. *HEDR* (1913), 41.

160. Wilhelm Ostwald, "Die chemische Reichsanstalt," *Zeitschrift für angewandte Chemie 19* (1906): 1025-27, on 1027.

161. *Verzeichnis der Veröffentlichungen aus der Physikalisch-Technischen Reichsanstalt 1887 bis 1900* (Berlin, 1901); [Friedrich Kohlrausch], "Denkschrift über die Tätigkeit der Physikalisch-Technischen Reichsanstalt von Anfang 1900 bis Ende 1903," *SBVR 206* (1903-4): 1139-59, on 1154-59; [Warburg], "Denkschrift von Anfang 1904 bis Ende 1906," 459-63; idem, "Denkschrift über die Tätigkeit der Physikalisch-Technischen Reichsanstalt von Anfang 1907 bis Ende 1911," *SBVR 298* (1912): 121-45, on 138- 45; and idem, "Denkschrift von Anfang 1912 bis Ende 1920," 898-911.

162. Friedrich Kohlrausch, "Vorwort zur 11. Auflage des Lehrbuchs der praktischen Physik [1910]," in *GAFK, 1,* 1084- 88, on 1086.

163. Board Session 2, 12 March 1908, BA 240, PTB-IB.

164. E. Warburg, "Werner Siemens und die Physikalisch-Technische Reichsanstalt," *NWN 4* (1916): 793-97, on 797; idem, "Denkschrift 1912 bis 1920," 880 (quote).

165. [Warburg], "Denkschrift 1912 bis 1920," 880.

166. [Emil Warburg], "Die Tätigkeit der Physikalisch-Technischen Reichsanstalt im Jahre 1915," *ZI 36* (1916): 84- 93, 116-30, 149-59, on 84.

167. Ibid., 84-85. For further details on the Reichsanstalt's war-related work see [Emil Warburg], "Die Tätigkeit der Physikalisch-Technischen Reichsanstalt im Jahre 1916," *ZI 37* (1917): 70-78, 91-103, 120-32, on 70; idem, "Die Tätigkeit der Physikalisch-Technischen Reichsanstalt im Jahre 1917," *ZI 38* (1918): 59-65, 81-88, 94-106, on 59; idem, "Die Tätigkeit der Physikalisch-Technischen Reichsanstalt im Jahre 1918," *ZI 39* (1919): 105-16, 137-45, 180-94, on 105; and "Die Tätigkeit der Physikalisch-Technischen Reichsanstalt im Kriege," *Zeitschrift für komprimierte und flüssige Gase sowie für die Pressluft-Industrie 19* (1917-18): 105-7.

168. [Warburg], "Denkschrift 1912 bis 1920," 880.

169. Board Session for 1917, BA 240, PTB-IB.

170. *Albert Einstein in Berlin,* ed. Christa Kirsten and Hans- Jürgen Treder, Teil 1, 50; V. Ya. Frenkel, "On the History of the Einstein-de Haas Effect," *Soviet Physics Uspekhi 22* (1979): 580-87; Dieter Hoffmann, "Albert Einstein und die Physikalisch-Technische Reichsanstalt," in *Wirkung von Albert Einstein und Max von Laue,* Akademie der Wissenschaften der DDR, Institut für Theorie, Geschichte und Organisation der Wissenschaft, Kolloquien, Heft 21 (Berlin, 1980), 90-102; Peter Galison, "Theoretical Predispositions in Experimental Physics: Einstein and the Gyromagnetic Experiments, 1915- 1925," *HSPS 12* (1982): 285-323; and Pais, '*Subtle is the Lord . . .',* 245-49.

171. [Warburg], "Denkschrift 1912 bis 1920," 881.

172. "Verwaltungsrat des Helmholtz-Fonds E.V., Sitzung am 10. März 1914," and Warburg to Arnold v. Siemens (copy), 20 February 1914, both in ZStAP, RI, Nr. 13187/1, Bl. 10-11, 14-15; and documents in "Helmholtz-Fond," SAA 63/Lh 599.

173. Emil Warburg, "Bericht über die Tätigkeit des Vereins Helmholtz-Fonds E.V. im Geschäftsjahr 1914 . . . " (copy), 10 August 1915, ZStAP, RI, Nr. 13187/1, Bl. 21.

174. Emil Warburg, "Bericht über die Tätigkeit des Vereins Helmholtz-Fonds E.V. im Geschäftsjahr 1915, laufend vom 1 April 1915 bis 31 März 1916," ibid., Bl. 23.

175. "Sitzung des Verwaltungsrats der Emil Rathenau-Stiftung," 14 März 1917, ZStAP, RI, Nr. 13188/1, Bl. 44- 45.

# Bibliography

## Manuscript Sources

Baltimore. The Milton S. Eisenhower Library, Special Collections, The Johns Hopkins University: Henry A. Rowland Papers, MS. 6.

Berkeley. The Bancroft Library, University of California, Berkeley: Emil Fischer Papers.

Berlin. Physikalisch-Technische Bundesanstalt Institut-Berlin: Reichsanstalt Papers, BA 239, 240, 265, 658, 816.

Berlin. Staatsbibliothek Preussischer Kulturbesitz, Handschriftenabteilung: Sammlung Darmstaedter (including correspondence and other materials of Ernst Hagen, Hermann von Helmholtz, Heinrich Kayser, and Otto Lummer); Nachlass Emil Warburg; Nachlass Wilhelm Wien.

Berlin. Zentrales Archiv der Akademie der Wissenschaften der DDR, Historische Abteilung: Nachlass Hermann von Helmholtz; *Verhandlungen [= Protokolle] Physikalisch-mathematische Klasse,* II, Sig. II-VI, 50, 60, 118.

Braunschweig. Physikalisch-Technische Bundesanstalt: Correspondence of Friedrich Kohlrausch.

Cambridge. Cambridge University Library: Ernest Rutherford Collection.

Cambridge, Massachusetts. Massachusetts Institute of Technology, the Libraries, Institute Archives and Special Collections: Samuel Wesley Stratton Papers.

Göttingen. Niedersächsische Staats- und Universitätsbibliothek Göttingen, Handschriftenabteilung: Sig. H. Wagner 29 (correspondence of Friedrich Kohlrausch).

Jerusalem. Hebrew University of Jerusalem, Department of Manuscripts and Archives: The Albert Einstein Papers, Emil Warburg File.

Koblenz. Bundesarchiv: Reichsanstalt Papers, R 2 12375; R 2 12376; R 43 F/2365, 3/8; R 43 F/2365, 4/8.

Lanaken, Belgium. Dr. Peter Meyer-Viol: Emil Warburg Nachlass (privately held).

Merseburg. Zentrales Staatsarchiv der DDR: Rep. 76 Vb Sekt. 1 Tit. X Nr. 4;
    Rep. 92 Althoff B Nr. 60; Rep. 92 Althoff C Nr. 20.
Munich. Deutsches Museum, Sondersammlungen: Correspondence of Her-
    mann von Helmholtz, Heinrich Hertz, Friedrich Kohlrausch, and Wil-
    helm Wien; Wilhelm Wien Nachlass and Papers.
Munich. Siemens Museum, Siemens-Archiv-Akten: Reichsanstalt Papers,
    4/Lk 28, 4/Lk 29, 61/Lc 973, 63/Lh 599.
New York. Niels Bohr Library, Center for History of Physics, American In-
    stitute of Physics: Archive for History of Quantum Physics (interviews
    with Walther Meissner by T. S. Kuhn, 8 February 1963; and with Richard
    Wichard Pohl by T. S. Kuhn and F. Hund, 25 June 1963); Bohr Scientific
    Correspondence; Interview with Sir James Chadwick by Charles Weiner,
    15-21 April 1969; Arnold Sommerfeld Correspondence.
Philadelphia. American Philosophical Society: MS of Heinrich Kayser, *Erin-
    nerungen aus meinem Leben* (n.p., 1936).
Potsdam. Zentrales Staatsarchiv der DDR: Reichsanstalt Papers, Auswärtiges
    Amt, Band 24, Sig. 37880-37884; Reichskanzlei, Vf, Band 1, Sig. 975;
    RI, Sig. 13144-13191, and Personalakten of Reichsanstalt employees;
    Reichsministerium für Wissenschaft, Erziehung und Volksbildung, Per-
    sonalakten of Reichsanstalt employees.
Providence. The John Hay Library, Archives, Brown University: Carl Barus
    Papers.
Remscheid. Deutsches Röntgen-Museum: Wilhelm Conrad Röntgen Corre-
    spondence and Papers.
Stockholm. The Nobel Archives, Kgl. Vetenskapsakademien: Emil Warburg
    Correspondence with Nobel Committee for Physics.
Stockholm. Universitetsbibliotek: Svante Arrhenius Collection.
Strasbourg. Archives du Bas-Rhin: Dossier personnel de Friedrich
    Kohlrausch, AL 103/102.
Washington, D.C. Carneigie Institution of Washington: George Ellery Hale
    Papers (held in the National Academy of Sciences).
Washington, D.C. National Archives: National Bureau of Standards, Entry
    64, Records of the Heat Division, Temperature Physics Section, 1888-
    1911, RG 167.

*Printed Sources*

Auerbach, Felix. *The Zeiss Works and the Carl-Zeiss Stiftung in Jena: Their
    Scientific, Technical, and Sociological Development Popularly De-
    scribed.* 2nd ed. Trans. Siegfried Paul and Frederic J. Cheshire. London,
    1904.
— *Ernst Abbe: Sein Leben, sein Wirken, seine Persönlichkeit.* 2nd ed. Leip-
    zig, 1922.

Badash, Lawrence, ed. *Rutherford and Boltwood: Letters on Radioactivity.* New Haven, 1969.

Baker, Ray Stannard. *Seen in Germany.* London, 1902.

Becherer, Gerhard. "Die Geschichte der Entwicklung des Physikalischen Instituts der Universität Rostock." *Wissenschaftliche Zeitschrift der Universität Rostock: Mathematisch-Naturwissenschaftliche Reihe 16* (1967): 825-37.

Ben-David, Joseph. *The Scientist's Role in Society: A Comparative Study.* Chicago, 1984.

"Berichte von Universitäten, Hochschulen und wissenschaftlichen Einrichtungen." *PB 2* (1946): 85-91.

Beuthe, Hermann. "Die Physikalisch-Technische Reichsanstalt." *Der Deutsche Verwaltungsbeamte 4* (1937): 561- 63.

Bezold, Wilhelm von, et al. "Nr. 13. Wahlvorschlag für Friedrich Kohlrausch (1840-1910) zum OM. Berlin, 30 Mai 1895." In *PPI,* 98.

Bild- und Filmsammlung Deutscher Physiker, ed. *Deutsche Senioren der Physik.* Leipzig, 1936.

Boettcher, A. "Nachruf auf Geheimen Regierungsrat Prof. Dr. H. F. Wiebe und Regierungsrat Dr. J. Domke." *DMZ,* Heft 20 (15 October 1913): 209-13, in *ZI 33* (1913).

Bortfeldt, J., W. Hauser, and H. Rechenberg, ed. *Forschen-Messen-Prüfen: 100 Jahre Physikalisch-Technische Reichsanstalt/Bundesanstalt 1887-1987.* Weinheim, 1987.

Bradbury, S. *The Evolution of the Microscope.* Oxford, 1967.

Brauer, Ludolph, Albrecht Mendelssohn Bartholdy, and Adolf Meyer, ed. *Forschungsinstitute: Ihre Geschichte, Organisation und Ziele.* 2 vols. Hamburg, 1930.

Braun-Artaria, Rosalie. *Anna von Helmholtz: Ein Erinnerungsblatt.* N.p., 1899.

Bright, Arthur A., Jr. *The Electric Lamp Industry: Technological Change and Economic Development from 1800 to 1947.* New York, 1949.

Brodhun, E. "Die Physikalisch-Technische Reichsanstalt. Fünfundzwanzig Jahre ihrer Tätigkeit. 4. Optik." *NWN 1* (1913): 321-25.

Brüche, Ernst. "Das Schicksal der PTR." *PB 6* (1950): 422-26.

— "Physikalisch-Technische Reichsanstalt, Berlin-Charlottenburg." *PB 8* (1952): 429.

— "Aus der Vergangenheit der Physikalischen Gesellschaft. 4. An der Schwelle des Jahrhunderts." *PB 17* (1961): 120-27.

— "Aus der Vergangenheit der Physikalischen Gesellschaft. 5. Die alten Gesellschaftszeitschriften." *PB 17* (1961): 225-32.

— "Aus der Vergangenheit der Physikalischen Gesellschaft. VI. Die Jahre zwischen Quanten- und Relativitätstheorie." *PB 17* (1961): 400-10.

— "Unser Geheimrat Scheel." *PB 22* (1966): 121-28.

Buchheim, Gisela. "Initiativen zur Gründung der Physikalisch-Technischen Reichsanstalt (1887)." *NTM 11* (1974): 33-43.

— "Reichstagsdebatten über die Gründung der Physikalisch-Technischen Reichsanstalt." *NTM 12* (1975): 1-13.

— "Die Wechselbeziehungen zwischen Industrie, Staat und Wissenschaft gezeigt am Beispiel der Gründung der Physikalisch-Technischen Reichsanstalt (1887)." *Acta Historiae Rerum Naturalium Necnon Technicarum*, special issue, 8 (1976): 189-97.

— "Die Entwicklung des elektrischen Messwesens und die Gründung der Physikalisch-Technischen Reichsanstalt." *NTM 14* (1977): 16-32.

— "Die Gründungsgeschichte der Physikalisch-Technischen Reichsanstalt von 1872 bis 1887." *Dresdener Beiträge zur Geschichte der Technikwissenschaften*, Heft 3 (1981); Heft 4 (1982).

— "Die Denkschrift vom 30.7.1872 von K.-H. Schellbach: Ein Nachtrag zur Vorgeschichte der Physikalisch-Technischen Reichsanstalt." *NTM 23* (1986): 99-101.

Buchmann, Edward. *Die Entwicklung der Grossen Berliner Strassenbahn und ihre Bedeutung für die Verkehrsentwickelung Berlins*. Berlin, 1910.

Buchwald, E. "Erinnerung an Otto Lummer." *PB 6* (1950): 313-16.

Bunsen, Marie von. *Zur Erinnerung an Frau Anna von Helmholtz*. N.p., 1899.

Burchardt, Lothar. *Wissenschaftspolitik im Wilhelminischen Deutschland: Vorgeschichte, Gründung und Aufbau der Kaiser-Wilhelm-Gesellschaft zur Förderung der Wissenschaften*. Göttingen, 1975.

— "Der Weg zur PTR: Die Projektierung und Gründung hochschulferner physikalischer Forschungsstätten im Kaiserreich." *PB 32* (1976): 289-97.

Byatt, I. C. R. "Electrical Products." In *The Development of British Industry and Foreign Competition, 1875-1914*, ed. Derek H. Aldcroft, 238-73. London, 1968.

Cahan, David. "The Physikalisch-Technische Reichsanstalt: A Study in the Relations of Science, Technology and Industry in Imperial Germany. Ph.D. diss., The Johns Hopkins University, 1980.

— "Werner Siemens and the Origin of the Physikalisch-Technische Reichsanstalt, 1872-1887." *HSPS 12* (1982): 253-83.

— "The Institutional Revolution in German Physics, 1865-1914." *HSPS 15:2* (1985): 1-65.

— "Die Physikalisch-Technische Reichsanstalt 1887 bis 1918." In *Forschen-Messen-Prüfen*, ed. J. Bortfeldt et al., 27-67.

— *Meister der Messung: Die Physikalisch-Technische Reichsanstalt im Deutschen Kaiserrich*. Weinheim, 1988..

— "Instruments, Institutes, and Scientific Innovation: Friedrich Kohlrausch's Route to a Law of Electrolytic Conductivity." *Osiris 5* (forthcoming 1989).

Carhart, Henry S., "The Imperial Physico-Technical Institution in Charlotten-
burg." *TAIEE 17* (1900): 555-83.
— "The Imperial Physico-Technical Institution of Charlottenburg." *Annual
Report of the Board of Regents of the Smithsonian Institution . . . for the
Year Ending June 30, 1900*, 403-15. Washington, D.C., 1901.
Chemisch-Technische Reichsanstalt. See "Frankfurter Bezirksverein."
Cochrane, Rexmond. *Measures for Progress: A History of the National
Bureau of Standards*. Washington, D.C., 1966.
Crawford, Elisabeth. *The Beginnings of the Nobel Institution: The Science
Prizes, 1901-1915*. Cambridge, 1984.
Crocker, Francis. "The Precision of Electrical Engineering." *TAIEE 14*
(1897): 237-49.
Cruse, Alfred. "Die elektrische Industrie." In *Handbuch der Wirtschaftskunde
Deutschlands, 3*, 949-1008.
Czada, Peter. *Die Berliner Elektroindustrie in der Weimarer Zeit: Eine re-
gionalstatistisch-wirtschafthistorische Untersuchung*. Berlin, 1969.
Dahl, Per. "Kamerlingh Onnes and the Discovery of Superconductivity: The
Leyden Years, 1911-1914." *HSPS 15:1* (1984): 1-37.
[Debray, Jamin, Daubrée, and Laboulaye]. "Discours prononcés par des
membres de l'Académie aux funérailles de M. Regnault." *Comptes ren-
dus hebdomadaires des séances de l'Académie des sciences 86* (1878):
131-43.
Denke, Rudolf. "Das Siemens-Grundstück: Ein Beitrag zu seiner
Geschichte." *Der Bär von Berlin: Jahrbuch des Vereines für die
Geschichte des Berlins*, 6th series (1956): 108-34.
Dettmar, Georg. "Die Elektrizitäts-Industrie." In *Deutschland unter Kaiser
Wilhelm II*, ed. P. Zorn and H. v. Berger, *2*, 559-78. 3 vols. Berlin, 1914.
— *Die Entwicklung der Starkstromtechnik in Deutschland*. Berlin, 1940.
Deutscher Verein von Gas- und Wasserfachmännern zu Dresden. "Bericht der
Lichtmess-Commission." *Journal für Gasbeleuchtung und Wasserver-
sorgung 36* (1893): 449-51.
Diesselhorst, H. "Wärmeleitung." In *Encyklopädie der mathematischen Wis-
senschaften mit Einschluss ihrer Anwendungen*, ed. A. Sommerfeld,
*Physik*, Teil 1, *5*, 208-31. Leipzig, 1903-1921.
Dorn, Friedrich Ernst. *See* Physikalisch-Technische Reichsanstalt. Sundry
Official Publications. *Vorschläge zu gesetzlichen Bestimmungen über
elektrische Masseinheiten.*
Dostrovsky, Sigalia. "Bothe, Walther Wilhelm (Georg)." *DSB 2* (1970): 337-
39.
Drennan, O. J. "Kohlrausch, Friedrich Wilhelm Georg." *DSB 7* (1973): 449-
50.
DuBois-Reymond, Emil. "Gedächtnisrede auf Hermann von Helmholtz."
*APAWB* (1896): Gedächtnisrede 2, 1-50. Reprinted in *PPII*, 68-99.

— *Reden*, 2nd enl. ed. Ed. Estelle DuBois-Reymond. 2 vols. Leipzig, 1912. (See esp. "Der deutsche Krieg," *1*, 393- 420, and "Das Kaiserreich und der Friede," *1*, 421-30.)

Dumas, J.-B. "Eloge historique de Henri-Victor Regnault, membre de l'Académie des sciences de l'Institut de France." *Mémoires de l'Académie des sciences de l'Institut de France 42* (1883): 37-72.

Durm, Joseph, Hermann Ende, Eduard Schmitt, and Heinrich Wagner, ed. *Handbuch der Architektur. Teil 4: Entwerfen, Anlage und Einrichtung der Gebäude.* Halbband 6: *Gebäude für Erziehung, Wissenschaft und Kunst.* Heft 2: *Hochschulen, zugehörige und verwandte wissenschaftliche Institute* . . . . Darmstadt, 1888.

Ebert, H. "Karl Scheel zum siebzigsten Geburtstag." *ZTP 17* (1936): 65-66.

— "Nachruf auf Karl Scheel." *Zeitschrift für Physik 104* (1936): i-iii.

— "50 Jahre Physikalisch-Technische Reichsanstalt: Ein Rückblick." *Glas und Apparat 18* (1937): 203-5.

Ebert, Hans. "Baugeschichte und Wissenschaftsentwicklung: Zur Geschichte der TH/TU Berlin." In *Technische Universität Berlin: Baugeschichte-Bauplanung. TUB Dokumentation aktuell 1/1977*, ed. Jürgen Dietrich Besch et al., 29-98. Berlin, 1977.

Ebert, Hermann. *Hermann von Helmholtz*. Stuttgart, 1949.

Ehrenberg, Richard. "Werner Siemens in seiner Bedeutung für die deutschen Volkswirtschaft." *NWN 4* (1916): 823-27.

Einstein, Albert. "Emil Warburg als Forscher." *NWN 10* (1922): 823-38.

Elektrotechnischer Verein, Berlin. "Vereins-Angelegenheiten." *EZ 8* (1887): 2-3.

Eversheim, Paul. "Die Physik im Kriege." In *Deutsche Naturwissenschaft, Technik und Erfindung im Weltkriege*, ed. Bastian Schmid, 57-79. Munich, 1919.

Farrer, David. *The Warburgs: The Story of a Family*. New York, 1975.

Faulhaber, C. "Die Optische Industrie." In *Handbuch der Wirtschaftskunde Deutschlands, 3*, 455-72.

Feussner, Karl. "Die Thätigkeit der physikalisch-technischen Reichsanstalt auf elektrotechnischem Gebiete." *EZ 15* (1894): 672-75.

— "Die Ziele der neueren elektrotechnischen Arbeiten der Physikalisch-Technischen Reichsanstalt." In *Sammlung elektrotechnischer Vorträge*, ed. Ernst Voit, *1*, 115-46. Stuttgart, 1899.

Feussner, K., and St. Lindeck, "Die elektrischen Normal-Drahtwiderstände der Physikalisch-Technischen Reichsanstalt." *WAPTR 2* (1895): 501-41.

Fischer, Emil. "Die Begründung einer chemischen Reichsanstalt." *Verhandlungen des Vereins zur Beförderung des Gewerbefleisses* (1906): 113-23.

— *Aus meinem Leben*. Berlin, 1922.

Fischer, Emil, and Ernst Beckmann. *Das Kaiser-Wilhelm-Institut für Chemie Berlin-Dahlem*. Braunschweig, 1918.

Fleischmann, R. "Walter Bothe und sein Beitrag zur Atomkernforschung." *NWN 44* (1957): 457-60.

Foerster, Wilhelm. *Die Physikalisch-Technische Reichsanstalt: Ein Beitrag zur Verständigung.* Berlin, 1887.

— *Lebenserinnerungen und Lebenshoffnungen (1832 bis 1910).* Berlin, 1911.

Forman, Paul. "Paschen, Louis Carl Heinrich Friedrich." *DSB 10* (1974): 345-50.

Forman, Paul, John L. Heilbron, and Spencer Weart. "Physics *circa* 1900: Personnel, Funding, and Productivity of the Academic Establishments." *HSPS 5* (1975): 1-185.

Fox, Robert. *The Caloric Theory of Gases from Lavoisier to Regnault.* Oxford, 1971.

Franck, James. "Emil Warburg zum Gedächtnis." *NWN 19* (1931): 993-97.

"Frankfurter Bezirksverein." *Zeitschrift für angewandte Chemie 20* (1907): 603-8.

Fraunberger, F. "Quincke, Georg Hermann." *DSB 11* (1975): 241-42.

Frenkel, V. Ya. "On the History of the Einstein-de Haas Effect." *Soviet Physics Uspekhi 22* (1979): 580-87.

Fuess, R. "Gedenkfeier für Dr. Leopold Loewenherz." *ZI 13* (1893): 177-91.

Galison, Peter. "Theoretical Predispositions in Experimental Physics: Einstein and the Gyromagnetic Experiments, 1915- 1925." *HSPS 12* (1982): 285-323.

Galton, Douglas. "President's Address." *Report of the Sixty-fifth Meeting of the British Association for the Advancement of Science Held at Ipswich in September 1985 65* (1895): 3-35.

— "On the Reichsanstalt, Charlottenburg, Berlin." *Report of the Sixty-fifth Meeting of the British Association for the Advancement of Science Held at Ipswich in September 1985 65* (1895): 606-8.

— "Physics at the British Association." *Nature 52* (1895): 532-36.

Gehloff, Georg. "Die Strahlungsgesetze und ihre Anwendungen." *Helios: Fachzeitschrift für Elektrotechnik 16* (1910): 137-41, 149-55, 175-76, 181-86, 381-86, 393-96, 405-10, 417-20.

— "Emil Warburg als Lehrer." *ZTP 3* (1922): 193-94.

Gehrcke, Ernst. "Warburg als Physiker." *ZTP 3* (1922): 186-92.

— "Technische Physik und Physikalisch-Technische Reichsanstalt." *ZTP 10* (1929): 226-28.

— "Eugen Brodhun † [Obituary]." *Das Licht 8* (1938): 260.

— "Erinnerungen an Lummer." *PB 11* (1955): 315-17.

Geiger, Hans. "Ueber eine einfache Methode zur Zählung von $\alpha$- und $\beta$-Strahlen." *VDPG 15* (1913): 534-39.

German Budget, Reichsamt des Innern. "Anlage IV. Reichsamt des Innern." In *Haushalts-Etat des Deutschen Reichs. Reichshaushalts-Etat für das*

*Etatsjahr [auf das Rechnungsjahr]* . . . *nebst Anlagen* (title varies). Berlin, 1872-1918.

German Bundesrat. *See* Physikalisch-Technische Reichsanstalt. Founding Documents.

German Reichstag. *Stenographische Berichte über die Verhandlungen des Reichstages* (title varies). Berlin, 1867-1932.

— "Entwurf eines Gesetzes, betreffend die elektrischen Masseinheiten." *SBVR 164* (1897-98): 1735-41.

— *See also* Physikalisch-Technische Reichsanstalt. Denkschriften.

Gesellschaft Deutscher Naturforscher und Aerzte. "62. Versammlung deutscher Naturforscher und Aerzte zu Heidelberg in den Tagen von 17. bis 23. September 1889." *ZI 9* (1889): 224.

Gillispie, Charles Coulston, ed. *Dictionary of Scientific Biography*. 16 vols. New York, 1970-1980.

Gladstone, J. H. ["Henri Victor Regnault."] *Journal of the Chemical Society 33* (1878): 235-39.

Glasser, Otto. *Dr. W. C. Röntgen*. Springfield, Ill., 1945.

Glazebrook, R. T. "The Aims of the National Physical Laboratory of Great Britain." *Annual Report of the Board of Regents of the Smithsonian Institution . . . for the Year Ending June 30, 1901*, 341-57. Washington, D.C., 1902.

Goldstein, Eugen. "Aus vergangenen Tagen der Berliner Physikalischen Gesellschaft." *NWN 13* (1925): 39-45.

Görges, Hans, ed. *50 Jahre Elektrotechnischer Verein: Festschrift* . . . . Berlin, 1929.

Griewank, Karl. *Staat und Wissenschaft im Deutschen Reich: Zur Geschichte und Organisation der Wissenschaftspflege in Deutschland*. Freiburg i. B., 1927.

— "Aus den Anfängen gesamtdeutscher Wissenschaftspflege." In *Volkstum und Kulturpolitik: Eine Sammlung von Aufsätzen Gewidmet Georg Schreiber zum fünfzigsten Geburtstage*, ed. H. Konen and J. P. Steffes, 208- 36. Cologne, 1932.

— "Wissenschaft und Kunst in der Politik Kaiser Wilhelms I. und Bismarcks." *Archiv für Kulturgeschichte 34* (1952): 288-307.

Grüneisen, E. "Emil Warburg zum achtzigsten Geburtstages." *NWN 14* (1926): 203-7.

— "Emil Warburg." *AP 403* (1931): 521-24.

Gumlich, E. "Nachruf für Ernst Hagen." *PZ 24* (1923): 145-49.

Günther, N. "Abbe, Ernst." *DSB 1* (1970): 6-9.

Güntherschulze, A. "Die Physikalisch-Technische Reichsanstalt." *Frankfurter Zeitung* (13 July 1914).

Guthnick, P. "Wilhelm Foerster." *Vierteljahrsschrift der Astronomischen Gesellschaft 50* (1924): 5-13.

Hackmann, Willem D. "Sonar Research and Naval Warfare, 1914-1954: A Case Study of a Twentieth-Century Establishment Science," *HSPS 16:1* (1986): 83-110.

Haemmerling, K. *Charlottenburg: Das Lebensbild einer Stadt, 1905-1955.* Berlin, 1955.

Hagen, Ernst. *See* Physikalisch-Technische Reichsanstalt. Tätigkeitsberichte.

Hagen, E. and H. Rubens. "Ueber Beziehungen zwischen dem Reflexionsvermögen der Metalle und ihrem elektrischen Leitvermögen." *SB* (1903): (I) 269-77.

Hagen, Ernst and Karl Scheel. "Die Physikalisch-Technische Reichsanstalt." In *Ingenieurwerke in und bei Berlin: Festschrift zum 50 Jährigen Bestehen des VDI,* ed. Verein Deutscher Ingenieure, 60-67. Berlin, 1906.

*Handbuch der Wirtschaftskunde Deutschlands.* 4 vols. Leipzig, 1901-4.

Hauser, Wilfried. "85 Jahre Physikalisch-Technische Reichsanstalt/Bundesanstalt." *PTB-Mitteilungen* (June 1972): 449-50.

Hauser, Wilfried, and Helmut Klages. "Die Entwicklung der PTR zum metrologischen Staatsinstitut." *PB 33* (1977): 457-64.

*Haushalts-Etat des Deutschen Reichs. See* German Budget, Reichsamt des Innern.

Hefner-Alteneck, F. v. "Vorschlag zur Beschaffung einer constanten Lichteinheit." *Journal für Gasbeleuchtung und Wasserversorgung 27* (1884): 73-79.

— "Zur Frage der Lichteinheit." *Journal für Gasbeleuchtung und Wasserversorgung 27* (1886): 3-8.

Heilbron, J. L. *The Dilemmas of an Upright Man: Max Planck as Spokesman for German Science.* Berkeley, Calif., 1986.

Heintzenberg, Friedrich. *See* Siemens, Werner.

Helmholtz, Anna von. *Anna von Helmholtz: Ein Lebensbild in Briefen.* Ed. Ellen von Siemens-Helmholtz. 2 vols. Berlin, 1929.

Helmholtz, Hermann von. "Ueber die Berathungen des Pariser Kongresses, betreffend die elektrischen Masseinheiten." *EZ 2* (1881): 482-89.

— "Vorrede." *WAPTR 1* (1894): n.p.

— *Vorträge und Reden.* 4th ed. 2 vols. Braunschweig, 1896. (See esp. "Ueber das Verhältniss der Naturwissenschaften zur Gesammtheit der Wissenschaft. Akademische Festrede gehalten zu Heidelberg beim Antritt des Prorectorats 1862," *1,* 159-85.)

— *See also* Physikalisch-Technische Reichsanstalt.

Helmholtz, Hermann von, et al. "Nr. 8. Wahlvorschlag für Werner Siemens (1816-1892) zum OM. Berlin, 20 Oktober 1873." In *PPI,* 84-86.

Henning, F. "Henri Victor Regnault." *PZ 11* (1910): 770-74.

— "Ferdinand Kurlbaum † [Obituary]." *ZTP 8* (1927): 525-27.

— "Ferdinand Kurlbaum." *PZ 29* (1928): 97-104.

Hermann, Armin. *Frühgeschichte der Quantentheorie (1899-1913)*. Mosbach in Baden, 1969.

— "Lummer, Otto Richard." *DSB 8* (1973): 551-52.

Hertz, Heinrich. *Erinnerungen, Briefe, Tagebücher*. Ed. Johanna Hertz. Leipzig, 1927.

Heydweiller, Adolf. "Friedrich Kohlrausch." In *GAFK*, 2, xxxv-lxviii.

Hoffmann, Dieter. "Albert Einstein und die Physikalisch-Technische Reichsanstalt." In *Wirkung von Albert Einstein und Max von Laue*, 90-102. Akademie der Wissenschaften der DDR, Institut für Theorie, Geschichte und Organisation der Wissenschaft. Kolloquien, Heft 21. Berlin, 1980.

Hoffmann, Friedrich. "Strahlungsmessungen der PTR um die Jahrhundertwende." *PB 4* (1948): 143-45.

Hoffmann, Friedrich, H. Korte, and H. Willenberg. "Der Werdegang der neuen Lichteinheit." *Das Licht 11* (1941): 107- 12.

Holborn, Hajo. "The Prusso-German School: Moltke and the Rise of the General Staff." In *Makers of Modern Strategy from Machiavelli to the Nuclear Age*, ed. Peter Paret (with Gordon A. Craig and Felix Gilbert), 281-95. Princeton, N.J., 1986.

Holborn, L. "Elektrische Strassenbahnen und physikalische Institute." *Preussische Jahrbücher 81* (1895): 177-84.

— "Friedrich Kohlrausch † [Obituary]." *EZ 31* (1910): 211-12.

— "Die Physikalisch-Technische Reichsanstalt: Fünfundzwanzig Jahre ihrer Tätigkeit. 2. Wärme." *NWN 1* (1913): 225-29.

Holborn, L., and F. Kurlbaum. "Ueber ein optisches Pyrometer." *SB* (1901): (II) 712-19.

Holborn, L., and W. Wien. "Ueber die Messung tiefer Temperaturen." *AP 295* (1896): 213-28.

Hörz, Herbert, and Andreas Laass. "Hermann von Helmholtz und die Physikalisch-Technische Reichsanstalt." In *Perspektiven interkultureller Wechselwirkung für den wissenschaftlichen Fortschritt*, 123-30. Akademie der Wissenschaften der DDR, Institut für Theorie, Geschichte und Organisation der Wissenschaft. Kolloquien, Heft 48. Berlin, 1985.

Howard, Michael. *The Franco-Prussian War: The German Invasion of France, 1870-1871*. New York, 1981.

Hughes, Thomas P. *Networks of Power: Electrification in Western Society, 1880-1930*. Baltimore, Md., 1983.

Institut für Radiumforschung, Vienna. "Festschrift des Instituts für Radiumforschung anlässlich seines 40-jahrigen Bestandes (1910-1950)." *Sitzungsberichte der Akademie der Wissenschaften in Wien: Mathematisch-naturwissenschaftliche Klasse 159:2a* (1950): 1-57.

"Instruments of Precision at the Paris Exhibition." *Nature 63* (1900): 61-62.

Jaeger, Wilhelm. "Die Quecksilber-Normale der Physikalisch-Technischen Reichsanstalt für das Ohm." *WAPTR 2* (1895): 379-500.

— "Die Physikalisch-Technische Reichsanstalt: Fünfundzwanzig Jahre ihrer Tätigkeit. Elektrizität." *NWN 1* (1913): 273-79.

— *Die Entstehung der internationalen Masse der Elektrotechnik (Geschichtliche Einzeldarstellungen aus der Elektrotechnik,* ed. Elektrotechnischer Verein, Band 4). Berlin, 1932.

Jaeger, W., and St. Lindeck. "Die Internationale Konferenz über elektrische Einheiten und Normale zu London im Oktober 1908." *EZ 30* (1909): 344-48.

Jakob, Max. "C. Lindes Lebenswerk." *NWN 5* (1917): 417-23.

Jarausch, Konrad. *Students, Society and Politics in Imperial Germany: The Rise of Academic Illiberalism.* Princeton, N.J., 1982.

Julius, W. H. *Das Licht- und Wärmestrahlung verbrannter Gase.* Berlin, 1890.

Jungnickel, Christa, and Russell McCormmach. *Intellectual Mastery of Nature: Theoretical Physics from Ohm to Einstein.* Vol. 1: *The Torch of Mathematics 1800-1870.* Vol. 2: *The Now Mighty Theoretical Physics 1870-1925.* Chicago, 1986.

Kangro, Hans. *Vorgeschichte des Planckschen Strahlungsgesetzes: Messungen und Theorien der spektralen Energieverteilung bis zur Begründung der Quantenhypothese.* Wiesbaden, 1970.

— "Ultrarotstrahlung bis zur Grenze elektrisch erzeugter Wellen: Das Lebenswerk von Heinrich Rubens." *Annals of Science 26* (1970): 235-59; *27* (1971): 165-200.

— "Pringsheim, Ernst." *DSB 11* (1975): 149-51.

— "Rubens, Heinrich (Henri Leopold)." *DSB 11* (1975): 581-85.

— "Wien, Wilhelm Carl Werner Otto Fritz Franz." *DSB 14* (1976): 337-42.

Kaufmann, W. Review of *Die Physikalisch Technischen Staatslaboratorien,* by H. Pellat. *PZ 1* (1899-1900): 486.

Kern, Ulrich. "Die Physikalisch-Technische Reichsanstalt 1918 bis 1945." In *Forschen-Messen-Prüfen,* ed. J. Bortfeldt et al., 68-112.

Kirchhoff, Gustav Robert, Hermann von Helmholtz, and Werner Siemens. "Nr. 12. Wahlvorschlag für Friedrich Kohlrausch (1840-1910) zum KM. Berlin, 19. Juni 1884." In *PPI,* 95-97.

Kirsten, Christa, and Hans-Günther Körber, ed. *Physiker über Physiker: Wahlvorschläge zur Aufnahme von Physikern in die Berliner Akademie 1870 bis 1929 von Hermann v. Helmholtz bis Erwin Schrödinger.* Berlin, 1975.

— *Physiker über Physiker II: Antrittsreden, Erwiderungen bei der Aufnahme von Physikern in die Berliner Akademie, Gedächtnisreden von 1870 bis 1929.* Berlin, 1979.

Kirsten, Christa, and Hans-Jürgen Treder, ed. *Albert Einstein in Berlin 1913-1933.* Teil 1: *Darstellung und Dokumente.* Teil 2: *Spezialinventar.* Berlin, 1979.

Kiyonobu, Itakura, and Eri Yagi. "The Japanese Research System and the Establishment of the Institute of Physical and Chemical Research." In *Science and Society in Modern Japan: Selected Historical Sources*, ed. Nakayama Shigeru, David L. Swain, and Eri Yagi, 158-201. Cambridge, Mass., 1974.

Klein, F. "Ueber die Gründung eines physikalisch-technischen Universitätsinstituts in Göttingen." *Zeitschrift des Vereins deutscher Ingenieure 40* (1896): 75-78.

— "Plan eines physikalisch-technischen Instituts an der Universität Göttingen." *Zeitschrift des Vereins deutscher Ingenieure 40* (1896): 102-7.

Klein, Martin. "Max Planck and the Beginnings of the Quantum Theory." *Archive for History of Exact Sciences 1* (1962): 459- 79.

Klemm, Friedrich. "Linde, Carl von." *DSB 8* (1973): 365- 66.

Knoblauch, O. "Carl von Linde zum 11. Juni 1932." *ZTP 13* (1932): 25-52.

Koch, Peter Paul. "Wilhelm Conrad Röntgen als Forscher und Mensch." *ZTP 4* (1923): 273-79.

Kocka, Jürgen. *Unternehmensverwaltung und Angestelltenschaft am Beispiel Siemens 1847-1914: Zum Verhältnis von Kapitalismus und Bürokratie in der deutschen Industrialisierung*. Stuttgart, 1969.

Koenigsberger, Leo. *Hermann von Helmholtz*. 3 vols. Braunschweig, 1902-3.

Kohlrausch, Friedrich. *Leitfaden der praktischen Physik, zunächst für das physikalische Prakticum in Göttingen*. Leipzig, 1870. 11th rev. ed., *Lehrbuch der praktischen Physik*. Leipzig, 1910.

— "Diskussion über die Frage der Störungen wissenschaftlicher Institute durch electrische Bahnen." *EZ 16* (1895): 427-29, 444-45. Reprinted in *GAFK, 1*, 886-901.

— "Antrittsrede." *SB* (1896): (II) 736-43. Reprinted in *GAFK, 1*, 1024-31; and in *PPII*, 172-76.

— "Gustav Wiedemann. Nachruf." *VDPG 1* (1899): 155- 67. Reprinted in *GAFK, 1*, 1064-67.

— "Vorwort zur 11. Auflage des Lehrbuchs der praktischen Physik." In Kohlrausch's *Lehrbuch der Praktischen Physik*, 11th ed. ix-xii. Reprinted in *GAFK, 1*, 1084-88.

— *Gesammelte Abhandlungen von Friedrich Kohlrausch*. Ed. Wilhelm Hallwachs, Adolf Heydweiller, Karl Strecker, and Otto Wiener. 2 vols. Leipzig, 1910-11.

— *See also* Physikalisch-Technische Reichsanstalt.

Körber, Hans-Günther. "Holborn, Ludwig Christian Friedrich." *DSB 6* (1972): 469-70.

Krebs, Hans. "Two Letters by Wilhelm Conrad Röntgen." *Notes and Records of the Royal Society of London 28* (1973): 83- 92.

— (with Roswitha Schmid). *Otto Warburg: Zellphysiologe, Biochemiker, Mediziner, 1883-1970*. Stuttgart, 1979.

Krüss, Hugo. "Stephan Lindeck." *DMZ*, Heft 22 (15 November 1911): 233-34, in *ZI 31* (1911).

Küchler, G. W. "Physical Laboratories in Germany." In *Occasional Reports by the Office of the Director-General of Education in India*, No. 4, 181-211. Calcutta, 1906.

Kuhn, Thomas S. "Energy Conservation as an Example of Simultaneous Discovery." In *Critical Problems in the History of Science*, ed. Marshall Clagett, 321-56. Madison, Wis., 1959. Reprinted in Kuhn's *The Essential Tension: Selected Studies in Scientific Tradition and Change*, 66-104. Chicago, 1977.

— *Black-Body Theory and the Quantum Discontinuity, 1894- 1912*. New York, 1978.

Kundt, August. "Gedächtnisrede auf Werner von Siemens." *APAWB* (1893): Gedächtnisrede 2, 1-21. Reprinted in *PPII*, 111-23.

"A. Kundt." Obituary. *ZI 14* (1894): 258.

Kurylo, Friedrich, and Charles Susskind. *Ferdinand Braun: A Life of the Nobel Prizewinner and Inventor of the Cathode-Ray Oscilloscope*. Cambridge, Mass., 1981.

Landes, David. *The Unbound Prometheus: Technological Change and Industrial Development in Western Europe from 1750 to the Present*. Cambridge, 1972.

Lehmann, Otto. *Geschichte des Physikalischen Instituts der Technischen Hochschule Karlsruhe: Festgabe der Fridericiana zur 83. Versammlung Deutscher Naturforscher und Aerzte*. Karlsruhe, 1911.

Lemmerich, Jost. *Zur Geschichte der Physik an der Technischen Hochschule Berlin-Charlottenburg* (= *Berliner Beiträge zur Geschichte der Naturwissenschaften und der Technik*, ed. Friedrich G. Rheingans and Edgar Swinne, Vol. 3). Berlin, 1986.

Leyden, F. *Gross-Berlin: Geographie der Weltstadt*. Breslau, 1933.

— "Charlottenburg um 1880." In *Berlin*, ed. J. J. Haesslin, 322-24. Munich, 1971.

Linde, Carl. *Aus meinem Leben und von meiner Arbeit: Aufzeichnungen für meine Kinder und meine Mitarbeiter*. Munich, 1916.

"Literatur: Beleuchtungswesen." *Journal für Gasbeleuchtung und Wasserversorgung 36* (1893): 703.

Lodge, Oliver. "Section A. Mathematics and Physics. Opening Address by Prof. Oliver J. Lodge . . . , President of the Section." *Nature 44* (1897): 383-97.

Loewe, Joseph. "Die elektrotechnische Industrie." In *Die Störungen im deutschen Wirtschaftsleben während der Jahre 1900 ff.*, ed. Verein für Sozialpolitik, *3*, 75-155. 8 vols. in 4. Leipzig, 1903.

Loewenherz, L. "Die Aufgaben der zweiten (technischen) Abteilung der physikalisch-technischen Reichsanstalt." *ZI 8* (1888): 153-57.

Lummer, Otto. "Ueber die Ziele und die Thätigkeit der Physikalisch-Technischen Reichsanstalt." *Verhandlungen des Vereines zur Beförderung des Gewerbfleisses 73* (1894): 151- 84.

— "Light and Its Artificial Production." *Annual Report of the Board of Regents of the Smithsonian Institution . . . to July, 1897*, 273-99. Washington, D.C., 1898.

— *Die Ziele der Leuchttechnik.* Munich, 1903.

— "Wissenschaftliche Grundlagen zur ökonomischen Lichterzeugung." *Zeitschrift für Beleuchtungswesen 10* (1904): 1-3.

— *Die Lehre von der Strahlenden Energie (Optik),* Vol. 2 of *Müller-Pouillets Lehrbuch der Physik und Meteorologie.* 10th enl. ed. 4 vols. Braunschweig, 1909.

Lummer, O., and E. Brodhun. "Photometrische Untersuchungen. I. Ueber ein neues Photometer." *ZI 9* (1889): 41-50.

— "Photometrische Untersuchungen. IV. Die photometrische Apparate der Reichsanstalt für den technischen Gebrauch." *ZI 12* (1892): 41-50.

Lummer, O., and E. Gehrcke. "Ueber den Bau der Quecksilberlinien; ein Beitrag zur Auflösung feinster Spektrallinien." *SB* (1902): (I) 11-17.

Lummer, O., and E. Jahnke. "Ueber die Spectralgleichung des schwarzen Körpers und des blanken Platins." *AP 308* (1900): 283-97.

Lummer, O., and F. Kurlbaum. "Bolometrische Untersuchungen." *AP 282* (1892): 204-24.

— "Ueber die Herstellung eines Flächenbolometers." *ZI 12* (1892): 81-89.

— "Bolometrische Untersuchung für eine Lichteinheit." *SB* (1894): (I) 229-38.

— "Ueber die neue Platinlichteinheit der Physikalisch-technischen Reichsanstalt." *Verhandlungen der Physikalischen Gesellschaft zu Berlin 14* (1895): 56-70.

— "Der electrisch geglühte 'absolute schwarze' Körper und seine Temperaturmessung." *Verhandlungen der Physikalischen Gesellschaft zu Berlin 17* (1898): 106-11.

Lummer, O., and E. Pringsheim. "Die Strahlung eines 'schwarzen' Körpers zwischen 100 und 1300 °C." *AP 299* (1897): 395-410.

— "Die Vertheilung der Energie im Spektrum des schwarzen Körpers." *VDPG 1* (1899): 23-41.

— "1. Die Vertheilung der Energie im Spectrum des schwarzen Körpers und des blanken Platins. 2. Temperaturbestimmung fester glühender Körper." *VDPG 1* (1899): 215-35.

— "Ueber die Strahlung des schwarzen Körpers für lange Wellen." *VDPG 2* (1900): 163-80.

— "Temperaturbestimmung hocherhitzter Körper (Glühlampe etc.) auf bolometrischen und photometrischen Wege." *VDPG 3* (1901): 36-46.

— "Kritisches zur schwarzen Strahlung." *AP 311* (1901): 192-210.

— "Temperaturbestimmung mit Hilfe der Strahlungsgesetze." *PZ 3* (1901-2): 97-100.

Lummer, O., and Max Thiesen. "Ueber das Gesetz der schwarzen Strahlung." *VDPG 2* (1900): 65-70.

Macfarlane, A. "On the Units of Light and Radiation." *TAIEE 12* (1895): 85-98.

Manegold, Karl-Heinz. "Felix Klein als Wissenschaftsorganisator: Ein Beitrag zum Verhältnis von Naturwissenschaft und Technik im 19. Jahrhundert." *Technikgeschichte 35* (1968): 177-204.

— "Zur Emanzipation der Technik im 19. Jahrhundert im Deutschland." In *Wissenschaft, Wirtschaft und Technik: Studien zur Geschichte. Wilhelm Treue zum 60. Geburtstag*, ed. Karl-Heinz Manegold, 379-402. Munich, 1969.

— *Universität, Technische Hochschule und Industrie: Ein Beitrag zur Emanzipation der Technik im 19. Jahrhundert unter besonderer Berücksichtigung der Bestrebung Felix Kleins.* Berlin, 1970.

Marienfeld, Wolfgang. *Wissenschaft und Schlachtflottenbau in Deutschland 1897-1906 (= Marine-Rundschau, 1957, Beiheft 2)*. Frankfurt, 1957.

Massey, Harrie, and N. Feather, "James Chadwick." *Biographical Memoirs of Fellows of the Royal Society 22* (1976): 11-70.

Masur, Gerhard. *Imperial Berlin.* New York, 1970.

Matschoss, Conrad, ed. *Werner Siemens: Ein kurzgefasstes Lebensbild nebst einer Auswahl seiner Briefe.* 2 vols. Berlin, 1916.

May, Kenneth O. "Gauss, Karl Friedrich." In *DSB 5* (1972): 298-315.

McCormmach, Russell. "Editor's Forward." *HSPS 3* (1971): ix-xxiv.

— *Night Thoughts of a Classical Physicist.* Cambridge, Mass., 1982.

McKay, John P. *Tramways and Trolleys: The Rise of Urban Mass Transport in Europe.* Princeton, N.J., 1976.

Medicus, F. A. *Das Reichsministerium des Innern.* Berlin, 1940.

Meissner, W. "Hermann F. Wiebe." *Petroleum: Zeitschrift für die gesamten Interessen der Petroleum-Industrie und des Petroleum-Handels 8* (1912): 69-71.

Mendelssohn, Kurt. *The World of Walther Nernst: The Rise and Fall of German Science 1864-1941.* Edinburgh, 1973.

Messerschmidt, Manfred. "Die politische Geschichte der preussisch-deutschen Armee." In *Handbuch zur deutschen Militärgeschichte 1648-1939*, Band 2, Abschnitt 4, *Militärgeschichte im 19. Jahrhundert 1814-1890*, ed. F. Forstmeier and Hans Meier-Welcker. Munich, 1975.

Mie, Gustav. "Werner Siemens als Physiker." *NWN 4* (1916): 771-76.

Morley, Edward W. "Visits to Scientific Institutions in Europe." *American Architect and Building News 59* (1898): 12- 13.

Moseley, Russell. "Science, Government, and Industrial Research: The

Origins and Development of the National Physical Laboratory, 1900-1975." Diss., University of Sussex, 1976.

— "The Origins and Early Years of the National Physical Laboratory: A Chapter in the Pre-History of British Science Policy." *Minerva 16* (1978): 222-50.

Moser, H., ed. *Forschung und Prüfung: 75 Jahre Physikalisch-Technische Reichsanstalt.* Braunschweig, 1962.

Müller, C. "Emil Warburg 80 Jahre." *EZ 47* (1926): 317.

Müller, Felix, ed. *Karl Schellbach: Rückblick auf sein wissenschaftliches Leben, nebst zwei Schriften aus seinem Nachlass und Briefen von Jacobi, Joachimstahl und Weierstrass.* Leipzig, 1905.

Naglo, Emil, ed. *Die ersten 25 Jahre des Elektrotechnischen Vereins.* Berlin, 1904.

National Bureau of Standards, Washington, D.C. "Scientific Notes and News." *Science 16* (1902): 437.

National Physical Laboratory, Teddington, England. "A National Physical Laboratory." *Nature 55* (1897): 368-69.

— "The National Physical Laboratory." *Nature 55* (1897): 385-86.

— "Report on a National Physical Laboratory." *Nature 58* (1898): 548-49.

— "The National Physical Laboratory." *Nature 58* (1898): 565-66.

— "The National Physical Laboratory." *Engineering 71* (1901): 707-8.

Nernst, W. "Ueber die Helligkeit glühender schwarzer Körper und über ein einfaches Pyrometer." *PZ 7* (1906): 380-83.

Neumayer, G. "Die Kriegsflotten und die wissenschaftliche Forschung." In *Beiträge zur Beleuchtung der Flottenfrage*, 4th series, 27-34. Munich, 1900.

Nichols, E. F., and Heinrich Rubens. "Ueber Wärmestrahlen von grosser Wellenlänge." *Naturwissenschaftliche Rundschau 11* (1896): 545-49.

— "Beobachtung elektrischer Resonanz an Wärmestrahlen von grosser Wellenlänge." *SB* (1896): (II) 1393-1400.

Nichols, Ed. L., Clayton Sharp, and Chas. Matthews. "Standards of Light: Preliminary Report of the Subcommittee of the Institute." *TAIEE 13* (1896): 135-207.

Nitske, W. Robert. *The Life of Wilhelm Conrad Röntgen: Discoverer of the X Ray.* Tucson, Az., 1971.

Orlich, E. "Stephan August Lindeck." *ZI 31* (1911): 329- 31.

Ostwald, Wilhelm. "Die chemische Reichsanstalt." *Zeitschrift für angewandte Chemie 19* (1906): 1025-27.

— *Lebenslinien: Eine Selbstbiographie.* 3 vols. Berlin, 1926-27.

— *Aus dem wissenschaftlichen Briefwechsel Wilhelm Ostwalds. Teil 2: Briefwechsel mit Svante Arrhenius und Jacobus Hendricus Van't Hoff.* Ed. Hans-Günther Körber. 2 vols. Berlin, 1969.

Pais, Abraham. *'Subtle is the Lord . . .' : The Science and the Life of Albert Einstein.* New York, 1982.

Paschen, F. "Ueber Gesetzmässigkeiten in den Spectren fester Körper." *AP 294* (1896): 455-92.

— "Ueber Gesetzmässigkeiten in den Spectren fester Körper." *AP 296* (1897): 662-73.

— "Ueber die Vertheilung der Energie im Spectrum des schwarzen Körpers bei niederen Temperaturen." *SB* (1899): (I), 405-20; (II) 959-76.

— "Ueber das Strahlungsgesetz des schwarzen Körpers: Entgegnung auf Ausführungen der Herrn O. Lummer und E. Pringsheim." *AP 311* (1901): 648-58.

— "Gedächtnisrede des Hrn. Paschen auf Emil Warburg." *SB* (1932): (I) cxv-cxxiii. Reprinted in *PPII*, 185- 91.

Passer, Harold C. *The Electrical Manufacturers, 1875-1900.* Cambridge, Mass., 1953.

Pellat, H. "Les laboratoires nationaux physico-techniques." In *Congrès international de physique réuni à Paris en 1900: Rapports*, ed. C. E. Guillaume and L. Poincaré, *1*, 101-7. 4 vols. Paris, 1900-1.

Pernet, Johannes. "Ueber die physikalisch-technische Reichsanstalt zu Charlottenburg und die daselbst ausgeführten elektrischen Arbeiten." *Schweizerische Bauzeitung 18* (1891): 1-6.

— "Ueber den Einfluss physikalischer Präcisionsmessungen auf die Förderung der Technik und des Mass- und Gewichtswesens." *Schweizerische Bauzeitung 24* (1894): 110-14.

— "Hermann von Helmholtz. 31. August 1821 bis 8. September 1894: Ein Nachruf." *Neujahrsblatt der Naturforschenden Gesellschaft in Zürich 97* (1895): 1-36.

Pernet, J., W. Jaeger, and E. Gumlich. "Thermometrische Arbeiten betreffend die Herstellung und Untersuchung der Quecksilber-Normalthermometer." *WAPTR 1* (1894): 1-105, 1-439.

Peters, Theodore. *Geschichte des Vereines deutscher Ingenieure in zeitlicher Auseinanderfolge.* Berlin, 1912.

Pfetsch, Frank. "Scientific Organisation and Science Policy in Imperial Germany, 1871-1914: The Foundation of the Imperial Institute of Physics and Technology." *Minerva 3* (1970): 557- 80.

— *Zur Entwicklung der Wissenschaftspolitik in Deutschland, 1750-1914.* Berlin, 1974.

Physikalisch-Technische Reichsanstalt. Denkschrifte (Official Reports to the Reichstag). [Hermann von Helmholtz]. "Denkschrift über die bisherige Thätigkeit der Physikalisch-Technischen Reichsanstalt." *SBVR 122* (1890-91): 1368-80.

— [Hermann von Helmholtz]. "Denkschrift über die Thätigkeit der Physikalisch-Technischen Reichsanstalt in den Jahren 1891 und 1892." *SBVR 130* (1892-93): 385-402.

— [Friedrich Kohlrausch]. "Denkschrift über die Thätigkeit der Physikalisch-Technischen Reichsanstalt vom Anfang des Jahres 1893 bis Ostern 1895." *SBVR 151* (1895-97): 373-91.

— [Friedrich Kohlrausch]. "Denkschrift über die Thätigkeit der Physikalisch-Technischen Reichsanstalt vom Frühjahr 1895 bis zum Sommer 1897." *SBVR 163* (1897-98): 870-87.

— [Friedrich Kohlrausch]. "Denkschrift über die Thätigkeit der Physikalisch-Technischen Reichsanstalt vom Sommer 1897 bis Ende 1899." *SBVR 176* (1898-1900): 3449-63.

— [Friedrich Kohlrausch]. "Denkschrift über die Tätigkeit der Physikalisch-Technischen Reichsanstalt von Anfang 1900 bis Ende 1903." *SBVR 206* (1903-4): 1139-59.

— [Emil Warburg]. "Denkschrift über die Tätigkeit der Physikalisch-Technischen Reichsanstalt von Anfang 1904 bis Ende 1906." *SBVR 239* (1907): 446-63.

— [Emil Warburg]. "Denkschrift über die Tätigkeit der Physikalisch-Technischen Reichsanstalt von Anfang 1907 bis Ende 1911." *SBVR 298* (1912): 121-45.

— [Emil Warburg]. "Denkschrift über die Tätigkeit der Physikalisch-Technischen Reichsanstalt von Anfang 1912 bis Ende 1920." *SBVR 365* (1920): 880-911.

Physikalisch-Technische Reichsanstalt. Founding Documents. "Begründung der Vorschläge zur Errichtung einer 'physikalisch-technischen Reichsanstalt' für die experimentelle Förderung der exakten Naturforschung und der Präzisionstechnik. 'Vorbemerkungen.' [1-6] 'Aufgaben der ersten (wissenschaftlichen) Abtheilung der physikalisch- technischen Reichsanstalt. (Ausgearbeitet von Dr. von Helmholtz).' [6-9] 'Aufgaben der zweiten (technischen) Abtheilung der physikalisch-technischen Reichsanstalt (Ausgearbeitet von Dr. Foerster).' [10-13] 'Organisationsplan.' [14-49, including 5 Anlagen]." Beilage: "Denkschrift betreffend die Begründung eines Instituts für die experimentelle Förderung der exakten Naturforschung und der Präzisionstechnik (Physikalisch-mechanisches Institut). Vom 16. Juni 1883." [1-22]. Beilage I. "Votum des Herrn Geheimen Regierungs-Raths Dr. Werner Siemens (April 1883)" [23-27]. Beilage II. "Votum des Chefs der trigonometrischen Abtheilung der königlichen Landesaufnahme, Herrn Oberstlieutenant Schreiber (Mai 1883)" [29-32]. Beilage III. "Votum des Herrn Geheimen Regierungs-Rathes, Prof. Dr. von Helmholtz (Juni 1883). Ueber die Aufgaben der wissenschaftlichen Abtheilung des in Aussicht genommenen Physikalisch-Mechanischen Instituts" [33-36]. *Drucksachen zu den Verhandlungen des Bundesraths des Deutschen Reichs*, 1-49 and 1-36, respectively.

1886. Band 1. Aktenstück Nr. 50. Partially reprinted in *SBVR. Sammlung sämtlicher Drucksachen*. 1887. Drucksache Nr. IV. Beilage B. Siemens's remarks ("Begründung," 2-4) also reprinted as "Ueber die Bedeutung und die Ziele einer zu begründenden physikalisch-technischen Reichsanstalt" in Siemens's *Wissenschaftliche und Technische Arbeiten*, 2, 576-80; and Siemens's "Votum" (Beilage I) as "Votum betreffend die Gründung eines Instituts für die experimentelle Förderung der exakten Naturforschung und der Präzisionstechnik," ibid., 2, 568-75.

— *See also* Prussian Landtag, Haus der Abgeordneten.

Physikalisch-Technische Reichsanstalt. Miscellaneous. "Errichtung einer Reichsanstalt für die Förderung der Naturforschung und der Präcisions-technik." *Centralblatt der Bauverwaltung 6* (1886): 157-58.

— "IV. Gebäude für die Verwaltungsbehörden des Deutschen Reiches. 11. Die Physikalisch-Technische Reichsanstalt in Charlottenburg." In *Berlin und seine Bauten*, ed. Architekten-Verein zu Berlin und Vereinigung Berliner Architekten, 2, 80-84. 3 vols. in 2. Berlin, 1896.

— "Die Tätigkeit der Physikalisch-Technischen Reichsanstalt im Kriege." *Zeitschrift für komprimierte und flüssige Gase sowie für die Pressluft-Industrie 19* (1917-18): 105-7.

Physikalisch-Technische Reichsanstalt. Sundry Official Publications. *Geschäftsordnung für die Physikalisch-Technische Reichsanstalt*. Berlin, 1888.

— *Vorschläge zu gesetzlichen Bestimmungen über elektrische Masseinheiten, entworfen durch das Curatorium der Physikalisch-Technischen Reichsanstalt: Nebst kritischem Bericht über den wahrscheinlichen Werth des Ohm nach dem bisherigen Messungen verfasst von Dr. Friedrich Ernst Dorn*. Berlin, 1893.

— "Die Betheiligung der Physikalisch-Technischen Reichsanstalt an der Weltausstellung in Chicago." *ZI 13* (1893): 157-64.

— [Hermann von Helmholtz]. "Die Beglaubigung der Hefnerlampe. (Mittheilung aus der Physikalisch-Technischen Reichsanstalt)." *ZI 13* (1893): 257-67.

— "Biographische Notizen." *WAPTR 1* (1894): n.p.

— *Wissenschaftliche Abhandlungen der Physikalisch-Technischen Reichsanstalt*. Berlin, 1894-.

— *Verzeichnis der Veröffentlichungen aus der Physikalisch-Technischen Reichsanstalt 1887 bis 1900*. Berlin, 1901.

— *Bestimmungen über die elektrischen Masseinheiten im Deutschen Reiche*. Berlin, 1901.

— [Friedrich Kohlrausch]. *Die bisherige Tätigkeit der Physikalisch-Technischen Reichsanstalt: Aus dem Reichstage am 19. Februar 1904 überreichten Denkschriften. Mit einem Verzeichnis der Veröffentlichungen aus den Jahren 1901-1903*. Braunschweig, 1904. Reprinted in *DMZ*, Heft

13 (1 July 1904): 121-25; Heft 14 (15 July 1904): 134-37, in *ZI 24* (1904);
and in *GAFK, 1*, 1032-47.

— "Denkschrift betreffend die Vorarbeiten und die Ausarbeitung des Projekts
zu einem Neubaue des elektrischen Laboratoriums der Physikalisch-
Technischen Reichsanstalt, II. Abteilung." In *HEDR* (1908), Beilage A,
Kapitel 3, Titel 8, pp. 60-61. Berlin, 1907.

— [Emil Warburg]. "Prüfungsbestimmungen der Physikalisch-Technischen
Reichsanstalt. Teil A: Allgemeine Bestimmungen." *DMZ*, Heft 8 (15
April 1910): 73-77, in *ZI 30* (1910).

— *Geschäftsordnung für die Physikalisch-Technische Reichsanstalt.* Berlin,
1917.

Physikalisch-Technische Reichsanstalt. Tätigkeitsberichte. (Annual Reports).
"Die Thätigkeit der Physikalisch-Technischen Reichsanstalt bis Ende
1890." *ZI 11* (1891): 149-70.

— "Die Thätigkeit der Physikalisch-Technischen Reichsanstalt in den Jahren
1891 und 1892." *ZI 13* (1893): 113-40.

— [Ernst Hagen]. "5ter Bericht über die Thätigkeit der Physikalisch-Technischen Reichsanstalt (Dezember 1892 bis Februar 1894)." *ZI 14* (1894):
261-79, 301-16.

— [Ernst Hagen]. "Die Thätigkeit der Physikalisch- Technischen Reichsanstalt in der Zeit vom 1. März 1894 bis 1. April 1895." *ZI 15* (1895): 283-
300, 324-43.

— [Friedrich Kohlrausch]. "Die Thätigkeit der Physikalisch-Technischen
Reichsanstalt in der Zeit vom 1. April 1895 bis 1. Februar 1896." *ZI 16*
(1896): 203-18, 233-49.

— [Friedrich Kohlrausch]. "Die Thätigkeit der Physikalisch-Technischen
Reichsanstalt in der Zeit vom 1. Februar 1896 bis 31. Januar 1897." *ZI 17*
(1897): 140-54, 172-86.

— [Friedrich Kohlrausch]. "Die Thätigkeit der Physikalisch-Technischen
Reichsanstalt in der Zeit vom 1. Februar 1897 bis 31. Januar 1898." *ZI 18*
(1898): 138-51, 181-91.

— [Friedrich Kohlrausch]. "Die Thätigkeit der Physikalisch-Technischen
Reichsanstalt in der Zeit vom 1. Februar 1898 bis 31. Januar 1899." *ZI 19*
(1899): 206-16, 240-56.

— [Friedrich Kohlrausch]. "Die Thätigkeit der Physikalisch-Technischen
Reichsanstalt in der Zeit vom Februar 1899 bis Februar 1900." *ZI 20*
(1900): 140-50, 172-86.

— [Friedrich Kohlrausch]. "Die Thätigkeit der Physikalisch-Technischen
Reichsanstalt im Jahre 1900." *ZI 21* (1901): 105-21, 138-55.

— [Friedrich Kohlrausch]. "Die Thätigkeit der Physikalisch-Technischen
Reichsanstalt im Jahre 1901." *ZI 22* (1902): 110-24, 143-60.

— [Friedrich Kohlrausch]. "Die Tätigkeit der Physikalisch-Technischen
Reichsanstalt im Jahre 1902." *ZI 23* (1903): 113-25, 150-57, 171-84.

— [Friedrich Kohlrausch]. "Die Tätigkeit der Physikalisch-Technischen Reichsanstalt im Jahre 1903." *ZI 24* (1904): 133-47, 167-80.

— [Friedrich Kohlrausch]. "Die Tätigkeit der Physikalisch-Technischen Reichsanstalt im Jahre 1904." *ZI 25* (1905): 102-16, 137-53.

— [Emil Warburg]. "Die Tätigkeit der Physikalisch-Technischen Reichsanstalt im Jahre 1905." *ZI 26* (1906): 109-25, 145-60, 185-94.

— [Emil Warburg]. "Die Tätigkeit der Physikalisch-Technischen Reichsanstalt im Jahre 1906." *ZI 27* (1907): 109-24, 147-60, 184-200.

— [Emil Warburg]. "Die Tätigkeit der Physikalisch-Technischen Reichsanstalt im Jahre 1907." *ZI 28* (1908): 101-16, 139-57, 172-89.

— [Emil Warburg]. "Die Tätigkeit der Physikalisch-Technischen Reichsanstalt im Jahre 1908." *ZI 29* (1909): 103-18, 143-63, 179-96.

— [Emil Warburg]. "Die Tätigkeit der Physikalisch-Technischen Reichsanstalt im Jahre 1909." *ZI 30* (1910): 106-20, 140-60, 174-95.

— [Emil Warburg]. "Die Tätigkeit der Physikalisch-Technischen Reichsanstalt im Jahre 1910." *ZI 31* (1911): 112-25, 148-63, 183-97.

— [Emil Warburg]. "Die Tätigkeit der Physikalisch-Technischen Reichsanstalt im Jahre 1911." *ZI 32* (1912): 119-35, 155-69, 195-210.

— [Emil Warburg]. "Die Tätigkeit der Physikalisch-Technischen Reichsanstalt im Jahre 1912." *ZI 33* (1913): 84-98, 111-30, 152-72.

— [Emil Warburg]. "Die Tätigkeit der Physikalisch-Technischen Reichsanstalt im Jahre 1913." *ZI 34* (1914): 113-30, 151-64, 184-200.

— [Emil Warburg]. "Die Tätigkeit der Physikalisch-Technischen Reichsanstalt im Jahre 1914." *ZI 35* (1915): 96-111, 131-51, 174-91.

— [Emil Warburg]. "Die Tätigkeit der Physikalisch-Technischen Reichsanstalt im Jahre 1915." *ZI 36* (1916): 84-93, 116-30, 149-59.

— [Emil Warburg]. "Die Tätigkeit der Physikalisch-Technischen Reichsanstalt im Jahre 1916." *ZI 37* (1917): 70-8, 91-103, 120-32.

— [Emil Warburg]. "Die Tätigkeit der Physikalisch-Technischen Reichsanstalt im Jahre 1917." *ZI 38* (1918): 59-65, 81-88, 94-106.

— [Emil Warburg]. "Die Tätigkeit der Physikalisch-Technischen Reichsanstalt im Jahre 1918." *ZI 39* (1919): 105-17, 137-45, 180-94.

Planck, Max. "Ueber irreversible Strahlungsvorgänge." *AP 306* (1900): 69-122.

— "Ueber eine Verbesserung der Wien'schen Spectralgleichung." *VDPG 2* (1900): 202-4.

— "Zur Theorie des Gesetzes der Energievertheilung im Normalspectrum." *VDPG 2* (1900): 237-45.

— "Ueber das Gesetz der Energieverteilung im Normalspectrum." *AP 309* (1901): 553-63.

— "Persönliche Erinnerungen." *VDPG 16* (1935): 11-16. Reprinted in Planck's *Physikalische Abhandlungen und Vorträge, 3*, 358-63.

— "Persönliche Erinnerungen aus alten Zeiten." *NWN 33* (1946): 230-35. Reprinted in Planck's *Vorträge und Erinnerungen*, 5th ed., 1-14.

— *Wissenschaftliche Selbstbiographie*. 2nd ed. Leipzig, 1948.

— *Vorträge und Erinnerungen*. 5th ed. Stuttgart, 1949.

— "Die Physikalische Gesellschaft um 1890." *PB 6* (1950): 433-34.

— *Physikalische Abhandlungen und Vorträge*. 3 vols. Braunschweig, 1958.

Planck, Max, et al. "Nr. 24. Wahlvorschlag für Emil Warburg (1846-1931) zum OM. Berlin, 30. Mai 1895." In *PPI*, 131-32.

— "Feier des 80. Geburtstages des Ehrenmitgliedes der Deutschen Physikalischen Gesellschaft Herrn Geheimrat Professor Dr. Max Planck . . . ." *VDPG 19* (1938): 57-76. Reprinted in Planck's *Physikalische Abhandlungen und Vorträge, 3*, 402-16.

J. C. Poggendorff's *Biographisch-Literarisches Handwörterbuch zur Geschichte der exacten Wissenschaften* (title varies). Leipzig, 1863-.

Pohl, Robert W. "Von den Studien- und Assistentenjahren James Francks: Erinnerungen an das Physikalische Institut der Berliner Universität." *PB 28* (1972): 542-44.

Pringsheim, E. "Die Strahlungsgesetze und ihre Anwendungen." *Naturwissenschaftliche Rundschau 15* (1900): 1- 2, 17-19.

Pritchett, Henry S. "The Story of the Establishment of the National Bureau of Standards." *Science 15* (1902): 281-84.

Prussian Landtag, Haus der Abgeordneten. "Denkschrift betreffend die Begründung eines mechanischen Instituts"; "Vorschläge zur Hebung der wissenschaftlichen Mechanik und Instrumentenkunde"; "Grundzüge der Organisation des mechanischen Instituts." *Anlagen zu den Stenographischen Berichten über die Verhandlungen des Hauses der Abgeordneten*, 531-36. 1876. Band 1. 12th Legislatur-Periode. 3rd Session. Aktenstück Nr. 53. Berlin, 1876.

Pupin, Michael. *From Immigrant to Inventor*. New York, 1960.

Pyenson, Lewis. "Physical Sense in Relativity: Max Planck Edits the *Annalen der Physik*, 1906-1918." In Pyenson's *The Young Einstein: The Advent of Relativity*, 194-214. Bristol, 1985.

Ramser, Hans. "Warburg, Emil Gabriel." *DSB 14* (1976): 170-72.

Rayleigh, Lord [John William Strutt]. "Remarks Upon the Law of Complete Radiation." *Philosophical Magazine 49* (1900): 539-40.

Reiche, Fritz. "Otto Lummer." *PZ 27* (1926): 459-67.

Reichsamt des Innern. *See* German Budget, Reichsamt des Innern.

Reiner, Julius. *Hermann von Helmholtz*. Leipzig, 1905.

Remané, H. "Die geschichtliche Entwickelung, die Herstellung, die physikalischen Eigenschaften und die Anwendungen der elektrischen Glühlampen." *DMZ*, Heft 23 (1 December 1899): 209-13, 221-27, in *ZI 19* (1899).

Richter, Werner. "Die Organisation der Wissenschaft in Deutschland." In

*Forschungsinstitute: Ihre Geschichte, Organisation und Ziele,* ed. Ludolph Brauer et al., *1*, 1-12.

Riecke, Eduard. "Friedrich Kohlrausch † [Obituary]." *PZ 11* (1910): 73-76.

— "Friedrich Kohlrausch." *Nachrichten von der Königlichen Gesellschaft der Wissenschaften zu Göttingen: Geschäftliche Mitteilungen aus dem Jahre 1910,* Heft 1 (1910), 71-85.

Röntgen, Wilhelm Conrad. *W. C. Röntgen: Briefe an L. Zehnder.* Ed. Ludwig Zehnder. Zurich, 1935.

Rothenberg, Gunter E. "Moltke, Schlieffen and the Doctrine of Strategic Envelopment." In *Makers of Modern Strategy from Machiavelli to the Nuclear Age,* ed. Peter Paret (with Gordon A. Craig and Felix Gilbert), 296-325. Princeton, N.J., 1986.

Rubens, Heinrich. "A. Paalzow, Gedächtnisrede." *VDPG 10* (1908): 451-62.

— "Gedächtnisrede auf Friedrich Kohlrausch." *APAWB: Physikalisch-Mathematische Klasse* (1910), Gedächtnisrede 1, 3-11. Reprinted in *PPII,* 177-82.

— "Adresse an Hrn. Emil Warburg zum 50. Doktorjubiläum." *SB* (1917): (I) 269-71.

Rubens, H., and F. Kurlbaum. "Ueber die Emission langwelliger Wärmestrahlen durch den schwarzen Körper bei verschiedenen Temperatur." *SB* (1900): (II) 929-41.

— "Anwendung der Methode der Reststrahlen zur Prüfung des Strahlungsgesetzes." *AP 309* (1901): 649-66.

Ruske, Walter. *Hundert Jahre Materialprüfung in Berlin: Ein Beitrag zur Technikgeschichte.* Berlin, 1971.

— "Reichs- und preussische Landesanstalten in Berlin: Ihre Entstehung und Entwicklung als ausseruniversitäre Forschungsanstalten und Beratungsorgane der politischen Instanzen." *Bundesanstalt für Materialprüfung-Berichte* (1973), Nr. 23.

Rutherford, Ernest. "The British Radium Standard." *Nature 92* (1913), 402-3.

Schaefer, Clemens. "Otto Lummer zum 100. Geburtstag." *PB 16* (1960): 373-81.

Scheel, Karl. "Friedrich Kohlrausch † [Obituary]. Nachruf." *Naturwissenschaftliche Rundschau 25* (1910): 153-54.

— "Die Physikalisch-Technische Reichsanstalt in Charlottenburg." *Akademische Rundschau 1* (1913): 221-27.

— "Die Physikalisch-Technische Reichsanstalt: Fünfundzwangzig Jahre ihrer Tätigkeit. 1. Allgemeines." *NWN 1* (1913): 177-80.

— "Die Physikalisch-Technische Reichsanstalt." *Anzeiger für Berg-, Hütten- und Maschinenwesen 38* (1916): 5498-99, 5569-70.

— "Werner von Siemens und die Physikalisch-Technische Reichsanstalt." *Dinglers Polytechnisches Journal 331* (1916): 405-8.

— "Die Tätigkeit der Physikalisch-Technischen Reichsanstalt im Jahre 1915." *NWN 4* (1916): 569-74.

— "Forschungsstätten." In *Forschungsinstitute: Ihre Geschichte, Organisation und Ziele,* ed. Ludolph Brauer et al., *1,* 175-208.

— "Ernst Hagen." *Deutsches Biographisches Jahrbuch 5* (1930): 146-48.

Schellbach, Karl. *Erinnerungen an den Kronprinzen Friedrich Wilhelm von Preussen.* Breslau, 1890.

Schering, H. "Emil Warburg und die Technik." *NWN 14* (1926): 208-11.

Schmid, G. "Die Verluste des ehemaligen Reichsarchivs im zweiten Weltkrieg." In *Archivar und Historiker: Studien zur Archiv- und Geschichtswissenschaft: Zum 65. Geburtstag von Heinrich Otto Meissner,* ed. Staatliche Archivverwaltung im Staatssekretariat für Innere Angelegenheiten, 176-207. Berlin, 1956.

Schmitt, Eduard, ed. (with Josef Durm and Hermann Ende). *Handbuch der Architektur.* Teil 4: *Entwerfen, Anlage und Einrichtung der Gebäude* Halband 6: *Gebäude für Erziehung, Wissenschaft und Kunst.* Heft 2a: *Hochschulen, zugehörige und verwandte wissenschaftliche Institute.* I: *Universitäten und Technische Hochschulen. . . .* 2nd ed. Stuttgart, 1905.

Schönrock, O. "Hans Heinrich Landolt." *ZI 30* (1910): 92-96.

Schubring, Gert. "Mathematics and Teacher Training: Plans for a Polytechnic in Berlin." *HSPS 12* (1981): 161-94.

Schulze, A. "50 Jahre Physikalisch-Technische Reichsanstalt." *Metallwirtschaft 16* (1937): 1369-72.

Schwarzschild, K. "Präzisionstechnik und wissenschaftliche Forschung." *DMZ,* Heft 14 (15 July 1914): 149-53; Heft 15 (1 August 1914): 162-65, in *ZI 34* (1914).

"Scientific Notes and News." *See* National Bureau of Standards.

Siemens, Georg. *Geschichte des Hauses Siemens.* 3 vols. Munich, 1947.

Siemens, Werner von. ["Antrittsrede"]. *Monatsberichte der Preussischen Akademie der Wissenschaften zu Berlin* (1874): 464- 67. Reprinted in *PPII,* 106-8.

— "Die Elektrizität im Dienste des Lebens." *EZ 1* (1880): 16-22.

— *Wissenschaftliche und Technische Arbeiten.* 2nd ed. 2 vols. Berlin, 1889 and 1891. (See esp. "Das naturwissenschaftliche Zeitalter," 2, 491-500.)

— *Personal Recollections of Werner Siemens.* Trans. W. C. Coupland. London, 1893.

— *Scientific and Technical Papers.* Trans. from 2nd German ed. 2 vols. London, 1895.

— *Lebens-Erinnerungen.* Leipzig, 1943.

— *Aus einem reichen Leben: Werner von Siemens in Briefen an seine Familie und an Freunde.* Ed. Friedrich Heintzenberg. Stuttgart, 1953.

— *Inventor and Entrepreneur: Recollections of Werner Siemens.* 2nd English ed. London, 1966.

— *See also* Matschoss, Conrad.

Siemens, Wilhelm. "Werner Siemens und sein Wirkungsfeld." *NWN 4* (1916): 759-71.

Sonnemann, Rolf, et al. *Geschichte der Technischen Universität Dresden: 1828-1978.* Berlin, 1978.

Spieker, Paul. "E. Sternwarten und andere Observatorien." In *Handbuch der Architektur,* ed. Joseph Durm et al., Teil 4, Halbband 6, Heft 2, 474-567.

Spielmann, Wilhelm. *Handbuch der Anstalten und Einrichtungen zur Pflege von Wissenschaft und Kunst in Berlin.* Berlin, 1897.

Stark, Johannes, ed. *Forschung und Prüfung: 50 Jahre Physikalisch-Technische Reichsanstalt.* Leipzig, 1937.

— *Fortschritte der Physik 1887 bis 1937* (= "Die Physikalisch-Technische Reichsanstalt in der Entwicklung der Physik während der letzten 50 Jahre"). Leipzig, 1938.

Steinhaus, W. "Ernst Gumlich zum Gedächtnis." *ZTP 11* (1930): 129-31.

Stenzel, Rudolf. "Zusammenfassungen der Sitzungsprotokolle der Plenar- bzw. Vollversammlungen der Normal-Eichungs-Kommission des Norddeutschen Bundes, der Kaiserlichen Normal-E[A]ichungs- Kommission, der Reichsanstalt für Mass und Gewicht, der Physikalisch-Technischen Reichsanstalt, Abteilung I und der Physikalisch-Technischen Bundesanstalt von 1869 bis 1970." *Physikalisch-Technische Bundesanstalt. Bericht* (1972), JB 2/72.

— "Begründung für die Verschmelzung der *Reichsanstalt für Mass und Gewicht* mit der *Physikalisch- Technischen Reichsanstalt* in Berlin im Jahre 1923." *Annals of Science 33* (1976): 289-306.

Stichweh, Rudolf. *Zur Entstehung des modernen Systems wissenschaftlicher Disziplinen: Physik in Deutschland, 1740-1890.* Frankfurt am Main, 1984.

Stille, Ulrich, and Hermann Siemens. "Festveranstaltung aus Anlass des 60jährigen Bestehens des Helmholtz-Fonds E. V." *PTB-Mitteilungen* (May 1973): 297-306.

Stine, W. M. "Macfarlane on Units of Light and Radiation." *TAIEE 12* (1895): 152-53.

Strutt, John William. *See* Rayleigh, Lord.

Taylor, A. J. P. *Bismarck: The Man and the Statesman.* New York, 1967.

Thiesen, Max. "Ueber das Gesetz der schwarzen Strahlung." *VDPG 2* (1900): 65-70.

Thiesen, M., K. Scheel, and L. Sell. "Thermometrische Arbeiten, betreffend die Vergleichungen von Quecksilberthermometern unter einander." *WAPTR 2* (1895): 1- 72.

— "Untersuchungen über die thermische Ausdehnung von festen und tropfbar flüssigen Körpern." *WAPTR 2* (1895): 73-184.

Trenn, Thaddeus J. "Geiger, Hans (Johannes) Wilhelm." *DSB 5* (1972): 330-33.

— "Die Erfindung des Geiger-Müller- Zählrohres." *Deutsches Museum, Abhandlungen und Berichte 44* (1976): 54-64.

— "The Geiger-Müller Counter of 1928." *Annals of Science 43* (1986): 111-35.

Tsuneishi, Kei-ichi. "On the Abbe Theory (1873)." *Japanese Studies in the History of Science 12* (1973): 79-91.

Turner, R. Steven. "Helmholtz, Hermann von." *DSB 6* (1972): 241-53.

U.S. Congress, House of Representatives Committee on Coinage, Weights, and Measurements. *National Standardizing Bureau.* 56th Cong., 1st sess., 1900, H, Document No. 625, Serial 3997.

U.S. Congress, Senate Subcommittee of the Committee on Commerce. *Hearing before the Subcommittee of the Committee on Commerce, United States Senate, upon the Bill (S. 4680) to Establish a National Standardizing Bureau.* 56th Cong., 2d sess., 1900, S. Document No. 70, Vol. 5.

U.S. Congress, Senate Committee on Commerce. *National Standar[d]izing Bureau.* 56th Cong., 2d sess., 1901. To accompany S. Report 4680.

Van't Hoff, J. H. "Gedächtnisrede auf Hans Heinrich Landolt." *APAWB: Physikalisch-mathematische Klasse* (1910): Gedächtnisrede 2, 1-13. Reprinted in *PPII,* 130-37.

Verband Deutscher Elektrotechniker. *Der Verband Deutscher Elektrotechniker, 1893-1918.* N.p., 1918 (?).

Verein Deutscher Ingenieure. "Bitte des Vorstandes des Vereines Deutscher Ingenieure, betreffend die Errichtung einer 'physikalisch-technischen Reichsanstalt.'" *Zeitschrift des Vereins deutscher Ingenieure 30* (1886): 890-91.

"Vermischtes: Einfluss elektrischer Strassenbahn auf Galvanometer." *Centralblatt der Bauverwaltung 13* (1893): 551.

"Vermischtes." *Centralblatt der Bauverwaltung 15* (1895): 276.

Virchow, Rudolf. ["Ueber die Aufgaben der Naturwissenschaften in dem neuen nationalen Leben Deutschlands"]. *Tageblatt der 44. Versammlung Deutscher Naturforscher und Aerzte in Rostock 1871* Nr. 5 (1871): 73-81.

Voigt, Woldemar. *Physikalische Forschung und Lehre in Deutschland während der letzten hundert Jahre: Festrede im Namen der Georg-August Universität zur Jahresfeier der Universität am 5. Juni 1912.* Göttingen, 1912.

— "Entgegnung." *PZ 13* (1912): 1093-95.

[Waetzmann, Erich, and Fritz Reiche]. *Akademische Reden: Zum Gedächtnis an Otto Lummer gehalten in Breslau 1925.* Braunschweig, 1928.

Waidner, C. W., and G. K. Burgess. "Note on the Primary Standard of Light." *Electrical World 52* (1908): 625-28.

Warburg, Emil. "Das Physikalische Institut." In *Die Universität Freiburg seit dem Regierungsantritt seiner königlichen Hoheit des Grossherzogs Friedrich von Baden*, 91- 96. Freiburg i. B., 1881.

— "Ueber den Aufschwung der modernen Naturwissenschaften." *Berichte der Naturforschenden Gesellschaft zu Freiburg i. B. 6* (1892): 18-29.

— "Antrittsrede." *SB* (1896): (II) 743-45. Reprinted in *PPII*, 183-84.

— "Die Physikalisch-Technische Reichsanstalt in Charlottenburg." *Internationale Wochenschrift für Wissenschaft, Kunst und Technik 1* (1907): 537-48.

— "Die Physikalisch-Technische Reichsanstalt in Charlottenburg." *Zeitschrift des Oesterreichischen Ingenieur- und Architekten-Vereines 60* (1908): 513-17, 529-31.

— "Friedrich Kohlrausch. Gedächtnisred . . . " *VDPG 12* (1910): 911-38.

— "Ueber die Ziele der Physikalisch-Technischen Reichsanstalt; zur Abwehr." *PZ 13* (1912): 1091-93.

— "Bemerkungen zu der Aufspaltung der Spektrallinien im elektrischen Feld." *VDPG 15* (1913): 1259-66.

— "Ueber den Entwicklungsgang der Starkstromtechnik und über deren Beziehungen zur Physikalisch-Technischen Reichsanstalt." *Internationale Monatsschrift für Wissenschaft, Kunst und Technik 8* (1914): 914-29.

— "Verhältnis der Präzisionsmessungen zu den allgemeinen Zielen der Physik." In *Physik*, ed. Emil Warburg, 653-60.

— "Werner Siemens und die Physikalisch-Technische Reichsanstalt." *NWN 4* (1916): 793-97.

— "Ueber eine rationelle Lichteinheit." *VDPG 19* (1917): 3-10.

— "Ein Ehrentag für Emil Warburg." *Stahl und Eisen 37* (1917): 385-86.

— "Die technische Physik und die Physikalisch-Technische Reichsanstalt." *ZTP 2* (1921): 225-27.

— "Helmholtz als Physiker." In Emil Warburg, M. Rubner, and M. Schlick, *Helmholtz als Physiker, Physiologe und Philosoph: Drei Vorträge*, 3-14.

— *See also* Physikalisch-Technische Reichsanstalt.

Warburg, Emil, ed. *Physik*. In *Die Kultur der Gegenwart: Ihre Entwicklung und ihre Ziele*, Teil 3, *Mathematik, Naturwissenschaften, Medizin*, Abteilung 3, *Anorganische Naturwissenschaften*. Ed. P. Hinneberg. Leipzig, 1915.

Warburg, Emil, M. Rubner, and M. Schlick. *Helmholtz als Physiker, Physiologe und Philosoph: Drei Vorträge*. Karlsruhe, 1922.

Webster, A. G. "A National Physical Laboratory." *Pedagogical Seminary 2* (1892): 90-101.

Weiher, Sigfrid von. "Gelenkte Forschung schon im 19. Jahrhundert: Die Entstehungsgeschichte der Physikalisch-Technischen Reichsanstalt." *VDI-Nachrichten 16* (1962): 9-10.

— "Zur Entstehungsgeschichte der Physikalisch-Technischen Reichsanstalt." *Siemens Zeitschrift 41* (1967): 856.

Wendel, Günter. *Die Kaiser-Wilhelm-Gesellschaft 1911-1914: Zur Anatomie einer imperialistischen Forschungsgesellschaft*. Berlin, 1975.

— "On the History of the Founding in 1887 in Germany of the Imperial Institute of Physics and Technology" (in Russian). *Voprosy istorii estestvoznaniia i tekhniki* No. 3 (52) (1976): 66-70.

Westphal, Wilhelm. "Heinrich Rubens." *NWN 10* (1922): 1017-20.

— "Dem Andenken an Heinrich Rubens." *PB 8* (1952): 323-25.

— "Das Physikalische Institut der TU Berlin." *PB 11* (1955): 554-58.

"H. F. Wiebe." Obituary, editorial. *DMZ*, Heft 19 (1 October 1912): 197, in *ZI 32* (1912).

Wiedemann, Gustav. "Hermann von Helmholtz, Wissenschaftliche Abhandlungen." *AP 290* (1895): i-xxiv.

Wien, Wilhelm. "Eine neue Beziehung der Strahlung schwarzer Körper zum zweiten Hauptsatz der Wärmetheorie." *SB* (1893): (I) 55-62.

— "Die obere Grenze der Wellenlängen, welche in der Wärmestrahlung fester Körper vorkommen können; Folgerungen aus dem zweiten Hauptsatze der Wärmetheorie." *AP 285* (1893): 633-41.

— "Temperatur und Entropie der Strahlung." *AP 288* (1894): 132-65.

— "Ueber die Energievertheilung im Emissionsspectrum eines schwarzen Körpers." *AP 294* (1896): 662-69.

— "Friedrich Kohlrausch † [Obituary]." *AP 336* (1910): 449-54.

— *Aus der Welt der Wissenschaft: Vorträge und Aufsätze*. Leipzig, 1921. (See esp. "Physik und Technik," 235-63.)

— "Das Physikalische Institut und das Physikalische Seminar." In *Die wissenschaftlichen Anstalten der Ludwig-Maximilians-Universität zu München: Chronik zur Jahrhundertfeier*, ed. Karl Alexander von Müller, 207-11. Munich, 1926.

— "Ein Rückblick." In *Wilhelm Wien: Aus dem Leben und Wirken eines Physikers*, ed. Karl Wien, 1-50. Leipzig, 1930.

Wien, W., and O. Lummer. "Methode zur Prüfung des Strahlungsgesetzes absolute schwarzer Körper." *AP 292* (1895): 451-56.

Wildner, Paul. "Die Glasindustrie." In *Handbuch der Wirtschaftskunde Deutschlands, 3*, 263-92.

Woodruff, A. E. "Weber, Wilhelm Eduard." *DSB 14* (1976): 203-9.

*Zeitschrift für Instrumentenkunde*. "An unsere Leser." *ZI 1* (1881): n.p.

— Das Redaktionskuratorium. "An unsere Leser!" *ZI 15* (1895): 1.

Zenneck, Johannes. "Werner v. Siemens und die Gründung der Physikalisch-Technischen Reichsanstalt." *Deutsches Museum, Abhandlungen und Berichte 3* (1931): 1-26.

— "Carl von Lindes Wesen und Wirken." *Technische Ueberwachung 3* (1942): 93-97.

# Index